THE NEOSTRIATUM

Sheastore wrote about the six blind men of Indostan when they approached an elephant for the first time.

First blind man, falling against the elephant's side:
'God bless me! but this elephant
Is very like a wall.'
Second blind man, feeling the tusk:
'This wonder of an elephant
Is very like a spear.'
Third blind man, grasping the squirming trunk:
'I see,' quoth he, 'the elephant
Is very like a snake.'
The fourth blind man, clasping the knee:
'Tis clear enough, the elephant
Is very like a tree.'
The fifth blind man, catching the ear:
'This marvel of an elephant
Is very like a fan.'
The sixth, seizing the swinging tail:
'I see,' quoth he, 'the elephant
Is very like a rope.'
And so these men of Indostan
Disputed loud and long,
Each in his own opinion
Exceedingly stiff and strong,
Though each was partly in the right,
And all were in the wrong.

THE NEOSTRIATUM

Proceedings of a Workshop sponsored by the European Brain and Behaviour Society, Denmark, 17-19 April 1978

Editors:

IVAN DIVAC

Institute of Neurophysiology, University of Copenhagen, Denmark

and

R. GUNILLA E. ÖBERG

Laboratory of Psychology, University of Copenhagen, Denmark

PERGAMON PRESS

OXFORD · NEW YORK · TORONTO · SYDNEY · PARIS · FRANKFURT

U.K.	Pergamon Press Ltd., Headington Hill Hall, Oxford OX3 0BW, England
U.S.A.	Pergamon Press Inc., Maxwell House, Fairview Park, Elmsford, New York 10523, U.S.A.
CANADA	Pergamon of Canada, Suite 104, 150 Consumers Road, Willowdale, Ontario M2J 1P9, Canada
AUSTRALIA	Pergamon Press (Aust.) Pty. Ltd., P.O. Box 544, Potts Point, N.S.W. 2011, Australia
FRANCE	Pergamon Press SARL, 24 rue des Ecoles, 75240 Paris, Cedex 05, France
FEDERAL REPUBLIC OF GERMANY	Pergamon Press GmbH, 6242 Kronberg-Taunus, Pferdstrasse 1, Federal Republic of Germany

First edition 1979

British Library Cataloguing in Publication Data

The neostriatum.
1. Corpus striatum - Congresses
I. Divac, Ivan II. Oberg, R Gunilla E
III. European Brain and Behaviour Society
612'.826 QP383.9 79-40067

ISBN 0 08 023174 8

In order to make this volume available as economically and as rapidly as possible the authors' typescripts have been reproduced in their original forms. This method unfortunately has its typographical limitations but it is hoped that they in no way distract the reader.

Printed in Great Britain by A. Wheaton & Co. Ltd., Exeter

Contents

NEUROBEHAVIORAL APPROACHES

FUNCTIONAL CONCEPTS

Preface

This book contains proceedings of a workshop organized by the European Brain and Behaviour Society in the Vingsted Center, close to Vejle on the Danish peninsula Jutland, in April 1978. The administrative burden of the meeting was carried by the Institute of Neurophysiology of the University of Copenhagen. Professor H. Pakkenberg helped raising most of the funds which made the meeting possible. Thanks are due to him and to the following contributors: Astragruppen A/S, Civilingeniør Hans Lønborg-Madsens Mindelegat, CIBA-GEIGY, Danish Medical Research Council, Ferrosan, Merk Sharp & Dohme, Roche A/S and Sandoz A/S.

The present meeting was first conceived in 1974. At that time we felt a need for a broad review of all the different fields of research on the neostriatum. Considerable progress was being made by researchers of different methodological and theoretical orientation, but it was apparent that the cross-talk between different research groups and traditions was insufficient. We were, in fact, sharing the Zeitgeist, as now evidenced by two recent volumes which present proceedings from other meetings on the neostriatum, and which also give broad reviews of the field (1, 2). However, the contents of these recent books only partially overlap with the present volume. Because of financial restrictions we had to curb our original ambitions and impost certain limitations on the program of the Vejle meeting. We chose to stress basic experimental work and to deemphasize the clinical aspects which have been so generously covered in the past decade. Furthermore, pharmacological research is not prominent in the present volume since this aspect has attracted much interest and has been frequently reviewed in recent years. On the other hand, this book contains chapters on several less commonly discussed topics which were not comprehensively treated in any previous volume.

Since the purpose of the present conference was to illuminate the puzzles of the neostriatum from many different angles, the reader will find a certain amount of repetition of the factual material in different sections of this book.

It is a pleasure to present the contributions of many distinguished scientists in the present volume. Our editorial role has been limited to making suggestions for changes in the first version of some chapters. The authors remained arbiters of the feasibility of these suggestions, and they are obviously responsible for the final texts.

Throughout the book, the abbreviation NS stands for "neostriatum" or "neostriatal".

I.D., R.G.E.O.

1. Cools, A.R. Lohman, A.H.M. and van den Bercken, J.H.L. (Eds.) Psychobiology of the Striatum, Elsevier/North Holland, Amsterdam, 1977.

2. Yahr, M.D. (Ed.) The Basal Ganglia, Res Publ. Assoc. Nerv. Ment. Dis., 55, Raven Press, New York, 1976.

Basic Organisation

Introduction

I. Divac and R. G. E. Öberg

The last decade has brought revolutionary improvements in our tools for investigations of the anatomy and chemistry of the nervous system. These new techniques have provided data which require substantial changes in our notions about the NS. Some of the most significant recent findings concern the architecture and neurotransmitters (established as well as putative) of the NS, and the properties of the cortico-striatal projections. Thus, it has been demonstrated that some axons and perikarya in the NS contain peptides, and that some transmitters and their enzymes show gradients of concentration within the NS. It has also been established, by accumulation of converging evidence, that many medium-sized NS neurons are not interneurons as earlier believed, but that these cell bodies give rise to axons which reach the globus pallidus and the reticular part of substantia nigra. Autoradiographic techniques have revealed that a given cortical area projects not only to one part of the NS, as previously demonstrated by silver-impregnation methods, but also to the NS projection-targets of those cortical areas to which the first area richly and directly projects. Within their target regions, the cortico-neostriatal fibers terminate in patches of different neuronal density. It appears also, that these projections originate in a particular subset of cortical efferent neurons, and that they use glutamate as transmitter.

In the present section, these and other recent findings are integrated with previously established knowledge about the basic organisation of the NS. The informative summary figures in these reviews illustrate an impressive complexity of intrinsic and extrinsic NS circuitry. We are remote from the times when NS was regarded as an intermediate step between the (artifactual) "suppressor strips" of the neocortex and the remainder of the "extrapyramidal motor system".

The Internal Organization of the Neostriatum in Mammals

Pedro Pasik, Tauba Pasik and Marian DiFiglia

Department of Neurology, Mount Sinai School of Medicine
CUNY, New York, N.Y. 10029, U.S.A.

INTRODUCTION

Biochemical and pharmacological investigations into the nature of extrapyramidal disorders have brought forward the role of the neostriatum as the nodal structure of the basal ganglia system, hence the paramount importance of unraveling the intricate organization of this nuclear mass. Ultimately, it is through the knowledge of the morphologic bases for the synaptic transactions occurring within the neostriatum that a complete understanding of the physiopathology of these diseases will be reached. The present article will review the various stages in the gathering of information concerning the internal structure of the mammalian neostriatum. As with other areas of the central nervous system, successive developments in techniques mark the evolution of concepts on internal organization. Thus classic views are based mostly on data derived from general histologic procedures including the Nissl method, and rather fragmentary accounts from Golgi material. This is followed by the great advances made in the 1950's as part of the renaissance in the use of the latter technique, and in turn, by the results of tracers methods, particularly those based on retrograde axonal transport inasmuch as they help to recognize efferent elements. Finally, electron microscopy provides synaptologic data which, when correlated with light microscopy findings may lead to the formulation of neuronal networks as functional units of the neostriatum. Each of these methodologies have contributed both qualitative and quantitative information, and although much more must be done to arrive at a realistic model of neostriatal architecture, the last section will attempt to synthesize the present morphologic knowledge with that culled from physiology, biochemistry and pharmacology, thus providing certain directions for future research. Only the caudate-putamen will be dealt with in the discussion. Other possible striatal structures such as the nucleus accumbens and the olfactory tubercle will not be considered.

RESULTS OBTAINED WITH GENERAL HISTOLOGIC TECHNIQUES AND WITH THE NISSL METHOD

One of the earliest accounts on the cytology of the neostriatum can be found in Luys' (1865) description of human material fixed with chromic acid and examined unstained or treated with ammoniated carmine. Although he recognized large and small cells, it is evident from the illustrations that only the former are neuronal elements whereas the latter represent the nuclei of glial cells. Since this description was made over 20 years before Ramón y Cajal's neuron doctrine, the interpretations were reticularist in nature: the processes of the large cells were said to be continuous with the three types of afferents, which this author believed to be spinal fascicles, superior cerebellar peduncles and corticostriate fibers. Similar observations were made by Marchi (1886) who also offered the first demonstrations with the Golgi method (see below). Mostly triangular neurons were seen in human fresh material, 20-50 µm in size and having numerous protoplasmic processes (dendrites) and only one nerve process (axon).

Ramón y Cajal (1911) gives a good account of the corpus striatum of man as seen with the Nissl method: numerous small (8-10 µm) cells, either spherical or polygonal, with scanty and poorly stained cytoplasm; larger cells with few chromatic granules; and rare giant neurons, stellate in shape, with a large nucleus and the cytoplasm filled with chromatic granules. The three types are also seen in lower animals (rabbit, cat) where the giant cells are less voluminous. Later investigations consider only two neuronal types in the neostriatum of man (Spiegel, 1919; Bielschowsky, 1919; Vogt & Vogt, 1920; Foix & Nicolesco, 1925). The small cells are 12 x 15 µm in size; their shape can be stellate, triangular ovoid or fusiform; the cytoplasm is scarce with almost no chromatophilic granules and no pigment accumulation. These neurons have a tendency to form groups near a large cell (Foix & Nicolesco, 1925). The large cells are 30 x 40 µm in size and polygonal or fusiform in shape. The nucleus is eccentric, particularly in the adult where apparently it is displaced by the accumulation of lipochromic pigment, and the cytoplasm contains irregular masses of large chromatophilic granules near the cell membrane in the pole opposite to that occupied by the nucleus. The proportion of large to small cells is given as 1:20 (Foix & Nicolesco, 1925).

The cytoarchitectonic homogeneity of the human neostriatum (Spiegel, 1919; Vogt & Vogt, 1920) was challenged by extensive studies on 25 mammalian species,from mouse to man (Gurewitsch, 1930). This investigation shows that although the differences between caudate nucleus and putamen are minimal, the nuclei are not uniform. The bases for the 7 cytoarchitectonic fields described are difficult to interpret, however, because of the lack of actual values for cell size and density. Yet, it is of interest that this author distinguishes between effector giant cells of either polygonal or pyramidal shape and of rather even distribution, and receptor cells of various shapes and sizes,from large to small, which are arranged in groups and sometimes as sheets of spiral appearance. This study was extended later in human material (Brockhaus, 1942). Again 7 fields are described, but they are somewhat different than those of the previous investigation, and again no numerical data is provided. The few large cells are described as polygonal or narrowly pyramidal in shape.

A major effort was undertaken by Namba (1957) to subdivide the neostriatum into cytoarchitectonic fields. This author distinguishes two types of small cells in Nissl stained human material, namely a larger type α and a smaller type β, in addition to the large cells. On the basis of regional variations in the proportion of α and β cells and in the density of small cells as a whole, 5 fields are described in the caudate-putamen in the form of concentric shells from dorsolateral to ventromedial regions: lateral caudate, medial caudate and putamen, lateral putamen, ventral putamen and putamen limitans. The small-cell density increases in progressive steps so that the putamen limitans has twice as many of these neurons per field. The numerical statements, however, have a limited value because they are only relative and obtained without correction factors. The number of large cells parallels that of the small cells so that the large/small ratio remains relatively constant at 1:53, except for the extreme ventral regions (ventral putamen and putamen limitans) where the ratio becomes 1:76. More recently, a similar ratio (1:51) was found in the medial caudate nucleus of man between large cells of uniform size and a heterogeneous group of small cells (Treff, 1964). Additional information was provided in this study on absolute density values: $690/mm^3$ for the large cells and $35,000/mm^3$ for the small cells. No correction for shrinkage was used, however, so that these values were probably higher than the actual densities. A drastically different large/small cell ratio was reported by Tabuchi (1969) who found it to be 1:270. The discrepancy with Namba's (1957) results were explained on the basis of a sampling error in the latter study where small cells were counted only in fields containing a large neuron. This method resulted in lower numbers of small cells since the distance between large cells was at least one and one half times the diameter of the field used. Again no correction for shrinkage or for the "split cell" error were considered in this study. Similar sources of error may have played a role in the study of Dom, Baro & Brucher (1973) who determined cell densities in the putamen of 4 normal human brains. The values given are: 198 large cells/mm^3 and 29,558 small cells/mm^3 with a ratio of 1:149, the most frequent size of small cells being 8.5 μm. A considerably lower total cell density (11,575/mm^3) was reported by Carman (1966). Most sources of error have been accounted for in a recent monograph (Bøttcher, 1975). Counts were made on 15 normal human brains but unfortunately the numerical results were not expressed clearly enough. The mean ratio of large to small cells was 1:60. The differential cell distribution discussed above was not observed in this material. On the other hand, a good measurement of the total striatal volume referred to the fresh, unfixed state, was given as a mean of 13·130 cm^3 (internal capsule excluded). Without having been corrected for shrinkage the volume was found to be 11·960 cm^3 (internal capsule included (Harman & Carpenter, 1950), and 8·050 cm^3 (Carman, 1966). In the rhesus monkey, the volume was reported to be 0·822 cm^3 (no correction, internal capsule included, Harman and Carpenter, 1950).

Most investigators have reported a homogenous cellular distribution in the neostriatum of man (Vogt & Vogt, 1920), monkey (Fox, Andrade, Hillman & Schwyn, 1971/72a), cat (Kemp & Powell, 1971a) and rat (Lu & Brown, 1977). In the mouse, however, there is an apparent regional variation with the central core of the head of the caudate-putamen containing most of the large cells (Mensah, 1977). The latter author also describes the occurrence of clusters of 10-15 medium size cells. Some of these clusters may contain a large cell which can be multipolar or fusiform. Pairing of cells is also frequently seen (Kemp & Powell, 1971a; Lu & Brown, 1977; Mensah, 1977). Some regional variation is present in the opossum neostriatum as well (Martin & Hamel, 1967).

Three types of neurons are present in the cat caudate nucleus, on the basis of size (Kemp & Powell, 1971a): (1) The medium size cells measuring 9-18 μm comprise over 98% of all neurons. They have a large round and pale nucleus and a narrow rim of pale cytoplasm. A somewhat larger version of this type contains Nissl bodies and the nucleus is indented. (2) The large cells are over 20 μm and represent less than 1% of the population. Their nucleus is indented and large amounts of Nissl bodies are present. (3) The small cells, less than 8 μm in diameter, and of similar frequency as the large cells, have rather dark cytoplasm and no Nissl bodies. The size of the numerous medium size cells has been considered to be somewhat larger in the caudate than in the putamen (Bielschowsky, 1919; Vogt & Vogt, 1920; Foix & Nicolesco, 1925; Brockhaus, 1942; Tabuchi, 1969). More recently, however, a statistical analysis of 1200 neostriatal cells showed a significantly larger size for putaminal neurons (Adinolfi, 1971). The latter investigation apparently included all cell types, small and large. If this were the explanation for the discrepancy, the few large cells would have to be considerably larger in the putamen to reverse the mean difference.

The preceding review reveals several significant issues. Although cursory examination of the neostriatum results in the recognition of numerous small cells and few large cells, such a classification is indeed oversimplified. *Various types of small cells* have been described on the basis of size and shape differences (Bielschowsky, 1919; Gurewitsch, 1930; Namba, 1957). Moreover, one of these types was characterized by the presence of well formed Nissl bodies as opposed to their general absence in small cells (Kemp & Powell, 1971a). Similarly, at least *two types of large cells* have been repeatedly reported (Spiegel, 1919; Gurewitsch, 1930; Brockhaus, 1942; Mensah, 1977), one of them being polygonal in shape, and the other either fusiform or narrowly pyramidal. Examples of these two classes of large neurons from our own material is illustrated in Fig. 1. The quantitative data derived from Nissl or hematoxylin-eosin stained sections is rather ambiguous and points to an important area of future research. This should be undertaken with adequate controls over changes in the tissues during the various technical manipulations (shrinkage factors), as well as with the use of stereologic principles thereby arriving at more realistic determinations of cell numbers and densities. The ratios of large to small cell numbers have varied roughly from 1:50 to 1:270. This great discrepancy may reflect different criteria for classifying the neuronal

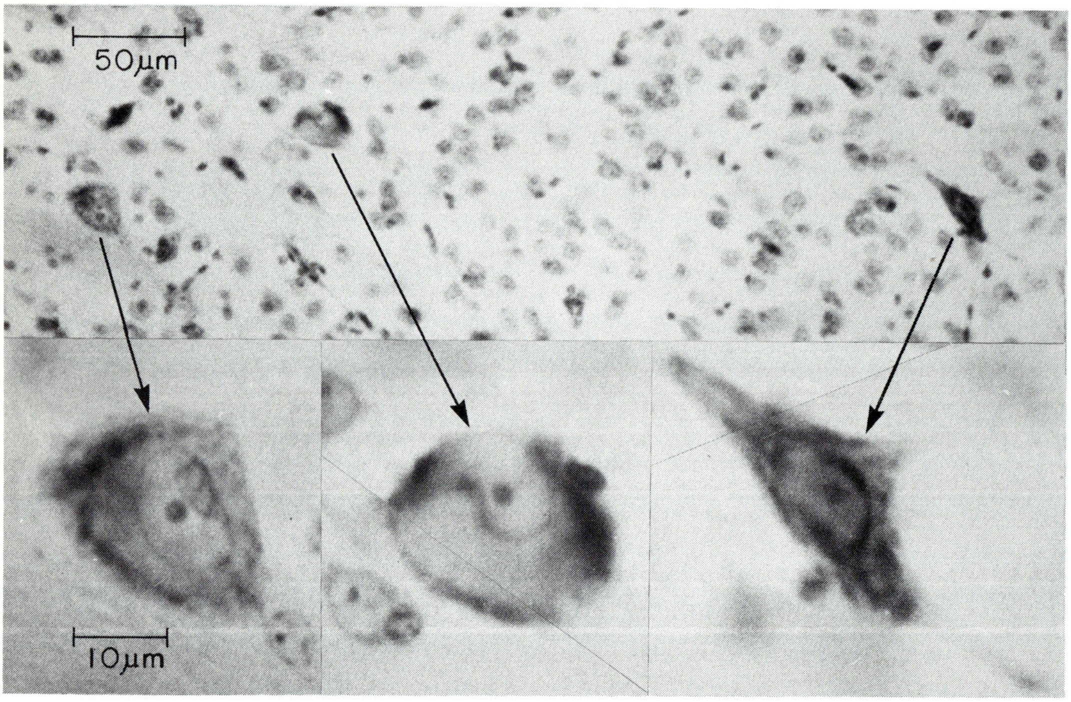

Fig. 1. Monkey Neostriatum. Klüver-Barrera stain. Arrows point to corresponding neurons at both magnifications. Note the high cell density and the relative distribution of the few large cells. The neurons at left and center belong to the polygonal or globular class, probably equivalent to the Aspiny II interneuron of the Golgi material. The large neuron on the right is long and rather narrow. It may represent the Spiny II long axon cell of Golgi impregnations (see next section).

elements. A step in the right direction is the measurement of cell nuclei (Bøttcher, 1975), but this karyometric approach was not pursued beyond the finding of a mean diameter (8 μm) for the nuclei of medium size cells. It is possible that full use of this technique (Palkovits & Fischer, 1968) will permit the segregation of various neuronal populations with adequate estimates of their relative occurrence.

RESULTS OBTAINED WITH THE GOLGI AND RELATED METHODS

The first description of the cellular types in the neostriatum as revealed by the "black reaction" comes from Golgi's laboratory itself (Marchi, 1886). Leaving aside the reticularist interpretation of the findings, this study shows both the existence of neurons of both Golgi types, namely the long axon type I and the short axon type II. It is also stated that in the caudate and lenticular nuclei there is a prevalence of Golgi type I cells, which is even more marked in the former nucleus (Marchi, 1886, p. 291). The article contains a dramatic illustration, reproduced in Fig. 2, of the abundance of long axon cells with the axons joining the internal capsule. These axons give off fine collaterals at acute angles within the striatum. In another figure, there is also an example of a large short axon cell. It is interesting that this investigator does not report on the presence of dendritic spines although the material is most probably from neonates or even fetuses as mentioned in his method section, and this alone may explain the relative absence of spines. Dendritic spines were first noted by Ramón y Cajal (1895) in the corpus striatum of an 8-day old rabbit. This was a general feature of all striatal neurons which included giant, medium and small cells with short axons, and large cells with long axons. These findings were extended (Ramón y Cajal, 1911) to include a dwarf or neurogliform neuron, 6-10 μm in diameter, with varicose dendrites and a short axon. It is important to note that most of these results were obtained from immature brains, and there is reason to believe that during early postnatal development, most neurons show spinelike processes which are lost during maturation in some types of cells (DiFiglia, Pasik & Pasik, 1978c). The best examples of more mature spiny neurons given by Ramón y Cajal are indeed long axon cells, some of large size as in his Fig. 325C, and some of medium size as in his Fig. 325B. The initial portion of these axons usually makes a long detour during which a number of collaterals are given off before entering fiber bundles. This investigator calls attention to the possibility of misinterpreting these long axons as short axons because in most cases their long tortuous course makes it difficult to identify their point of entrance into the fascicles (Ramón y Cajal, 1895, p. 60). In addition to the account of cell types, this author also gives a description of two types of afferent fibers: (1) an ascending thick axon which divides several times resulting in a vast arborization contacting both short and long axon cells, and (2) fine collaterals of passing corticofugal fibers which terminate in complex arborizations coming in contact with long axon cells.

Additional information was given by Kölliker (1896) and Dejerine (1901). The former recognizes 3 types of striatal neurons, namely a stellate medium size spiny cell, and a fusiform large spiny cell, both with long axons; and for the first time, a large polygonal neuron with varicose dendrites and a short axon. Dejerine (1901) describes a large less spiny cell with a long axon, but apparently misinterprets some of the findings of Ramón y Cajal and of Kölliker, erroneously stating that these authors consider the majority of striatal neurons as being of the Golgi II type. Dejerine in fact illustrates a medium and a large spiny neuron with short axons which could very well exemplify the mistaken identification commented upon by Ramón y Cajal (see above). Using methylene blue and hydrogen peroxide, Turner (1903) succeeded in staining the human caudate nucleus and found that the great majority of neurons were of medium size with an occasional very large one. The medium cells had either smooth, short dendrites, or long dendrites covered with spines. Only the initial portion of the axon could be followed with this method.

A long hiatus then occurs, extending to the 1950's. This is only briefly interrupted by Bielschowsky's (1919) report of an overwhelming number of small cells in the striatum, which he considers of Golgi II type, together with scarce large cells of either triangular or polygonal shape. These latter neurons comprise elements of Golgi I type exhibiting few collaterals, and also of Golgi II type with widely arborizing short axons which are interpreted as the associative cells of the striatum. The fact that there is no mention of dendritic spines is most probably due to the general use of his own neurofibrillary technique instead of the classic Golgi impregnation. This study is followed shortly by the extensive account on the diseases of the striatal systems by Vogt and Vogt (1920). Although their own work is based on the Nissl method, already reviewed above, they utilize Ramón y Cajal's (1911) findings and particularly his Fig. 325 to describe the "synaptology

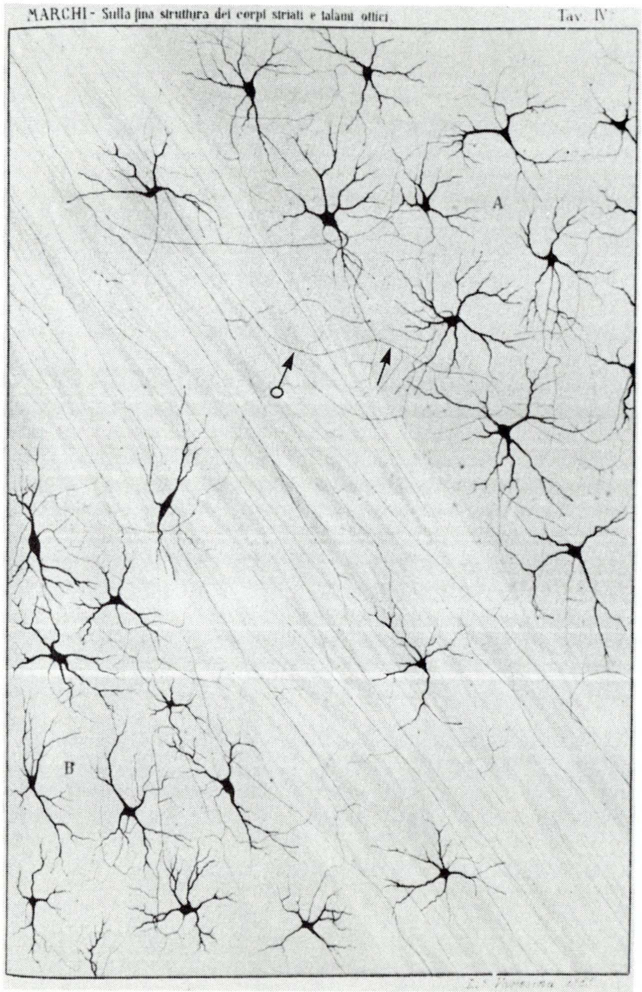

Fig. 2. Reproduction of Marchi's plate IV on corpus striatum (1886). Note the abundance of similar neurons both in the caudate nucleus (A) and in the putamen (B). The very fine processes represent the axons of these cells which are of the Golgi I type. The axons can be seen giving off collaterals (arrow) and then joining the internal capsule (ringed-arrow). In the original, the background is in beige, the cell bodies and dendrites in black, and the axons in red. Arrows have been added.

of striatal neurons." These authors essentially identify the four examples of short axon cells found in said figure as the numerous small neurons of the adult striatum. The two long axon cells represented in the figure are equated with the large neurons of the striatum in spite of the fact that one is a medium size cell. They speculate that the long axons of these "smaller" cells terminate within the striatum, and that those of the "larger" cells constitute the striopallidal fibers. The Vogts then ennunciate their theory of transmission mechanisms in the striatum as follows: the striopetal fibers, i.e. the afferents, terminate on the short axon cells and on the smaller long axon cells; these two groups of neurons then activate the larger long axon cells which finally transmit the impulses to the pallidum. This theory is still embraced by most students of the pathophysiology of extra-pyramidal diseases.

The first modern attempt to evaluate the neuronal types of the neostriatum is that of Leontovich. Her dissertation (1952) as a student of Polyakov, later published in full (1954), is based on material from hedgehog, rabbit, mouse, dog, monkey and man. Although the magnification used in the illustrations is too low to evaluate details, the descriptions of each cellular type are quite vivid. This investigator challenged the Vogts' theory on two basic findings: (1) many small neurons have long axons which join bundles going to the pallidum, (2) some large (giant) cells have short axons. She found that all long axon neurons have dendrites bearing spines and are the most numerous in the neostriatum. They comprise two varieties: one of small size with many richly branched, densely spiny dendrites, and an axon making complex loops and giving off collaterals before entering bundles; the other is of large size (giant) with fewer, sparsely branched, less spiny, thicker and straighter dendrites, and an axon which is goal oriented giving rise to many branched collaterals. The spines of these cells are thin, short and with a head at the end, and are not present on the soma or proximal dendrites (Leontovich, 1958). They all can be either stellate or fusiform in shape. The dendritic spines of these neurons, particularly the richly branched type, are frequently followed in their course by axons which establish contacts "en passant" with many spines of the same dendrite. The short axon cells are less numerous and their size varies from small to giant. The dendrites are smooth, with a rare spine, and relatively straight (stellate cells), or recurving around the body ("spiral" cells), or densely branching over a small territory (spidery cells). The axon of the small neurons arborizes near the cell body, and in the case of the giant elements of this kind, it covers a considerable volume of the nucleus.

In spite of these findings, the Vogts' theory still permeates the thinking of more recent investigations. An extensive study on the caudate nucleus of the cat (Kemp & Powell, 1971a) reports that the overwhelming majority of neurons is represented by a medium size cell with dendrites densely covered with spines except for their proximal 20 μm. The number of spines reaches 26 per 20 μm of dendrite length. Although the axon of these cells is described as a smooth process that loops on itself, gives off fine collaterals and finally terminates in a profuse arborization of beaded terminals within the dendritic field, the illustration provided (Kemp & Powell, 1971a, Fig. 1A) shows a rather poorly impregnated axon which does not allow a proper classification. These authors also describe 3 other types of much less frequent medium size cells: (1) Slightly larger neurons with larger, thicker dendrites exhibiting less spines which are distributed evenly along their course, and a long axon giving off varicose collaterals in its proximal portion. (2) Neurons with slender dendrites displaying distal varicosities and a few distal spines, with a short axon developing into a beaded arborization. (3) Neurons with complex branching of very varicose dendrites without spines, and with a short axon difficult to differentiate from the dendrites due to the beaded character of its collaterals. Finally very infrequent large and small cell types are described. The large neuron is fusiform with large straight dendrites extending for up to 1 mm, bearing few scattered spines, and a long axon giving off few collaterals with varicosities. The small cell has varicose dendrites which branch in a complex pattern and exhibit short twigs, and no identifiable axon. This study gives additional information on 3 types of possible afferent fibers: bundles of fine, wavy axons which do not branch; moderately fine axons dividing into fine beaded branches; and thicker axons branching less profusely. In an earlier publication, great importance was given to the axon collaterals of the medium size spiny neuron which were interpreted as a main component of the dense axonal plexus characteristic of the striatal neuropil (Kemp, 1968). These collaterals were observed approaching spines of other spiny neurons.

At about the same time, a series of articles by Fox and collaborators give a detailed account of their findings in the neostriatum of monkeys. They classify the neurons according to the presence or relative absence of dendritic spines, namely spiny and aspiny neurons respectively. The spiny neurons, present in great numbers, are equated with the achromatic cells of Nissl preparations (Fox *et al.*, 1971a). They are of various shapes, medium size, with spineless primary and initial secondary dendrites which are thereafter densely covered with sessile or pedunculated spines. Fox considers these spines as the most robust in the entire nervous system. The pioneering work of this author with the Golgi method gives substantial credence to this statement. The axon of the spiny neurons is interpreted as terminating near the soma after giving off collaterals. The illustrations provided (Fox *et al.*, 1971a, Figs. 5 & 23), however, are not convincing in this respect. In fact the original cautionary note of Ramón y Cajal (1895) applies to this finding just as it does to Dejerine's (1901) and Kemp and Powell's (1971a). The aspiny neurons are of 3 kinds (Fox, Andrade, Schwyn & Rafols, 1971/72b): (1) A large, elongated bulbous cell, representing the large chromatic cells shown by the Nissl method, with long thick, straight dendrites and a long myelinated axon. It is interesting to note that the example offered seems to exhibit some somatic and dendritic spines (Fox *et al.*, 1971/72b, Fig. 16). (2) An intermediate size, round cell with circling dendrites and no identified axon, designated as a spidery neuron of presumably short axon; (3) A small element with many dendrites corresponding to the dwarf or neurogliform cell of Ramón y Cajal which the authors designate as "aspiny neuron with somatic spines" (Fox, Lu Qui & Rafols, 1974).

TABLE 1 Neuronal Types in Monkey Neostriatum

Type	Size	Dendrites			Axon	
		Length	Spines	Varicosities	Length	Collaterals
Spiny I	medium	200	+++	0	long	+++
Spiny II	medium large	600	+	0	long	++
Aspiny I	medium	150	0	+++	short	arborization
Aspiny II	large	250	0	+++	short	arborization
Aspiny III	medium	150	rare	rare	short	arborization
Neurogliform	small	70	twigs	++	?	?

Length of dendrites in μm. Abundant, regular, sparse and absent indicated by +++, ++, + and 0 respectively.

TABLE 2 Differences Between Two Types of Spiny Neurons

	Type I	Type II
Mean maximal cross-sectional area	221 μm2 [52]	249 μm2 [104]
Dendritic field radius	200 μm	600 μm
Mean distance of first spine from soma	22 μm [3]	0
Mean overall spine density per 10 μm	15 [1]	5 [1.4]
Thin spines	62%	52%
Stubby spines	29%	46%
Mushroom-shaped spines	9%	2%

Figures in brackets indicate standard deviations. From Pasik *et al.*, 1976.

In the mid 1970's our laboratory started a program designed to unravel the synaptology of the monkey neostriatum. A thorough analysis of well impregnated Golgi-Kopsch series revealed at least 6 neuronal types (DiFiglia, Pasik & Pasik, 1976). Table 1 gives a summary of the results. It is apparent from these data that: (1) There are two classes of spiny neurons *(Spiny I and II)* and that both have long axons. (2) There are at least 3 types of true Golgi type II interneurons *(Aspiny I, II and III)*; (3) There are two distinct classes of large cells, one with spines and a long axon (large version of *Spiny II* and another without spines and a short axon *(Aspiny II)*. Table 2 depicts the differences between the two types of spiny neurons. The findings are self explanatory but attention is called to the large standard deviation of the maximal crossectional area of *Spiny II* neurons which indicates a broad variation in the size of this cell class ranging from 18 to 60 μm in its largest dimension. The long axon nature of spiny neurons is clearly seen in Fig. 3, which also illustrates the extensive fine collateral system originating mostly within 100 μm from the soma. Although the main axon has not been followed outside the neostriatum, it has been seen taking a parallel course to other similar axons particularly in sagittal sections (Fig. 4). The aspiny neurons, all of Golgi type II, are illustrated in Fig. 5. In addition to the characteristics listed in Table 1, it is worth mentioning that the *Aspiny I* neuron with varicose and recurving dendrites is the most frequently impregnated neostriatal cell in some of our sections (Fig. 6). The *Aspiny II* correlates well with the typical large chromatic cell of Nissl preparations (Fig. 1), and as shown recently (Pasik, Pasik & DiFiglia, 1977), it is beyond doubt a short axon cell. This finding, therefore, supports the previous controversial observations of Kölliker (1896) and of Leontovich (1954), and has been further confirmed by Rafols (1978). The *Aspiny III* neuron has rather straight, smooth dendrites occasionally exhibiting long thin processes and bulbous appendages, and a very dense axonal arborization of fine beaded branches. We have not taken issue with the *Neurogliform* neuron beyond its small size and dendritic characteristics since no axons have been found impregnated in our series.

In a most recent study (DiFiglia, Pasik & Pasik, 1978b), four types of afferent fibers are recognized in both the caudate nucleus and the putamen. They are depicted in Fig. 7. The most frequently impregnated axons are thick (1.5 μm) and have been followed for long distances within the structure; they branch at acute angles and the distal segments exhibit varicosities (Fig. 7a). They have been seen entering the putamen from the external capsule. These fibers match well with

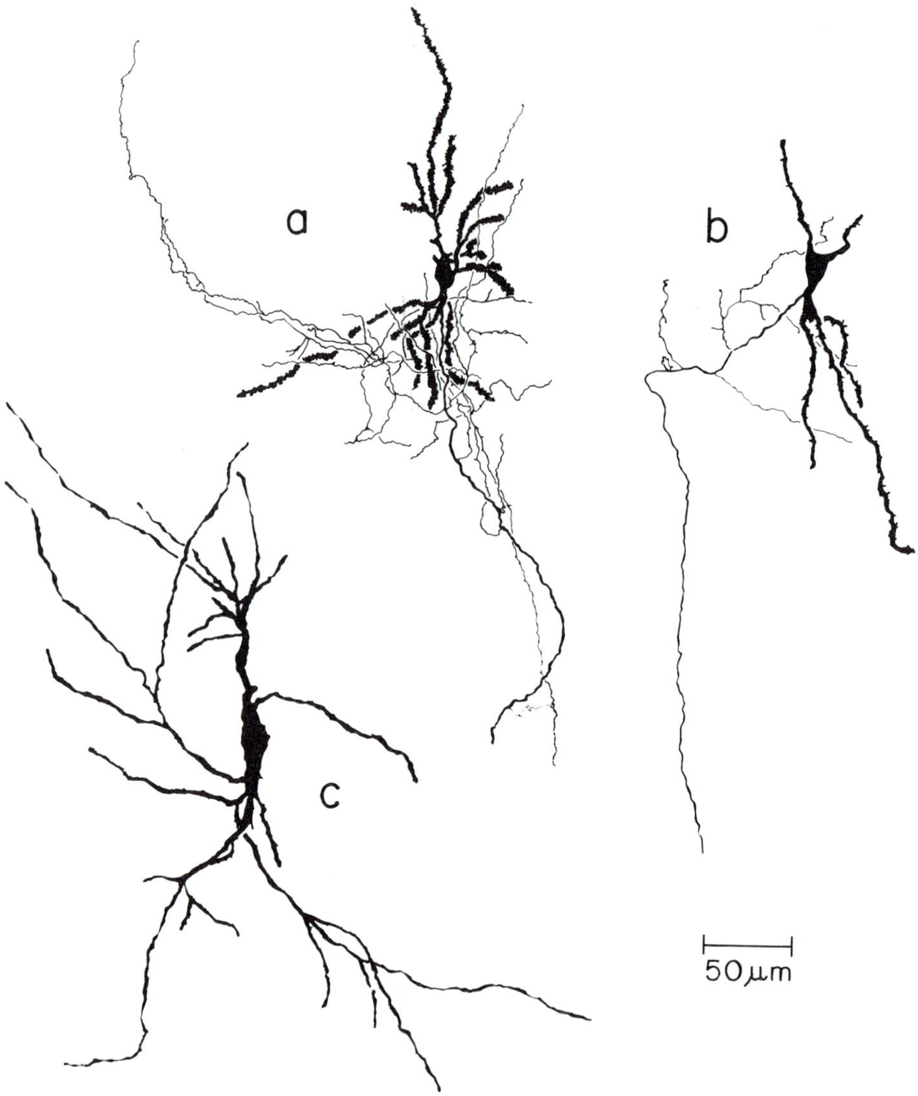

Fig. 3. Spiny Neurons of Monkey Neostriatum. Golgi-Kopsch method. Camera lucida. a, *Spiny I* neuron with densely spiny dendrites, and long axon with abundant collateral branches. b, *Spiny II* neuron with sparsely spiny dendrites and a long axon with some collaterals. c, Large version of *Spiny II* neuron with spines on the somatic and dendritic surfaces; the axon is not impregnated. Redrawn from Pasik et al., 1976, 1977.

the "ascending fiber" of Ramón y Cajal (1911) and may originate in the diencephalon or brainstem. Another type of thick axon (Fig. 7b) branches sparsely into thin processes and all along its course gives rise to grape-like appendages, some of which may reach over 4 µm in size. These are the largest boutons found in the striatum and as such they may represent the terminals of thalamic origin (Hassler, 1975). A third type is a thin fiber (0.8 µm) which gives rise to collaterals and terminal branches, both arborizing profusely into very delicate beaded processes (Fig. 7c). This pattern is similar to the fine mesh of beaded catecholaminergic terminals revealed by fluorescence microscopy in the neostriatum and, therefore, may represent these particular afferents. Finally, the thinnest axons (0.6 µm) take a long straight course within the structure giving off numerous side twigs with occasional terminal bulbs (Fig. 7d). They have been observed entering from the external capsule. They resemble very closely those identified in other regions of the central nervous system as fibers of cortical origin (Guillery, 1967). Golgi impregnations have also been used to study the mode of termination of afferents from the cerebral cortex and thalamus after lesions of these structures. A statistical analysis of a reduction in the number of spines in these cases led to the conclusion that the two inputs terminate on different spines of the same spiny neuron particularly in a region 40-120 µm from the soma (Kemp & Powell, 1971d).

The preceding review allows the comparison of the results obtained in the various

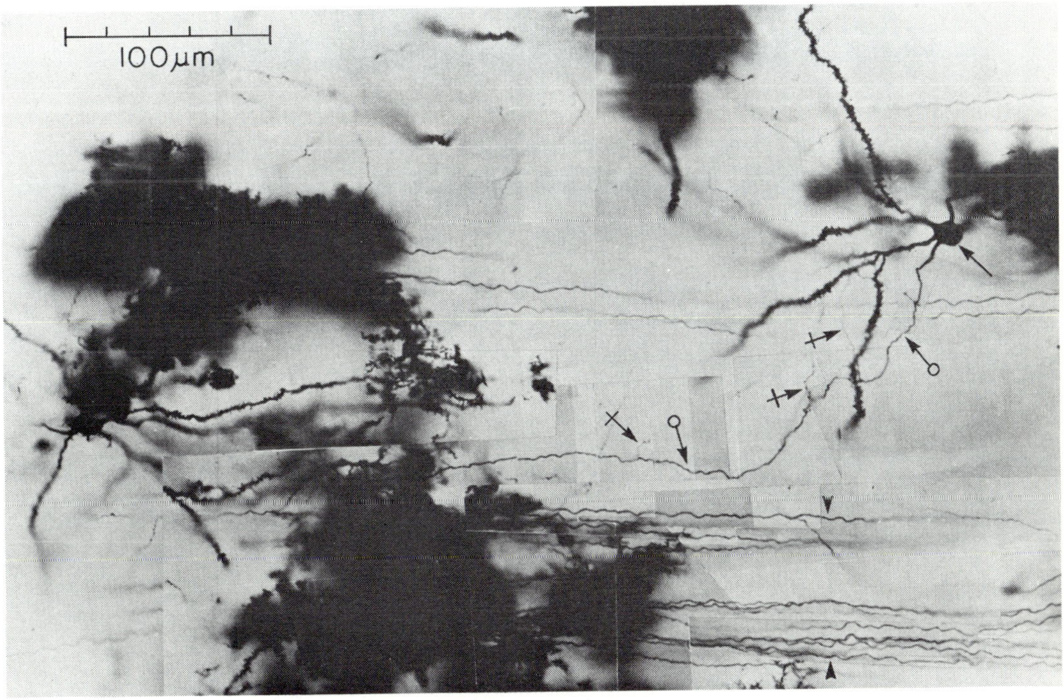

Fig. 4. *Spiny I* Neuron in Monkey Putamen. Golgi-Kopsch method. Photomontage. Sagittal section. A characteristic *Spiny I* cell (arrow) with long axon (ringed-arrow) joining a bundle of similar axons (between arrowheads) coursing in the direction of the pallidum. Some of the axon collaterals are also visible (crossed arrows).

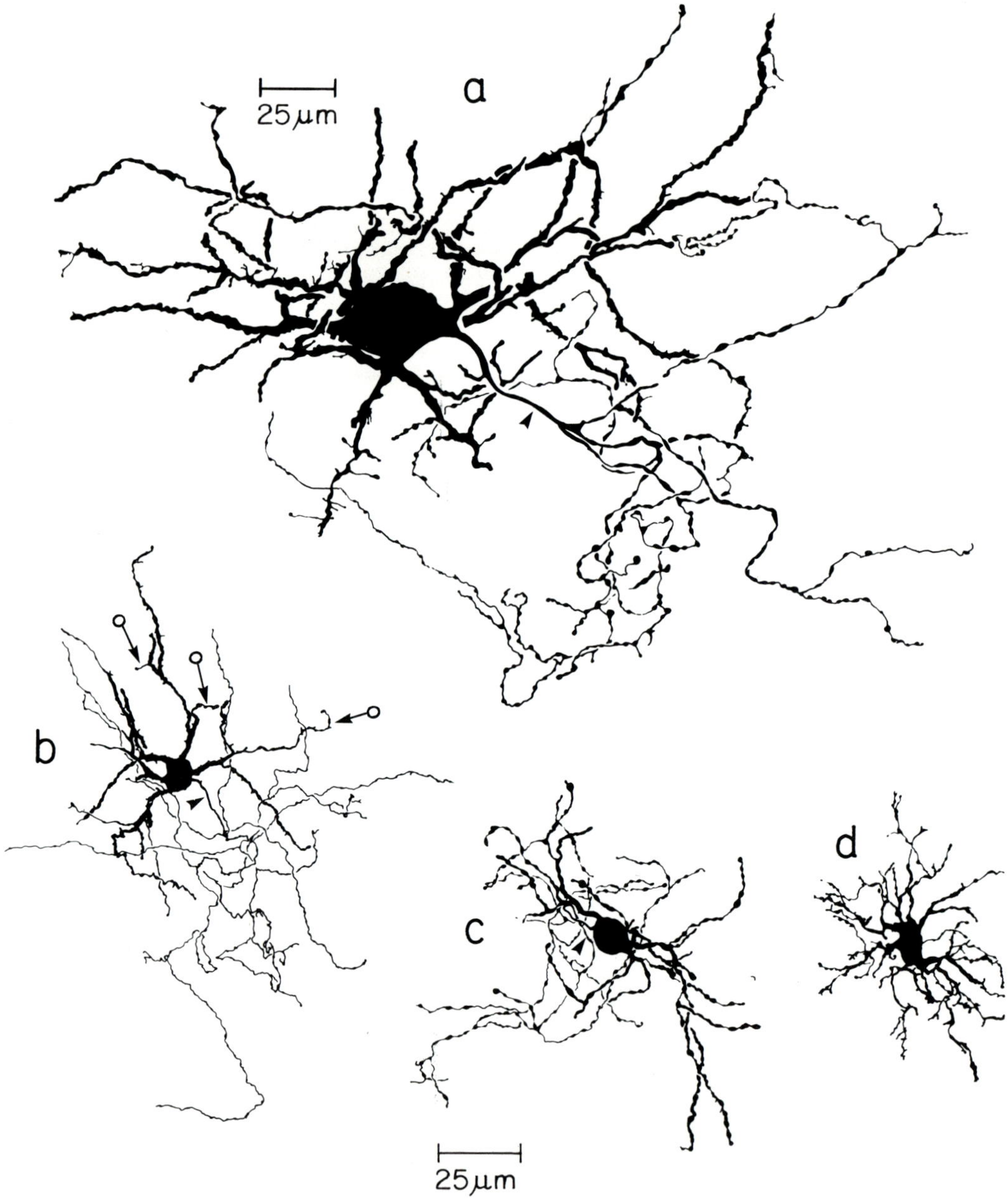

Fig. 5. Aspiny Neurons of Monkey Neostriatum. Golgi-Kopsch method. Camera lucida. a, *Aspiny II* neuron. b, *Aspiny III* neuron. Note the spine-like processes and dendritic appendages (ringed arrows). c, *Aspiny I* neuron. d, *Neurogliform* neuron. Short axons indicated by arrowheads. The *Aspiny II* cell is at a slightly smaller magnification. Redrawn from Pasik *et al.*, 1976, 1977.

Golgi studies, and this is attempted in Table 3. In establishing these correlations, however, we have followed certain principles of utmost importance in the interpretation of data derived from this kind of material. A fundamental characteristic of a Golgi impregnation is that it outlines only 5-10% of the neurons present in a given structure. This limitation, which is basic for a successful preparation, restricts considerably its value regarding the quantitative evaluation of the relative proportion of various neuronal types. Such a limitation is even more strict since it is possible that certain classes of cells are more susceptible to impregnation than others (Fox *et al.*, 1971a). It is, therefore, of little value to state that 95% of neostriatal neurons are of the spiny type (Kemp & Powell, 1971a). Another relevant issue is that positive findings with this technique supersede any negative result. This aspect of the method applies not only to particular cells but also to parts of cells. For instance, the discovery of a proven axon in a previously considered amacrine cell, such as the interneuron in the lateral geniculate nucleus, immediately changes the cellular category to a true axonal one (Hámori, Pasik & Pasik, 1978). Similarly, the rule should apply to the problem of whether a particular neuronal class belongs to the Golgi type I or type II. The observation of a long axon overules the designation of short axon cell based on a previous observation. It is always possible to interpret as a short axon the proximal segment of a long axon emitting many collaterals when the more distal portions fail to impregnate properly or do not impregnate at all. This may be the explanation for the controversial issue of whether the spiny cell in the neostriatum is of Golgi type I or II. Finally, the degree of maturity of the material is of paramount importance. Most of the earlier studies were made with the rapid Golgi method which requires the use of very young, neonate or even

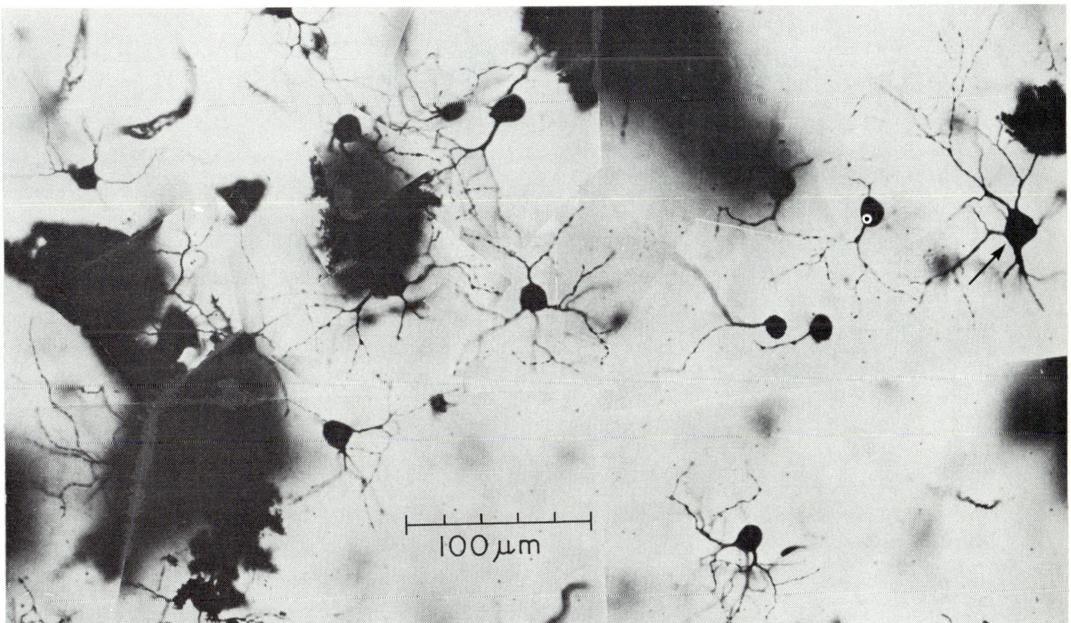

Fig. 6. Aspiny Neurons. Golgi-Kopsch method. Photomontage. Note the almost exclusive impregnation of *Aspiny I* neurons. There is also an *Aspiny II* cell (arrow).

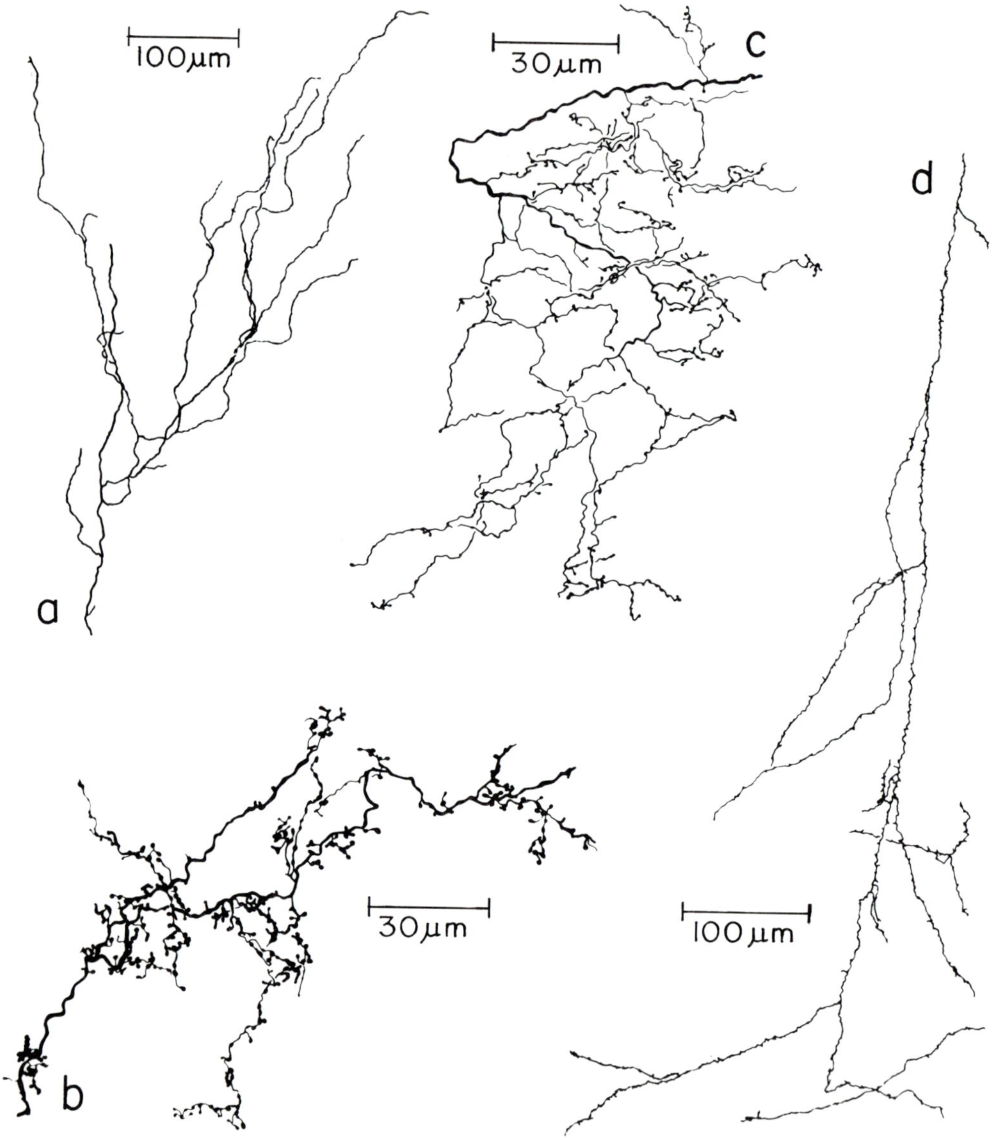

Fig. 7. Afferent Axons in Monkey Neostriatum. Golgi-Kopsch method. Camera lucida. Note the different scales used. a, Thick axon with varicose branches. b, Thick axon with large, grape-like appendages. c, Axon with thin beaded branches. d, Thin axon with side twigs. Redrawn from DiFiglia *et al.*, 1978b.

TABLE 3 Correlation of Neuronal Types

Our Findings	Ramón y Cajal (1911)	Kölliker (1896)	Dejerine (1901)	Leontovich (1954)	Kemp & Powell (1971a)	Fox et al. (A, B, C)*	Lu & Brown (1977)
monkey	human, mouse rat, rabbit, dog	human, cat	rabbit	human, monkey, dog, hedgehog rabbit, mouse	cat	monkey	rat
Spiny I long axon	long axon Figs. 325B 327A	long axon Fig. 713	short axon Fig. 279	rich dendrites long axon, Fig. 1	medium spiny short axon Fig. 1A	spiny short axon (A) Figs. 5 23	medium, long axon Fig. 7c, medium spidery Fig. 7a
Spiny II long axon					medium, long axon, Fig. 1B		medium spindle Fig. 7b
Spiny II (large)	long axon Fig. 325C	long axon Fig. 712	long axon Fig. 279	poor dendrites long axon Fig. 3 b_1	large, long axon, Fig. 4A	large aspiny long axon (B)Figs. 3, 16	large Fig. 7f
Aspiny I short axon	short axon Figs. 325A 326C			"spiral" short axon Fig. 3 B_2	medium, varic. dendrites short axon, Fig. 2B	spidery aspiny (B) Figs. 5, 6	medium, varic. dendrites, Fig. 7d
Aspiny II short axon		short axon Fig. 714		giant, short axon, Fig. 2			
Aspiny III short axon	short axon Fig. 325D			stellate, short axon Fig. 4 B_1	medium, smooth dendr., short axon, Fig. 2A	aspiny with somatic spines (C) Fig. 1A	medium, long slender dendr. Fig. 7e
Neurogliform	neurogliform Fig. 325E			spidery Fig. 4 B_3	small Fig. 4B	Oligodendroc. (C) Fig. 1G	small Fig. 7g

* References of Fox et al.: A, 1971/72a; B, 1971/72b; C, 1974.
Figure numbers in table are as found in the original references.

fetal material (Marchi, 1886; Ramón y Cajal, 1895, 1911). It is a conspicuous characteristic of the early postnatal development of the neostriatum that most neurons show spines or spinelike processes in the dendrites and even the surface of their somata. This is true even for the short axon cells which will eventually become aspiny neurons (DiFiglia *et al.*, 1978c). The loss of spines during development has been documented in spinal cord neurons in very early application of the Golgi technique (Geier, 1903). This fact must be taken into consideration in interpreting the figures offered by Ramón y Cajal (1911) and his very succinct description of the entire corpus striatum which only takes 15 pages of his *magnum opus*.

Table 3 gives the correlations between the neuronal types found in our laboratory with those of the other investigators reviewed above and one other recent study in the rat (Lu & Brown, 1977). Our *Spiny I* neuron can be found in every study. It has been assigned a short axon by Dejerine (1901), Kemp and Powell (1971a) and Fox *et al.*, (1971a). More recently, however, the latter group has recognized its long axon character (Fox & Rafols, personal communication). The *Spiny II* medium size cell has been recognized by Kemp and Powell (1971a) and in part also by Lu and Brown (1977). The large version of the *Spiny II* is present with similar characteristics in every study. The *Aspiny I* and *Aspiny III* interneurons are also found in most investigations except for those of Kölliker (1896) and Dejerine (1901). The *Aspiny II*, the large interneuron, has been recognized in the past only by Kölliker (1896) and Leontovich (1954), and presently also by Rafols (1978). The *Neurogliform* neuron (Ramón y Cajal, 1911, Fig. 325E) has been observed by all except Kölliker (1896) and Dejerine (1901), and Fox and Rafols (1971/72) consider it a form of oligodendrocyte. In addition to all of these cellular types, a large, spiny, short axon neuron has been observed only by Ramón y Cajal (1895, 1911) and by Dejerine (1901).

RESULTS OBTAINED WITH TRACER METHODS AND INTRACELLULAR INJECTIONS

The use of techniques based on retrograde axonal transport of various macromolecules have helped in identifying at least some neostriatal efferent cells. In the first such study, horseradish peroxidase (HRP) was injected into the substantia nigra of cats and numerous medium size cells were found labelled in most of the striatum but particularly concentrated in the head of the caudate nucleus (Grofová, 1975). No loaded giant cells were noted. Figure 8 is taken from a current study by this author (Grofová, 1978) where she describes two types of neurons marked with HRP: a medium size cell when the injection of the enzyme is restricted to the substantia nigra, and a larger elongated cell of up to 45 μm when HRP is deposited into the retrorubral nucleus. According to our findings with the Golgi method (see above) they may represent the *Spiny I* and large version of *Spiny II* neurons respectively. These results were confirmed in similar experiments on rats (Bunney & Aghajanian, 1976; Kanasawa, Marshall & Kelly, 1976). In the former study, medium size labelled neurons were present in high proportion (30-50%) in the entire ipsilateral caudate nucleus except for the medial core which as reviewed earlier contains most of the large cells in another rodent (Mensah, 1977). More recently, the findings were replicated in the cat and extended to include HRP injection into the pallidum (Szabo, 1977, 1978). This author postulates 3 types of efferent neostriatal cells: the most numerous are medium size with scant cytoplasm and HRP granules forming only a ring around the nucleus; fewer in number are other medium size cells, multipolar or elongated with heavier label; only occasionally seen is a larger cell (30-50 μm) with heavy label.

A new method for marking neurons by retrograde axonal transport has been introduced by Bak, Markham, Cook and Stevens (1977). It consists in injecting herpes simplex virus (HSV) which apparently is taken up by axon terminals, thus infecting the parent cell. Injections in the neostriatum of rats result in the appearance of subviral particles within the nucleus of large cells (about 30 μm) which comprise only 2% of the neuronal population. Although this finding is of interest in terms of marking by local injection large striatal cells which, as discussed above, are in fact interneurons *(Aspiny II)*, it is puzzling that medium size cells are not

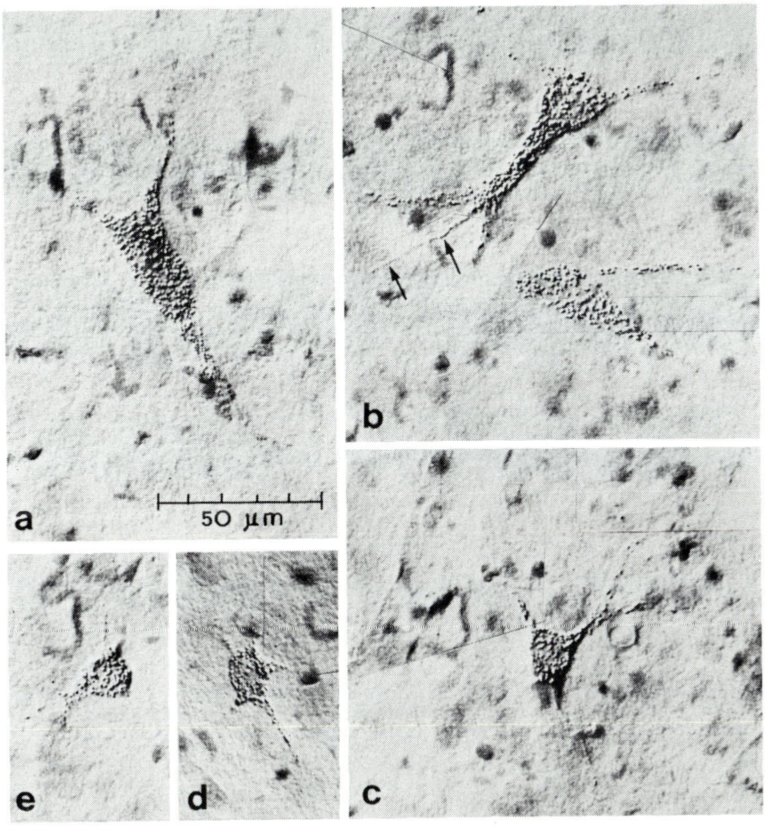

Fig. 8. Neurons of Cat Neostriatum. Interference light microscopy. a, b, Large striatal cells labelled after HRP injection in the retrorubral nucleus. c, d, e, Medium size labelled cells after HRP injection in the substantia nigra, pars reticulata. Arrows point to a presumed axon. From Grofová, 1978.

labeled with the virus. It should be recalled that some of them are of short axon as well *(Aspiny I and III)* and that the numerous long axon *Spiny I* neurons have intrinsic collaterals terminating within the striatum. In a more recent experiment HSV injections into the substantia nigra resulted in a high proportion (14-47%) of infected medium size cells and of only few large cells in the neostriatum of the rat (Bak *et al.*, 1978), thus confirming the results of the previous studies with HRP (Grofová, 1975, 1978; Szabo, 1977, 1978).

HRP has also been used to identify neurons from which intracellular recordings have been obtained. The enzyme is ejected through the same recording pipette. Thus, it has been shown in the cat that caudate neurons receiving a convergence of excitatory inputs from the substantia nigra, cerebral cortex and centromedianus-parafascicular complex are identical to our *Spiny I* neuron in their somatic and dendritic characteristics (Kitai, Kocsis, Preston & Sugimori, 1976). Moreover, the axon of such a cell was followed for up to 2·2 mm without evidence of a terminal arborization (Kocsis, Sugimori & Kitai, 1977a). Apparently, neostriatal neurons of similar morphology have been activated antidromically by stimulation of the substantia nigra (Kocsis, Van der Maelen & Kitai, 1977b).

The results of all of these studies coincide in showing that many medium size spiny neurons are efferent cells of the neostriatum, and probably represent the *Spiny I* and *II* neurons of our classification. The larger size output cells appear to be the large version of the *Spiny II* neuron. The essential agreement between the findings just reviewed and those derived from Golgi material (Leontovich, 1954; DiFiglia *et al.*, 1976) makes untenable the Vogts' (1920) theory of transmission mechanisms in the striatum.

RESULTS OBTAINED WITH ELECTRON MICROSCOPY

Normal Ultrastructure

Electron microscopy has contributed much to the understanding of the internal organization of the neostriatum. There have been a number of investigations on the normal ultrastructure at the descriptive level. The first thorough examination was made in the rat (Mori, 1966). In this study it is shown that there are: (1) small cells (10-15 μm) either with scant or with abundant cytoplasm, both containing small, round mitochondria; and (2) two distinct classes of large cells which as a whole are few in number: one is fusiform or triangular, with a nucleus without indentations and many microtubules, and the other is polygonal with an eccentric nucleus showing marked indentations, more abundant cytoplasm richer in all organelles except for microtubules. This author correlates the two classes of large cells with the Golgi type I and type II respectively. He also emphasizes the prevalence of axospinous contacts and the frequent occurrence of synapses between an axon terminal and two or more dendritic elements. Most of these findings were confirmed in the cat neostriatum (Adinolfi & Pappas, 1968; Adinolfi, 1971; Bak, Choi, Hassler, Usunoff & Wagner, 1975). The axon terminals which are present in the striatal neuropil have been described in the rat (Kawana, Akert & Bruppacher, 1971), the cat (Kemp & Powell, 1971b; Chung, Hassler & Wagner, 1977; Hassler, Chung, Rinne & Wagner, 1978, and the monkey (Rafols and Fox, 1971/72; Pasik, Pasik & DiFiglia, 1976), using a variety of criteria which has resulted in a rather confusing picture. There are, however, several similar findings which are noteworthy. The most frequent profile appears to contain small vesicles and makes an asymmetric contact predominantly with spines which usually exhibit an evagination, a so-called spinule, into the axonal bag (asymmetrical round vesicles of Kemp & Powell; small round vesicles of Pasik *et al.*; types I, III and IV of Chung *et al.*). Among this type of terminals there is one which is unusually large (Kemp & Powell, 1971b; type IV of Chung *et al.*, 1977, and Hassler *et al.*, 1978). Kemp and Powell (1971b), however, very properly avoid the segregation of this element in their quantitative estimates because small endings may be part of large ones. This caution has been recently substantiated by the full reconstruction of such terminals in thin serial sections (DiFiglia, Pasik & Pasik, unpublished observations). Symmetric synapses on dendritic trunks, somata and axon initial segments are made by terminals containing pleomorphic or large round vesicles (symmetrical large vesicles of Kemp & Powell; large vesicles of Rafols & Fox; pleomorphic and large vesicles of Pasik *et al.*; Types V and IX of Chung *et al.*, & Hassler *et al.*). In addition, a small number of boutons containing flat vesicles and making symmetric contacts is generally observed. An important finding has been the frequent occurrence of axon terminals containing pleomorphic or large round vesicles making synapses on the initial segment of other axons (Kemp & Powell, 1971b; Rafols and Fox 1971/72; Pasik *et al.*, 1976). Figure 9 illustrates an example taken from serial sections with at least 4 types of these contacts present. The occurrence of dendrodendritic synapses is rare. In the cat, the "presynaptic" dendrites have been seen to contain either large pleomorphic or flat vescicles, and to contact a spiny or a varicose dendrite (Kemp & Powell, 1971b). Examples of this type of arrangement have also been given from the neostriatum of the rat (Bak *et al.*,1975) and the monkey (Pasik *et al.*, 1976). It is possible that these presynaptic dendritic profiles correspond to the fine processes and appendages coming off dendrites of *Aspiny III* cells (Fig. 5).

The apparent lack of correlation among other findings of the various studies may be due to the use of different criteria, some of which are open to question. For

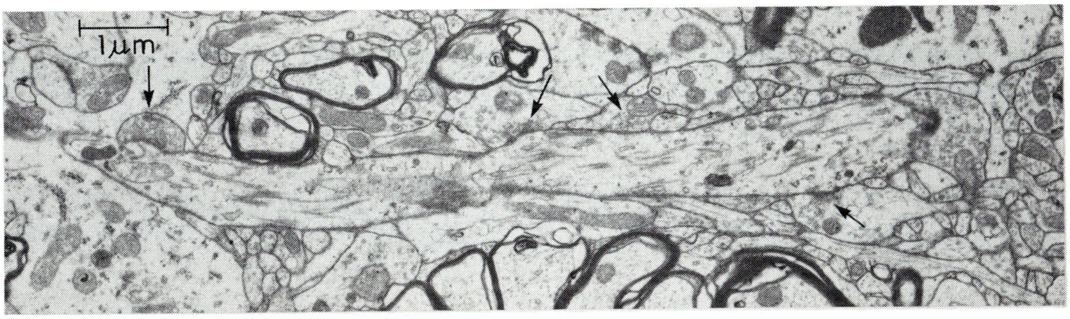

Fig. 9. Axon Initial Segment in Monkey Neostriatum. Note the characteristic undercoating of the membrane and the fasciculation of microtubules. This initial segment is postsynaptic to at least four axonal profiles (arrows).

instance, to establish a separate category of terminal on the basis of its large size can be confounded by the possibility that some of the smaller profiles are in fact smaller sections of the same type (Kemp & Powell, 1971b). Similarly, the segregation of terminals according to the postsynaptic element (Chung *et al.'s* type VII on dendritic trunks and somata, as opposed to types I, III and IV on spines) is not warranted in view of the very common occurrence of synapses en passant made by the same axon on both spines and dendrites or somata (Kemp & Powell, 1971b; Pasik *et al.*, 1976). The classification of synaptic vesicles involves also a large subjective component, particularly when dealing with the various sizes and degrees of pleomorphism. Very few studies indeed have approached this problem with painstaking objective measurements (Kawana *et al.*, 1971; Rafols & Fox, 1971/72) and have resulted in only 3 or 4 categories. Another debatable issue is the distribution of axosomatic synapses. The relative paucity of these contacts is generally recognized, but the differential occurrence on medium size and large cells is controversial. Some have noted their almost total absence on medium size neurons and their abundance on large cells (Fox *et al.*, 1971/72a; Chung *et al.*, 1977; Hassler *et al.*, 1978), whereas others have reported a considerable number on small cells (Rafols & Fox 1971/72; Pasik *et al.*, 1976), and their scarcity on large ones (Pasik *et al.*, 1976).

Few attempts have been made to determine quantitative relationships among the various components of the neostriatal neuropil. In fact, these studies have been limited to the axon terminal fraction. A statistical analysis in the monkey

Pasik *et al.*, 1976) revealed that axospinous synapses represent 82% of the total, and that 92% of those are asymmetric. Contrariwise, from among the axodendritic synapses which are only 18% of the total, 88% are symmetric. All comparisons give significant mean differences. The majority of axospinous synapses include terminals with small round vesicles (69%) and with pleomorphic vesicles (24%). The majority of axodendritic synapses comprise axons with pleomorphic vesicles (34%) and with large round vesicles (37%). The relative distribution of various axon terminals is also given for the cat (Chung *et al.*, 1977; Hassler *et al.*, 1978). Although sampling methods and statistics are not given in the latter studies, the data are interesting in suggesting possible differences between caudate and putamen.

The preceding review of normal ultrastructure points to the need for more sophisticated investigations which will enhance the value of the purely morphologic descriptions. They must be geared to three dimensional reconstruction of serial sections as well as to more refined stereologic quantitations. Hopefully, these types of studies will be forthcoming.

Correlation with Light Microscopy

The usefulness of ultrastructural data for the understanding of internal organization is considerably increased by their correspondence with light microscopy information particularly that derived from Golgi material. Only few studies have attempted such a correlation in the cat (Kemp & Powell, 1971a) and the monkey (Fox *et al.*, 1971/72 a, b); Pasik *et al.*, 1976). It should be emphasized that the evidence provided for the comparisons between light and electron microscopic data in all of these studies is mostly circumstantial and the conclusions drawn are purely inferential. With respect to neuronal types in the monkey neostriatum, we have offered a correlation which implicitly contains all of these reservations (Pasik *et al.*, 1976). For instance, a *Spiny I* neuron is recognized on the basis of similar mitochondria present in dendrites exhibiting spines; a *Spiny II* can be identified through the observation of a spine at the base of an emerging dendrite; an *Aspiny I* because of an emerging dendrite with strictures and dilations; an *Aspiny II* due to its large, globular soma with eccentric lobulated nucleus; an *Aspiny III* on account of smooth dendrites followed for over 40 μm from the cell body; even a *Neurogliform* cell because of its minute size. Some of these observations had been made earlier in rhesus and squirrel monkeys (Fox *et al.*, 1971/72 a, b). Other investigators have used the presence of an indentation in the nucleus of the cell to recognize neuronal types (Kemp & Powell, 1971a). Similar limitations are valid in the case of dendritic profiles found in the neuropil unless a high density of spines is seen in a relatively long stretch of longitudinally cut dendrite. Again on an inferential basis, we proposed (Pasik *et al.*, 1976) that cross sections of dendrites containing: (1) numerous mitochondria and few synapses on them belong to spiny dendrites *(Spiny* neuron); (2) few mitochondria and many synaptic contacts are from varicose dendrites *(Aspiny I and II* neurons); (3) few mitochondria, many microtubules and few synapses are from smooth surface dendrites *(Aspiny III* neuron). All of these criteria which have been used so far in every investigation are at the most, good educated guesses. Fortunately, a new technique has become available making possible a positive identification at the electron microscopy level of properly recognized elements at the light microscopy level. The procedure allows the examination of cytological detail of Golgi impregnated material after gold toning and subsequent deimpregnation (Fairén, Peters & Saldanha, 1977). A modification of this method is being extensively used in our laboratory, and recently the unequivocal recognition of *Spiny I* and *Aspiny I* neurons in the monkey neostriatum was accomplished (DiFiglia, Pasik & Pasik, 1978a). Figure 10a is a photomicrograph of a gold-toned neuron typical of the *Spiny I* category embedded in a thin sheet of Epon. The nucleus of the cell and also the nucleolus have become visible. Ultrathin sections of the material containing only this gold-toned Golgi impregnated element

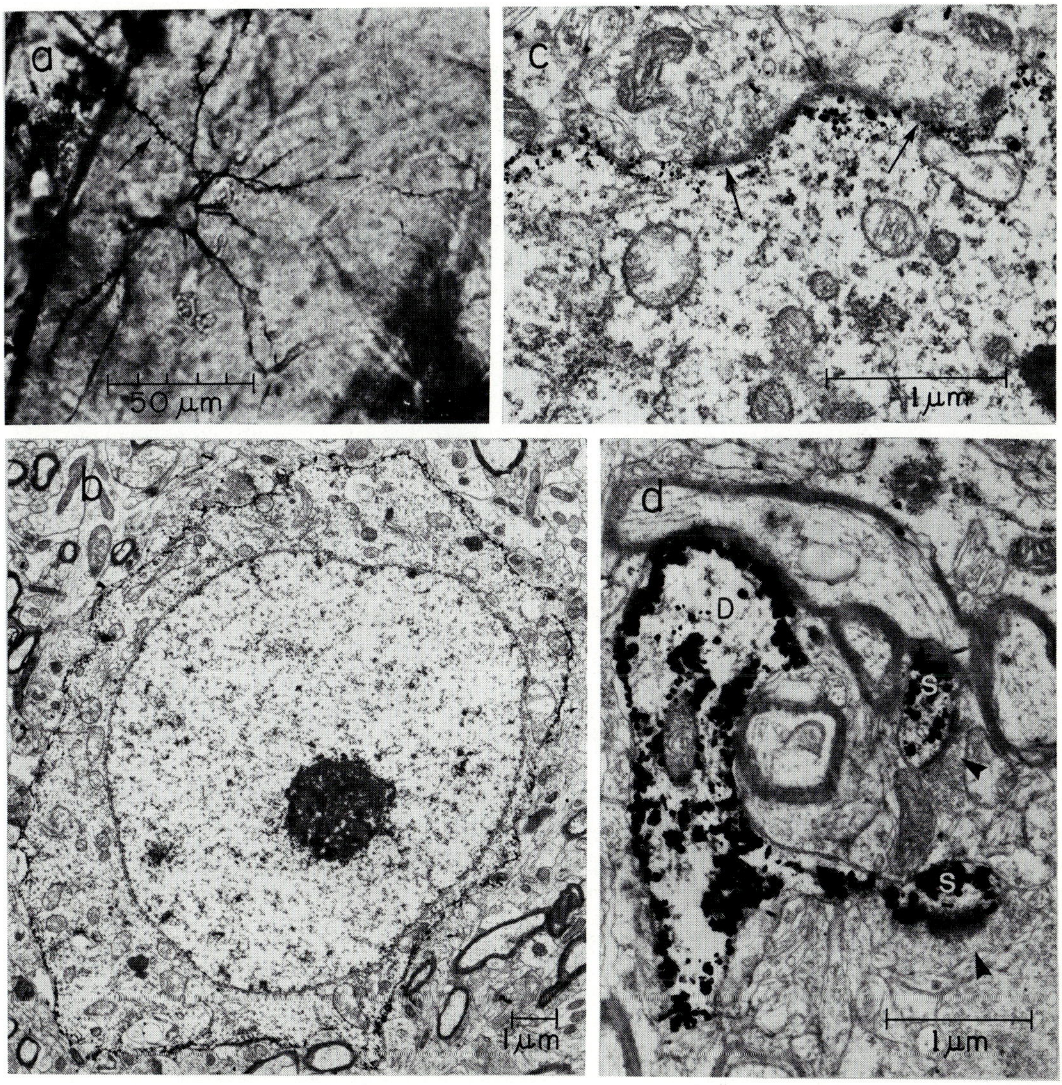

Fig. 10. Gold-toned *Spiny I* Neuron. a, Photomicrograph showing neuron after gold-toning and deimpregnation; the nucleus and nucleolus have become visible; note the presence of spines in characteristic location (arrow). b, Electron micrograph of the soma of the same cell sharply outlined by the deposit of electron dense gold particles beneath the membrane; note the paucity of organelles. c, A portion of the cell, 10 sections away, showing two axosomatic synapses (arrows); note again the submembranous granules. d, A dendrite of the same neuron (D) with two spines (S), one in continuity with the shaft, receiving axospinous synapses (arrowheads).

are illustrated in the electron micrographs of Fig. 10 b, c, d. Profiles belonging to this cell are identified by a discrete deposit of electron dense gold particles under the cell membrane and a heavier accumulation in the dendrites and the axon. It is apparent that the soma has the characteristics inferred in previous studies (Pasik *et al.*, 1976), i.e. scanty cytoplasm, poor in organelles with no stacks of rough endoplasmic reticulum. Serial sections of this neuron shows without question the occurrence of several axosomatic synapses (Fig. 10c). The latter finding clears the controversy discussed earlier: the presence of axosomatic synapses is in fact characteristic of *Spiny I* cells. Serial sections also show the axospinous complexes belonging to this neuron (Fig. 10d). Numerous synapses are also present on the axon initial segment. On the other hand, the *Aspiny I* neuron is of similar size but it is much richer in organelles particularly small stacks of rough endoplasmic reticulum. Axosomatic synapses are very rare which explains the confusion in the literature where medium size neurons showing no such contacts have been misinterpreted as spiny neurons on the basis of their size alone (Fox *et al.*, 1971/72 a; Chung *et al.*, 1977; Hassler *et al.*, 1978). An additional piece of information in the analysis of these serial sections has been the observation of axonal endings of at least two types (small round or pleomorphic vesicles) synapsing on a spine probably belonging to a spiny dendrite and on a dendrite of an *Aspiny I* neuron. It is clear that this method bridges the gap between light and electron microscopy and will contribute greatly to unraveling the synaptology of the neostriatum.

Nature of Axonal Elements

A last aspect of ultrastructural investigations deals with the identification of the nature of the axonal profiles which are present in the neostriatal neuropil. These studies essentially provide a marker for specific elements and have utilized a variety of techniques with various degrees of success. There have been 3 basic approaches: (1) to damage the source of the axons and examine for boutons in the process of degeneration, (2) to follow the antegrade axonal transport of tritiated amino acids injected at the source of the axons; (3) the immunocytochemical recognition of various neurotransmitters located in specific axon terminals. The terminal degeneration method is useful to determine the type and location of synapses made by axon terminals according to their origin. It is less valuable for correctly identifying axon profiles present in normal material (see above) because of the marked morphologic changes imposed by the degeneration process. Moreover, although cortical lesions can be rather restricted, electrolysis of deep structures (thalamus, midbrain) is more difficult to control with the implicit possibility of unintended damage to structures other than the selected target, and to fibers passing through it. Some of these drawbacks have been circumvented by the use of neurotoxic agents which specifically destroy certain classes of cells, as for instance 6-OH-Dopamine for catecholaminergic neurons (Hökfelt & Ungerstedt, 1973). Extensive ablations of cerebral cortex in the cat (Kemp & Powell, 1971c; Hassler *et al.*, 1978)have resulted in the degeneration of axon terminals which make asymmetric synapses mainly with dendritic spines, less frequently with the shaft of dendrites of various types, and very rarely with main stem dendrites or somata. These profiles can also be seen contacting elements "en passant." They are probably the small boutons with round vesicles seen in the normal structure. We have observed similar degeneration in both the ipsilateral and contralateral neostriatum of the rhesus monkey after cerebral hemidecortication (Fig. 11). Thalamic lesions were placed in the cat to either interrupt the major efferents from the intralaminar nuclei (Kemp & Powell, 1971c) or destroy the centromedianus-parafascicular complex (Chung *et al.*, 1977). The axonal endings in the process of degeneration found in the striatum resemble those seen after cortical ablations, although the synapses on dendritic shafts are less frequent. They have been interpreted either as the same type of profile (Kemp & Powell, 1971c) or the equivalent of the large terminals seen in the normal (Chung,*et al.*,

1977). The illustrations provided in the latter study, however, show profiles of about 1 μm and it is difficult to decide about the original size of such elements in the normal state. Lesions have also been placed in the midbrain involving both the substantia nigra and the tegmentum (Kemp & Powell, 1971c). The location of degenerating axon profiles is similar to that resulting from cortical or thalamic destruction except that they are not seen on main stem dendrites or somata. Moreover, they are much sparser than the other two. These findings were essentially confirmed in cat and rat, where the synaptic junction on spines was found

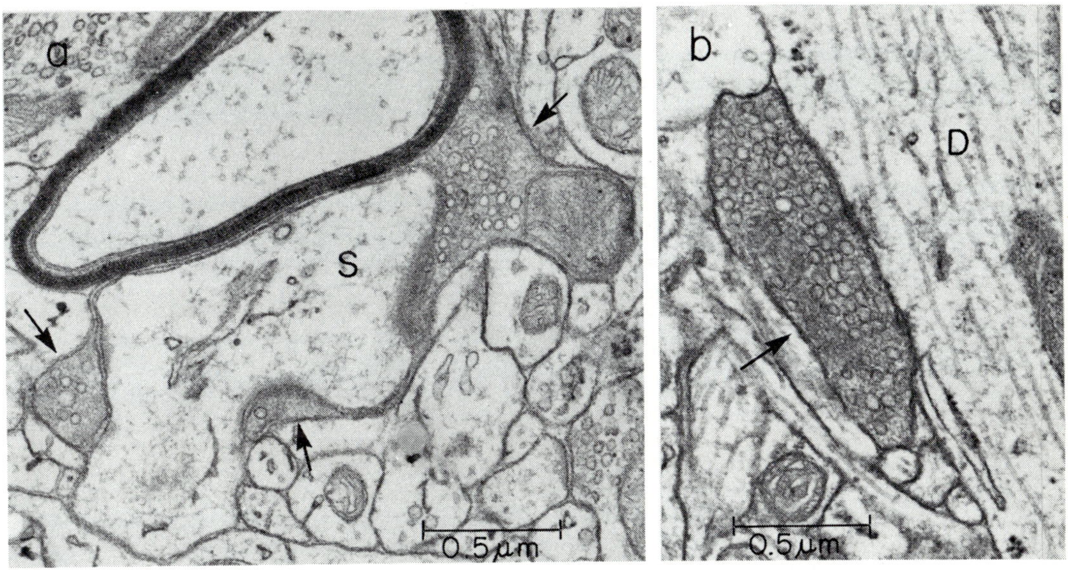

Fig. 11. Degeneration of Axon Terminals in Monkey Neostriatum after Contralateral Cerebral Hemidecortication. a, Degenerated terminal (arrows) making asymmetric synapses with a spine (S); 6 days survival. b, Degenerated terminal (arrow) making an asymmetric synapse with a dendritic trunk (D) containing many microtubules, ribosomes and a cistern; 4 days survival.

to be interrupted by a spinule (Bak *et al.*, 1975; Hassler, Bak, Usunoff & Choi, 1975). Axon terminals with degenerative changes and the same location were observed after intraventricular injection of 6-OH-Dopamine in the rat (Hattori, Fibiger, McGeer & Maler, 1973a). Another neurotoxic agent which can be used in this type of study is kainic acid. This compound when injected into the neostriatum destroys nearly all neurons with preservation of afferent axons (Schwarcz & Coyle, 1977). This approach, however, has not been followed at the electron microscopic level to further characterize such axon terminals in the absence of interneuron axons and collaterals of long axon cells. The only ultrastructural examination of rat neostriatum subjected to this treatment revealed the frequent retention of postsynaptic membranes attached to presumably afferent axons for at least three weeks postinjection (Olney & de Gubareff, 1978). Finally, the lesion technique has been used to investigate neostriatal intrinsic terminals, i.e. those derived from short axon cells and from collaterals of long axon cells. The examination of neostriatal tissue within 1 mm of a small lesion in the head of the caudate nucleus of cat (Kemp & Powell, 1971c) reveals a large number of degenerated terminals making asymmetric contacts. Perhaps more importantly, boutons making symmetric synapses with somata, main stem dendrites, shafts of dendrites, dendritic spines and the initial segment of other axons, are also affected. The latter terminals appear only within 450 μm from the lesion. Using a different approach, a surgically isolated portion of the caudate nucleus has been studied for remaining viable terminals. The results are somewhat discrepant. In one study, both axospinous asymmetric synapses with interrupted contacts and round vesicles are observed together with axodendritic asymmetric and symmetric junctions, and round or ovoid vesicles (Tennyson & Marco, 1973). In the other, only endings with large round or pleomorphic vesicles are found synapsing on dendritic shafts (Hassler, Chung, Wagner & Rinne, 1977).

Autoradiographic studies are few and limited to the injection of labelled amino acids in the substantia nigra and midbrain tegmentum of the rat (Hattori *et al.*, 1973a; McGeer, Hattori & McGeer, 1975). Tritiated leucine, and also norepinephrine,are effectively transported and mark axon terminals in the striatum containing slightly pleomorphic vesicles and establishing asymmetric synapses with spines. These boutons degenerate after treatment with 6-OH-Dopamine and are interpreted, therefore, as terminals of the nigrostriatal dopaminergic pathway. The isotopes also label a smaller number of another type of ending with highly pleomorphic vesicles and in symmetric contact with dendrites, suggesting the existence of a separate input to the striatum from the same nigrotegmental region. Studies utilizing specific uptake of putative neurotransmitters and their ultrastructural localization by autoradiography, have added knowledge to the relative distribution and the changes produced by various pharmacologic or lesion methods. Since the techniques call for incubation periods before fixation, the quality of the material does not usually permit the observation of precise synaptic arrangements and, therefore, has been of little help in understanding the intimate organization of the striatum. Thus, examination of rat neostriatum after incubation with tritiated GABA shows the uptake of this transmitter by profiles of various sizes, containing dark mitochondria and round vesicles, but no synapses can be recognized (Hattori, *et al.*, 1973b). On purely biochemical grounds these authors suggest that the labelled terminals may belong to collaterals of some efferent cells and/or to a special type of short axon neuron.

Finally, immunohistochemical methods have succeeded in localizing synthetizing enzymes for neurotransmitters in certain neostriatal elements and since the techniques can be applied after adequate fixation, the quality of the material allows the localization of the labelled elements within the neuronal circuitry. Choline-acetyltransferase can be visualized in guinea pig neostriatum within small and medium size neurons, dendrites, and small axonal boutons containing round vesicles and making axospinous asymmetric synapses (Hattori, Singh, McGeer & McGeer, 1976). Although the investigators state that spines are also labelled, thereby implying that there are cholinergic spiny neurons, the illustrations provided are not convincing in this respect. It is noteworthy that many medium size neurons are not labelled, and that axon profiles, degenerated by way of 6-OH-Dopamine treatment, contact both labelled and non-labelled dendrites. These findings suggest that dopaminergic terminals synapse on both cholinergic and non-cholinergic medium size cells, and that in turn the axons of cholinergic Golgi type II neurons establish synaptic contacts with the spines of spiny neurons. Some of these spiny neurons apparently are gabaergic since glutamic acid decarboxylase can be localized by similar immunocytochemical methods in their somata, dendrites and spines (Ribak, 1978).

SYNAPTOLOGY OF THE NEOSTRIATUM

An attempt can now be made to synthesize the morphologic information reviewed above into a working hypothesis on the possible neuronal networks which operate in the neostriatum. Figure 12 is a diagramatic representation of such circuitry which has been developed with additional data borrowed from electrophysiology and the biochemistry of neurotransmitters. The known afferents are represented at the top with their general morphologic characteristics and their transmitters if known or suspected. The classic inputs from the cerebral cortex, substantia nigra and thalamus with possible transmitters for the first two being glutamic acid (Spencer, 1976; Kim, Hassler, Haug and Paik, 1977) and dopamine (Andén, Carlsson, Dahlstrom, Fuxe, Hillarp & Larsson,1964) respectively, have been shown to be predominantly

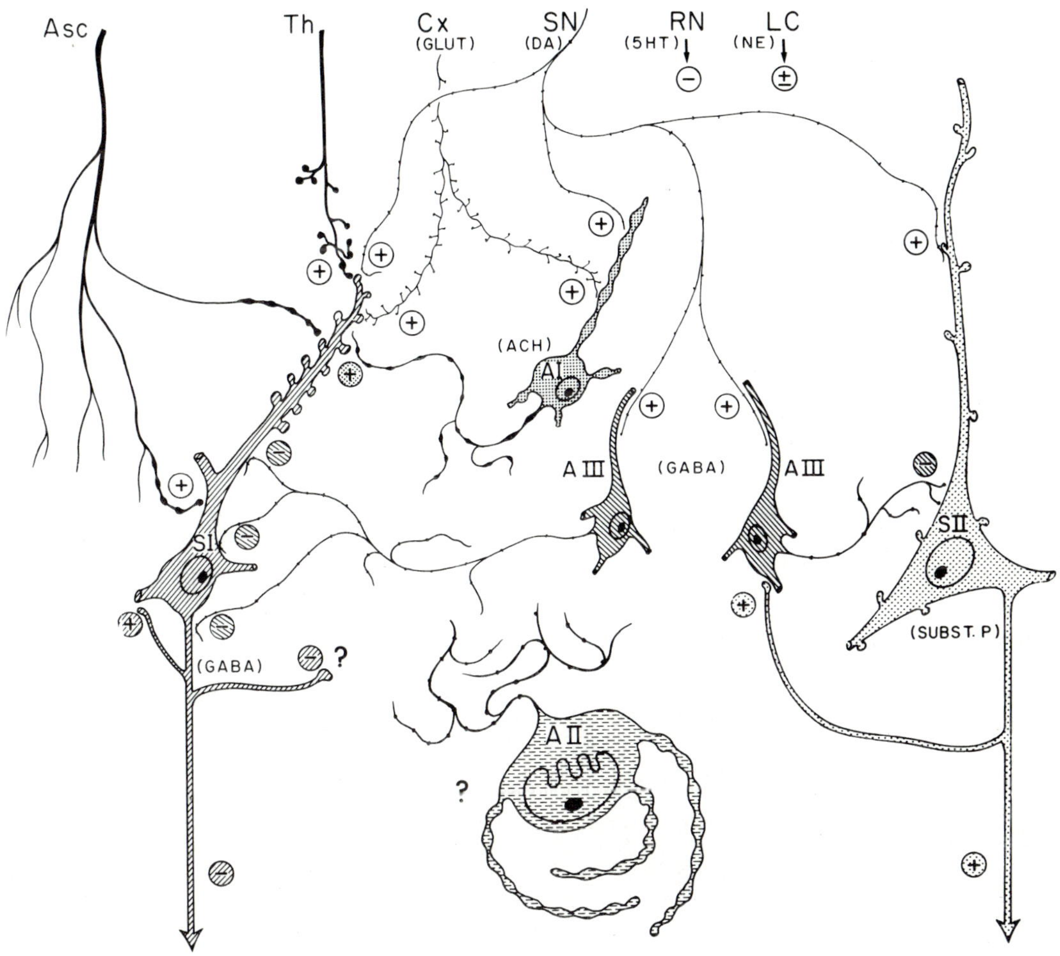

Fig. 12. Tentative Neuronal Circuits in the Neostriatum. Afferents from: brain stem or diencephalon (Asc), thalamus (Th),cerebral cortex (Cx), substantia nigra (SN), raphe nuclei (RN), locus ceruleus (LC). Interneurons: *Aspiny I* (A I), *Aspiny II* (A II), *Aspiny III* (A III). Output neurons: *Spiny I* (S I), *Spiny II* (S II). Excitatory and inhibitory synapses indicated by + and - respectively. Neurotransmitter designated in parenthesis near the element which contains it. Full explanation in the text.

excitatory in nature (Rocha-Miranda, 1965; Purpura & Malliani, 1967; Frigyesi & Purpura, 1967; Feltz & Albé-Fessard, 1972; Buchwald, Price, Vernon & Hull, 1973; Buchwald, Hull & Levine, 1975; Kitai, Sugimori and Kocsis, 1976). It has been recently demonstrated that these three inputs converge on spiny neurons (Kocsis *et al.*, 1977a), and as the preceding review has shown at least for those of nigral origin, they make synapses also with interneurons. It is possible that short latency excitatory responses are related to direct input to output cells *(Spiny I and II)*, long latency excitatory responses to the input to excitatory interneurons *(Aspiny I)* in turn exciting output cells, and EPSP-IPSP sequences to input to inhibitory interneurons *(Aspiny III)* in turn inhibiting output cells. It may be that both types of spiny neurons share these inputs (shown only partially in the diagram). Less known inputs are represented by the thick ascending afferent axons which could be responsible for the additional excitatory influence on proximal portions of spiny dendrites from stimulation in the region of the substantia nigra (Kocsis & Kitai, 1977). These afferents may originate in the brain stem and course through the substantia nigra or neighboring tegmental areas (Nauta & Kuypers, 1957; Hedreen & Chalmers, 1972; Robertson & Travers, 1975). Finally, the input from the raphe nuclei and its putative neurotransmitter serotonin, exerting an inhibitory effect on striatal neurons (Miller, Richardson, Fibiger & McLennan, 1975) has been demonstrated. Less clear are possible projections from the locus ceruleus providing noradrenergic terminals to the striatal neuropil (Couch & Goldstein, 1977), since apparently the small norepinephrine content of the neostriatum can represent adrenergic fibers of passage (Fonnum, this volume).

The interneurons of the neostriatum are of at least three types. One is cholinergic (McGeer, McGeer, Fibiger & Wickson, 1971) and probably corresponds to the *Aspiny I* cell since its frequent occurrence matches well with the high neostriatal content of acetylcholine. Another is gabaergic (McGeer & McGeer, 1975) and it may be represented by the *Aspiny III* cell. Both of these interneurons are contacted by afferent fibers and exert their excitatory and inhibitory effects respectively on spiny cells. Excitation by the *Aspiny I* neuron would be through asymmetric synapses on spines (Hattori, *et al.*, 1976). Inhibition by the *Aspiny III* neuron would occur at symmetric synapses on main stem dendrites, somata and axon initial segments (Figs. 9, 10c, and Kemp & Powell, 1971c). The latter contact on initial segments is a powerful mechanism to shut off the cell, whatever the magnitude of excitatory input since it feeds inhibition at the site of generation of the nerve impulse. A third type of interneuron is the giant *Aspiny II* cell; we know nothing of its connectivity. It is hoped that some answers will become available through the electron microscopic examination of properly isolated cells in gold-toned Golgi impregnated material. In the meantime, the only other giant interneuron that comes to mind for comparison is the Golgi cell of the cerebellar cortex (Golgi, 1883). This neuron is gabaergic (Wilkin, Wilson, Balász, Schon & Kelly, 1974) and has a potent inhibitory influence on granule cells (Eccles, Ito and Szentágothai, 1967). Perhaps the *Aspiny II* neuron of the neostriatum has a similar role on excitatory interneurons.

The direct contact of afferents with neostriatal output neurons, as opposed to linking through an interneuron as the Vogts' theory mandated, was first suggested on physiological grounds (Fuller, Hull & Buchwald, 1975) and recently confirmed by intracellular injection of HRP (Kocsis *et al.*, 1977 a, b). The candidates for output cells according to the preceding review are the spiny neurons. The fact that there are two categories of spiny neurons agrees with the electrophysiological data since it has been repeatedly shown that there are at least two different types of neostriatal outputs (Malliani & Purpura, 1967; Levine, Hull & Buchwald, 1974; Frigyesi & Purpura, 1967; Frigyesi & Szabo, 1975; Féger & Ohye, 1975). The most frequently encountered is essentially inhibitory and may be mediated by the *Spiny I* cell. GABA is the neurotransmitter responsible for the inhibitory action

(Kim, Bak, Hassler & Ikada, 1971; Yoshida & Precht, 1971; Kataoka, Bak, Hassler, Kim & Wagner, 1974; Fonnum, Gottesfeld and Grofová, 1978). It is possible, therefore, that the collaterals of the *Spiny I* neuron contribute to the intrinsic GABA pool of the neostriatum. However, no information is available on which elements are contacted by these collaterals. Similar branches of Purkinje cells in the cerebellar cortex are known to contact other Purkinje cells as well as interneurons (Eccles, Ito & Szentágothai, 1967). It should be noted that in our Golgi material some recurrent collaterals of *Spiny I* neuron axons appear to contact spiny dendrites of the same parent cell resembling the "autapses" seen in other regions of the central nervous system (van der Loos, 1972). Finally, the excitatory output is probably mediated by *Spiny II* neurons and there is evidence that substance P may be the transmitter involved (Hong, Jang, Racagni & Costa, 1977; Kanasawa, Emson & Cuello, 1977; Gale, Hong & Guidotti, 1977). Physiological data suggest that collaterals of excitatory neostriatal output neurons contact inhibitory interneurons which in turn inhibit the original output cell in a Renshaw cell type of arrangement (Richardson, Miller & McLennan, 1977).

It is important to realize that these proposed neuronal circuits for the neostriatum are incomplete by their very hypothetical nature. A notable absence is the recent discovery of relatively high concentrations of opioid peptides in the neostriatum (Hughes, 1975; Yang, Hong & Costa, 1977). Although there is some evidence of the existence of receptors for these peptides, both on cholinergic interneurons and on dopaminergic terminals (Vizi, Hársing & Knoll, 1977), the localization of the peptides themselves remains unknown.

ACKNOWLEDGMENTS

This work was supported in part by Grant # NS-11631 from the National Institute of Neurological and Communicative Disorders and Stroke, U.S.P.H.S. The authors are indebted to Marilyn Ilvento, Lilia Pasik, Victor Rodriguez and Minerva Travieso for their skillful assistance. Special thanks are due to Dr. Irena Grofová for supplying the photomicrograph reproduced in Fig. 8.

REFERENCES

Adinolfi A. M., The organization of synaptic junctions in cat putamen, Brain Res. 32, 53-67 (1971).

Adinolfi A. M. & G. D. Pappas, The fine structure of the caudate nucleus of the cat, J. comp. Neurol. 133, 167-184 (1968).

Andén N. -E., A. Carlsson, A. Dahlström, K. Fuxe, N. -A. Hillarp & K. Larsson, Demonstration and mapping out of nigro-neostriatal dopamine neurons, Life Sci. 3, 523-530 (1964).

Bak I. J., W.B. Choi, R. Hassler, K. G. Usunoff & A. Wagner, Fine structural synaptic organization of the corpus striatum and substantia nigra in rat and cat, In D. B. Calne, T. N. Chase & A. Barbeau (eds.) (1975) Dopaminergic Mechanisms, Adv. Neurol., Vol. 9, pp. 25-41, Raven Press, New York.

Bak I. J., C. H. Markham, M. L. Cook & J. G. Stevens, Intraaxonal transport of herpes simplex virus in the rat central nervous system, Brain Res. 136, 415-429 (1977).

Bak I. J., C. H. Markham, M. L. Cook & J. G. Stevens, Ultrastructural and immunoperoxidase study of striatonigral neurons by means of retrograde axonal transport of herpes simplex virus, Brain Res. 143, 361-368 (1978).

Bielschowsky M., Einige Bemerkungen zur normalen und pathologischen Histologie des Schweif- und Linsenkerns, J. Psychol. Neurol. (Leipzig) 25, 1-11 (1919).

Bøttcher J., Morphology of the basal ganglia in Parkinson's disease, Acta Neurol. Scand. 52 (Suppl. 62), 1-87 (1975)

Brockhaus H., Zur feineren Anatomie des Septum und des Striatum, J. Psychol. Neurol. (Leipzig) 51, 1-56 (1942).

Buchwald N. A., C. D. Hull & M. S. Levine, Neurophysiological and anatomical interrelationships of basal ganglia, In N. A. Buchwald and M. A. B. Brazier (eds.) (1975) Brain Mechanisms in Mental Retardation, pp. 187-203, Academic Press, New York.

Buchwald N.A., D. C. Price, L. Vernon & C. D. Hull, Caudate intracellular response to thalamic and cortical inputs, Exp. Neurol. 38, 311-323 (1973).

Bunney B.S. & G. K. Aghajanian, The precise localization of nigral afferents in the rat as determined by a retrograde tracing technique, Brain Res. 117, 423-435 (1976).

Carman J.B., The volume and neurone content of the striatum in man, Proc. Univ. Otago Med. Sch. 44, 49-51 (1966).

Chung J.W., R. Hassler & A. Wagner, Degeneration of two of nine types of synapses in the putamen after center median coagulation in the cat, Exp. Brain Res. 28, 345-361 (1977).

Couch J. R. & G. D. Goldstein, Projections of locus coeruleus and raphe nuclei to neostriatum, Neurosc. Abstr. 3, 36 (1977).

Dejerine J. (1901) Anatomie des Centres Nerveux, Vol. 2, pp. 305-344, Rueff, Paris.

DiFiglia M., P. Pasik & T. Pasik, A Golgi study of neuronal types in the neostriatum of monkeys, Brain Res. 114, 245-256 (1976).

DiFiglia M., P. Pasik & T. Pasik, Electron microscopy of Golgi impregnated spiny and aspiny neurons in monkey neostriatum, Neurosc. Abstr., 4, in press (1978a).

DiFiglia M., T. Pasik & P. Pasik, A Golgi study of afferent fibers in the neostriatum of monkeys, Brain Res. 153, in press (1978b).

DiFiglia M., T. Pasik & P. Pasik, Developmental aspects of neostriatal organization in monkeys, J. Appl. Neurophysiol., in press (1978c).

Dom R., F. Baro & J. M. Brucher, A cytometric study of the putamen in different types of Huntington's chorea, In A. Barbeau, T. N. Chase & G. W. Paulson (eds.) (1973) Huntington's Chorea, 1872-1972, Adv. Neurol., Vol. 1, pp. 369-385, Raven Press, New York.

Eccles J.C., M. Ito & J. Szentágothai (1967) The Cerebellum as a Neuronal Machine, Springer-Verlag, New York.

Fairén A., A. Peters & J. Saldanha, A new procedure for examining Golgi impregnated neurons by light and electron microscopy, J. Neurocytol. 6, 311-337 (1977).

Féger J. & C. Ohye, The unitary activity of the substantia nigra following stimulation of the striatum in the awake monkey, Brain Res. 89, 155-159 (1975).

Feltz P. & D. Albé-Fessard, The study of an ascending nigro-caudate pathway, Electroencephalogr. Clin. Neurophysiol. 33, 179-193 (1972).

Foix C. & J. Nicolesco (1925) Les Noyaux Gris Centraux et la Région Mésencéphalo-sous-optique, pp. 296-298, Masson, Paris.

Fonnum F. (this volume).

Fonnum F., Z. Gottesfeld & I. Grofová, Distribition of glutamate decarboxylase, choline acetyltransferase and aromatic amino acid decarboxylase in the basal ganglia of normal and operated rats. Evidence for striatopallidal, striato-entopeduncular and striatonigral gabaergic fibres, Brain Res. 143, 125-138 (1978).

Fox C.A., A. N. Andrade, D. E. Hillman & R. C. Schwyn, The spiny neurons in the primate striatum: A Golgi and electron microscopic study, J. Hirnforsch. 13, 181-201 (1971/72a).

Fox C.A., A. N. Andrade, R. C. Schwyn & J. A. Rafols, The aspiny neurons and the glia in the primate striatum: A Golgi and electron microscopic study, J. Hirnforsch. 13, 341-362 (1971/72b).

Fox C.A., I. J. Lu Qui & J. A. Rafols, Further observations on Ramón y Cajal's "dwarf" or "neurogliaform" neurons and the oligodendroglia in the primate striatum. J. Hirnforsch. 15, 517-527 (1974).

Fox C. A. & J. A. Rafols, Observations on the oligodendroglia in the primate striatum. Are they Ramón y Cajal's "dwarf" or "neurogliaform" neurons?, J. Hirnforsch. 13, 331-340 (1971/72).

Frigyesi T. L. & D. P. Purpura, Electrophysiological analysis of reciprocal caudato-nigral relations, Brain Res. 6, 440-456 (1967).

Frigyesi T. L. & J. Szabo, Caudate evoked synaptic activities in nigral neurons, Exp. Neurol. 49, 123-139 (1975).

Fuller D. R. G., C. D. Hull & N. A. Buchwald, Intracellular responses of caudate output neurons to orthodromic stimulation, Brain Res. 96, 337-341 (1975).

Gale K., J. -S. Hong & A. Guidotti, Presence of substance P and GABA in separate striatonigral neurons, Brain Res. 136, 371-375 (1977).

Geier T., Sur la forme et le développement des prolongements protoplasmatiques des cellules spinales chez les vertébrés superieurs, Névraxe 4, 233-249 (1903).

Golgi C., Sulla fina anatomia degli organi centrali del sistema nervoso. IV. Sulla fina anatomia delle circonvoluzioni cerebellari, Riv. sper. Freniat. 9, 1-17 (1883).

Grofová I., The identification of striatal and pallidal neurons projecting to substantia nigra. An experimental study by means of retrograde axonal transport of horseradish peroxidase, Brain Res. 91, 286-291 (1975).

Grofová I., Types of striato-nigral neurons labeled by retrograde transport of horseradish peroxidase, J. Appl. Neurophysiol. in press (1978).

Guillery R. W., Patterns of fiber degeneration in the dorsal lateral geniculate nucleus of the cat following lesions in the visual cortex, J. comp. Neurol. 130, 197-222 (1967).

Gurewitsch M., Cytoarchitektonische gliederung des Neostriatum der Säugetiere, Z. Anat. Enwickl. Gesch. 93, 723-742 (1930).

Hámori J., T. Pasik & P. Pasik, Electron microscopic identification of axonal initial segments belonging to interneurons in the dorsal lateral geniculate nucleus of the monkey, Neuroscience 3, 403-412 (1978).

Harman P. J. & M. B. Carpenter, Volumetric comparisons of the basal ganglia of various primates including man, J. comp. Neurol. 93, 125-137 (1950).

Hassler R., A cholinergic centro-thalamic input to the strionigral and strio-pallidal systems, Proc. 10th Int. Congr. Anat., Tokyo, p. 136 (1975).

Hassler, R., I. J. Bak, K. G. Usunoff & W. B. Choi, Synaptic organization of the descending and ascending connections between the striatum and the substantia nigra in the cat, In J. R. Boissier, H. Hippius and P. Pichot (eds.) (1975) Neuropsychopharmacology, pp. 397-411, Excerpta Medica, Amsterdam.

Hassler, R., J. W. Chung, U. Rinne & A. Wagner, Selective degeneration of two out of the nine types of synapses in cat caudate after cortical lesions. Exp. Brain Res. 31, 67-80 (1978).

Hassler, R., J. W. Chung, A. Wagner & U. Rinne, Experimental demonstration of intrinsic synapses in cat's caudate nucleus, Neurosci. Letters 5, 117-121 (1977).

Hattori T., H. C. Fibiger, P. L. McGeer & L. Maler, Analysis of the fine structure of the dopaminergic nigrostriatal projection by electron microscopic autoradiography, Exp. Neurol. 41, 599-611 (1973a).

Hattori T., P. L. McGeer, H. C. Fibiger & E. G. McGeer, On the source of GABA-containing terminals in the substantia nigra. Electron microscopic autoradiographic and biochemical studies, Brain Res. 54, 103-114 (1973b).

Hattori T., V. K. Singh, E. G. McGeer & P. L. McGeer, Immunohistochemical localization of choline acetyltransferase containing neostriatal neurons and their relationship with dopaminergic synapses, Brain Res. 102, 164-173 (1976).

Hedreen J.C. & J. P. Chalmers, Neuronal degeneration in rat brain induced by 6-hydroxydopamine: a histological and biochemical study, Brain Res. 47, 1-36 (1972).

Hökfelt T. & U. Ungerstedt, Specificity of 6-hydroxydopamine induced degeneration of central monoamine neurones: an electron and fluorescence microscopic study with special reference to intracerebral injection on the nigro-striatal dopamine system, Brain Res. 60, 269-297 (1973).

Hong J.S., H. -Y. T. Yang, G. Racagni & E. Costa, Projections of substance P containing neurons from neostriatum to substantia nigra, Brain Res. 122, 541-544 (1977).

Hughes I., Isolation of an endogenous compound from the brain with pharmacological properties similar to morphine, Brain Res. 88, 295-308 (1975).

Kanazawa I., P. C. Emson & A. C. Cuello, Evidence for the existence of substance P-containing fibres in striato-nigral and pallido-nigral pathways in rat brain, Brain Res. 119, 447-453 (1977).

Kanazawa I., G. R. Marshall & J. S. Kelly, Afferents to the rat substantia nigra studied with horseradish peroxidase, with special reference to fibres from the subthalamic nucleus, Brain Res. 115, 485-491 (1976).

Kataoka K., I. J. Bak, R. Hassler, J. S. Kim & A. Wagner, L-glutamate decarboxylase and cholinacetyltransferase activity in the substantia nigra and the striatum after surgical interruption of the strio-nigral fibers of the baboon, Exp. Brain Res. 19, 217-227 (1974).

Kawana E., K. Akert & H. Bruppacher, Enlargement of synaptic vesicles as an early sign of terminal degeneration in the rat caudate nucleus, J. comp. Neurol. 142, 297-308 (1971).

Kemp J. M., An electron microscopic study of the termination of afferent fibres in the caudate nucleus, Brain Res. 11, 464-467 (1968).

Kemp J. M. & T. P. S. Powell, The structure of the caudate nucleus of the cat: light and electron microscopy, Phil. Trans. R. Soc. Lond. B. 262, 383-401, (1971a).

Kemp J. M. & T. P. S. Powell, The synaptic organization of the caudate nucleus, Phil. Trans. R. Soc. Lond. B. 262, 403-412 (1971b).

Kemp J. M. & T. P. S. Powell, The site of termination of afferent fibres in the caudate nucleus, Phil. Trans. R. Soc. Lond. B. 262, 413-427 (1971c).

Kemp J. M. & T. P. S. Powell, The termination of fibres from the cerebral cortex and thalamus upon dendritic spines in the caudate nucleus: a study with the Golgi method, Phil. Trans. R. Soc. Lond. B. 262, 429-439 (1971d).

Kim J. S., I. J. Bak, R. Hassler & Y. Okada, Role of γ-Aminobutyric Acid (GABA) in the extrapyramidal motor system. 2. Some evidence for the existence of a type of GABA-rich strio-nigral neurons, Exp. Brain Res. 14, 95-104 (1971.

Kim J. S., R. Hassler, P. Haug & K. -S. Paik, Effect of frontal cortex ablation on striatal glutamic acid level in rat, Brain Res. 132, 370-374 (1971).

Kitai S. T., J. D. Kocsis, R. J. Preston & M. Sugimori, Monosynaptic inputs to caudate neurons identified by intracellular injection of horseradish peroxidase, Brain Res. 109,601-606 (1976).

Kitai S. T., M. Sugimori and J. Kocsis, Excitatory nature of dopamine in the nigro-caudate pathway, Exp. Brain Res. 24, 351-363 (1976).

Kocsis J. D. & S. T. Kitai, Dual excitatory inputs to caudate spiny neurons from substantia nigra stimulation, Brain Res. 138, 271-283 (1977).

Kocsis J. D., M. Sugimori & S. T. Kitai, Convergence of excitatory synaptic input to caudate spiny neurons, Brain Res. 124, 403-413 (1977a).

Kocsis J. D., C. P. Van der Maelen & S. T. Kitai, Conduction properties of caudate efferent axons in the cat, Neurosc. Abstr. 3, 40 (1977b).

Kölliker A. (1896) Handbuch der Gewebelehre des Menschen, Vol. 2, pp. 612-623, Engelmann, Leipzig.

Leontovich T.A. (1952) The neuronal structure of the striopallidum of certain mammals (in Russian), Abstract of dissertation, Moscow.

Leontovich T.A., Fine structure of subcortical ganglia (in Russian), Z. Nevropat. Psikh. 54,168-178 (1954).

Leontovich T.A., Features of interneuronal relations in subcortical ganglia of mammals (in Russian), Arkh. Anat. 35, 16-25 (1958).

Levine M.S., C. D. Hull & N. A. Buchwald, Pallidal and entopeduncular intracellular responses to striatal, cortical, thalamic and sensory inputs, Exp. Neurol. 44, 448-460 (1974).

Lu E. J. & W. J. Brown, The developing caudate nucleus in the euthyroid and hypothyroid rat, J. comp. Neurol. 171, 261-284 (1977).

Luys J. (1865) Recherches Sur Le Système Nerveux Cérébro-spinal, Sa Structure, Ses Fonctions Et Ses Maladies, Baillière, Paris.

Malliani A. & D. P. Purpura, Intracellular studies of the corpus striatum. II. Patterns of synaptic activities in lenticular and entopeduncular neurons, Brain Res. 6, 341-354 (1967).

Marchi V., Sulla fine struttura dei corpi striati e dei talami ottici, Riv. sper. Freniat. 12, 285-306 (1886).

Martin G. F., Jr. & E. G. Hamel, Jr., The striatum of the oppossum, Didelphis virginiana. Description and experimental studies, J. comp. Neurol. 131, 491-516 (1967).

McGeer E. G., T. Hattori & P. L. McGeer, Electron microscopic localization of labeled norepinephrine transported in nigro-striatal neurons, Brain Res. 86, 478-482 (1975).

McGeer P. L. & E. G. McGeer, Evidence for glutamic acid decarboxylase-containing interneurons in the neostriatum, Brain Res. 91, 331-335 (1975).

McGeer P. L., E. G. McGeer, H. C. Fibiger & V. Wickson, Neostriatal choline acetylase and acetylcholinesterase following selective brain lesions, Brain Res. 35, 308-314 (1971).

Mensah P. L., The internal organization of the mouse caudate nucleus: Evidence for cell clustering and regional variation, Brain Res. 137, 53-66 (1977).

Miller J. J., T. L. Richardson, H. C. Fibiger & H. McLennan, Anatomical and electrophysiological identification of a projection from the mesencephalic raphe to the caudate-putamen in the rat, Brain Res. 97, 133-138 (1975).

Mori S., Some observations on the fine structure of the corpus striatum of the rat brain, Z. Zellforsch. 70, 461-488 (1966).

Namba M., Cytoarchitektonische Untersuchungen am Striatum, J. Hirnforsch. 3, 24-48 (1957).

Nauta W. J. H. & H. G. J. M. Kuypers, Some ascending pathways in the brain stem reticular formation, In H. H. Jasper, L. D. Proctor, R. S. Knighton, W. C. Noshay & R. T. Costello (eds.) (1957) Reticular Formation of the Brain, pp. 3-30, Little Brown, Boston

Olney J. W. & T. de Gubareff, The fate of synaptic receptors in the Kainate-lesioned striatum, Brain Res. 140, 340-343 (1978).

Palkovits M. & J. Fischer (1968) Karyometric Investigations, Akadémai Kiadó, Budapest.

Pasik P., T. Pasik & M. DiFiglia, Quantitative aspects of neuronal organization in the neostriatum of the macaque monkey, In M. D. Yahr (ed.) (1976) The Basal Ganglia, Res. Publ. Assoc. Res. Nerv. Ment. Dis., Vol. 55, pp. 57-90, Raven Press, New York.

Pasik T., P. Pasik & M. DiFiglia, Interneurons in the neostriatum of monkeys, In J. Szentágothai, J. Hámori and E. S. Vizi (eds.) (1977) The Neuron Concept Today, pp. 153-162, Akadémai Kiadó, Budapest.

Purpura D. P. & A. Malliani, Intracellular studies of the corpus striatum. I. Synaptic potentials and discharge characteristics of caudate neurons activated by thalamic stimulation, Brain Res. 6, 325-340 (1967).

Rafols J. A., Local circuit neurons in the primate corpus striatum, Anat. Rec. 190, 516-517 (1978).

Rafols J. A. & C. A. Fox, Further observations on the spiny neurons and synaptic endings in the striatum of the monkey (Saimiri sciureus), J. Hirnforsch. 13, 299-308 (1971/72).

Ramón y Cajal, S., Corps strié, Bibliographie Anatomique 3, 58-62 (1895).

Ramón y Cajal, S. (1911) Histologie du Système Nerveux de l'Homme et des Vertébrés (transl. by L. Azoulay), Vol. 2, pp. 504-518, Maloine, Paris.

Ribak C. E., Immunocytochemical localization of glutamic acid decarboxylase (GAD) in the rat corpus striatum, Anat. Rec. 190, 521 (1978).

Richardson T. L., J. J. Miller & H. McLennan, Mechanisms of excitation and inhibition in the nigrostriatal system, Brain Res. 127, 219-234 (1977).

Robertson R. T. & J. B. Travers, Brainstem projections to the striatum:

Experimental morphological studies in the rat, Exp. Neurol. 48, 447-459 (1975).

Rocha-Miranda C. E., Single unit analysis of cortex-caudate connections, Electroenceph. clin. Neurophysiol. 19, 237-247 (1965).

Schwarcz R. & J. T. Coyle, Jr., Striatal lesions with kainic acid: neurochemical characteristics, Brain Res. 127, 235-249 (1977).

Spencer H. J., Antagonism of cortical excitation of striatal neurons by glutamic acid diethyl ester: Evidence for glutamic acid as an excitatory transmitter in the rat striatum, Brain Res. 102, 91-101 (1976).

Spiegel E., Die Kerne im Vorderhirn der Säuger, Arb. Neurol. Inst. Wiener Univ. 22, 418-497 (1919).

Szabo J., Horseradish peroxidase study of the striatal and nigral efferents, Anat. Rec. 187, 726 (1977).

Szabo J., Strio-nigral and nigro-striatal connections: Anatomical studies, J. Appl. Neurophysiol. in press (1978).

Tabuchi K., Zytoarchitektonische Untersuchung des Corpus striatum beim Menschen, J. Hirnforsch. 11, 325-332 (1969).

Tennyson V. M. & L. A. Marco, Intrinsic connections of caudate neurons. II. Fluorescence and electron microscopy following chronic isolation, Brain Res. 53, 307-317 (1973).

Treff W. M., Numerische und Volumenzelldichte im Caudatum Mediale: Mit besonderer Berücksichtigung des quantitativen Auswertungsfehlers bei Zellzählung, Prog. Brain Res. 6, 139-146 (1964).

Turner J., Notes on the minute structure of the human caudate nucleus and optic thalamus, Brain 26, 400-411 (1903).

van der Loos H. & E. M. Glaser, Autapses in neocortex cerebri: Synapses between a pyramidal cell's axon and its own dendrites, Brain Res. 48, 355-360 (1972).

Vizi E. S., L. G. Hársing, Jr. & J. Knoll, Presynaptic inhibition leading to disinhibition of acetylcholine release from interneurons of the caudate nucleus: Effects of dopamine, β-endorphine and D-Ala^{2}-Pro^{5}-Enkephalinamide, Neuroscience 2, 953-961 (1977).

Vogt C. & O. Vogt, Zur Lehre der Erkrankungen des striären Systems, J. Psychol. Neurol. (Leipzig) 25, 627-846 (1920).

Wilkin G. P., J. E. Wilson, R. Balázs, F. Schon & J. Kelly, A comparison of GABA uptake into the excitatory and inhibitory nerve terminals of the isolated cerebellar glomerulus, Nature (London) 252, 397-399 (1974).

Yang H. Y. T., J. S. Hong & E. Costa, Regional distribution of Leu and Met enkephalin in rat brain, Neuropharmacology 16, 303-307 (1977).

Yoshida M. & W. Precht, Monosynaptic inhibition of neurons of the substantia nigra by caudato-nigral fibers, Brain Res. 32, 225-228 (1971).

Extrinsic Connections of the Neostriatum

Irena Grofová

Anatomical Institute, University of Oslo, Oslo 1, Norway

The neostriatum (NS) or striatum consists of the nucleus caudatus and putamen which in higher mammals are separated by the fibers of the internal capsule. In all mammals, these nuclei are entirely covered by the cerebral cortex and subcortical white matter and they are contiguous rostroventrally with the nucleus accumbens septi or the fundus striati (50). They harbour radial bundles of very thin fibers which account for their "striatal" appearance.

The striatal projections have for a long time attracted interest but in the absence of suitable surgical and histological techniques they remained largely unknown until the fifties when studies of retrograde cell degeneration following striatal lesions disclosed the thalamostriatal connections (25,31,32,106,107,112). In the sixties, silver impregnation techniques for degenerating fibers revealed the existence and basic features of the organization of the corticostriatal (14,15,65, 70,83,99,100,129,130,131), striatopallidal and striatonigral fibers(26,70,92,95,114, 115,116,118,128). Finally, biochemical and fluorescent histochemical techniques provided convincing evidence for the existence of an important nigrostriatal dopaminergic connection (3,4,8,44,56,57,105,122) and a serotoninergic projection from the mesencephalic raphe (54,73,74,75,103,104). Modern autoradiographic (AR) and horseradish peroxidase (HRP) tracing techniques are presently providing further valuable information on the exact origin and detailed organization of these connections.

This review is intended to summarize experimental anatomical findings on the origin and organization of neostriatal connections with special emphasis on more recent observations. The fiber connections of the nucleus accumbens septi are not reviewed. Readers interested in the morphology and organization of other basal ganglia are referred to recent reviews on this topic (16,17,86).

AFFERENT CONNECTIONS

NS receives profuse and extensive topographically organized inputs from the cerebral cortex, thalamus and brain stem. These are schematically summarized in Fig. 1.

Corticostriatal Fibers

The corticostriatal connection was demonstrated experimentally with silver impregnation degeneration techniques in the opossum (83), rat (15,70,129), rabbit (14,15), cat (15,130) and monkey (65,99,100,131). In general, these fibers arise from nearly the entire ipsilateral neocortex and distribute in an organized manner to all parts of NS. The heaviest contribution originates from the sensorimotor cortex (14,65,99,129,130), the smallest projection from the visual cortex (65). This difference is especially pronounced in lower species (rat, rabbit, cat) in which the frontal cortex seems to be related to a considerably greater volume of NS than the posterolateral cortex (65). The cortical fibers reach the caudate nucleus via the internal capsule and subcallosal fasciculus, and the putamen through both internal and external capsules. In addition to ipsilateral fibers there exists also a modest contralateral projection. The origin of the latter projection seems to vary in different species. In the opossum it originates in the frontal, orbital and parietal regions and decussates in the anterior commissure (83). In the rat, rabbit and cat it arises exclusively from the sensorimotor cortex and crosses the midline in the corpus callosum (15). In the monkey the crossed corticostriatal fibers were described from the sensorimotor cortex and parietal area 5 (65). However, it appears from a recent HRP study that their origin is actually limited to the cortical areas 6, 8 and 4 (61).

Topographical organization and distribution within NS. The principles of the topographical organization of the corticostriatal afferents were clearly established from the degeneration studies (14,15,65,70,99,100,129,130). In the opossum (83), rabbit (14) and cat (130) fibers from every cortical area terminate in both the putamen and the caudate. In all species studied, the cortical fibers show a distinct mediolateral, dorsoventral and anteroposterior organization, although there is a considerable degree of overlap. It thus seemed that within the gross subdivisions of the striatum there are regions which are specifically influenced by functionally different cortical areas. Recent findings in the monkey with the AR technique, however, reveal a somewhat different distribution of corticostriatal fibers. Thus, fibers from the primary motor area 4 (77) and primary sensory areas 3, 1 and 2(61,78) project to nearly the entire anteroposterior extent of the putamen only. The patch-like terminal fields are topographically similarly organized suggesting a high degree of convergence between primary motor and primary somatosensory projections. Both these projections are somatotopically organized (61,77,78), fibers from the leg-tail area terminating most dorsally and rostrally, those from the face area most ventrally and caudally (Fig. 2). Premotor area 6 projects to a large anteroposterior extent of both the caudate and the putamen (79). The frontal eye field (area 8) projects prevalently to the body of the caudate nucleus (80) and frontal area 9 to the entire anteroposterior extent of the caudate (45,79). Similarly, association areas in the frontal, parietal, occipital and temporal lobes project to widespread and partially overlapping regions of the caudate (134). In addition, the latter study (134) reveals a new principle of the organization of the corticostriatal fibers from pre- and post-Rolandic association areas. It appears that cortical areas which have strong reciprocal cortico-cortical connections project, in part, to the same region of the caudate. This region thus receives input not only from a particular area of cortex but also from all other cortical areas reciprocally interconnected with that area. Whether this is true also for the specific cortical areas, and also for other animal species, remains to be established.

All autoradiographic studies stress a peculiar strip-like and patchy appearance of the distribution of the corticostriatal terminals originating from one cortical area. Strips and clusters of terminals emanating from several cortical regions lie in intimate relation to one another and may possibly interdigitate. There is some evidence that each cluster of terminals probably represents termination of fibers arising from a localized patch of cortex (61). This would imply that cortical

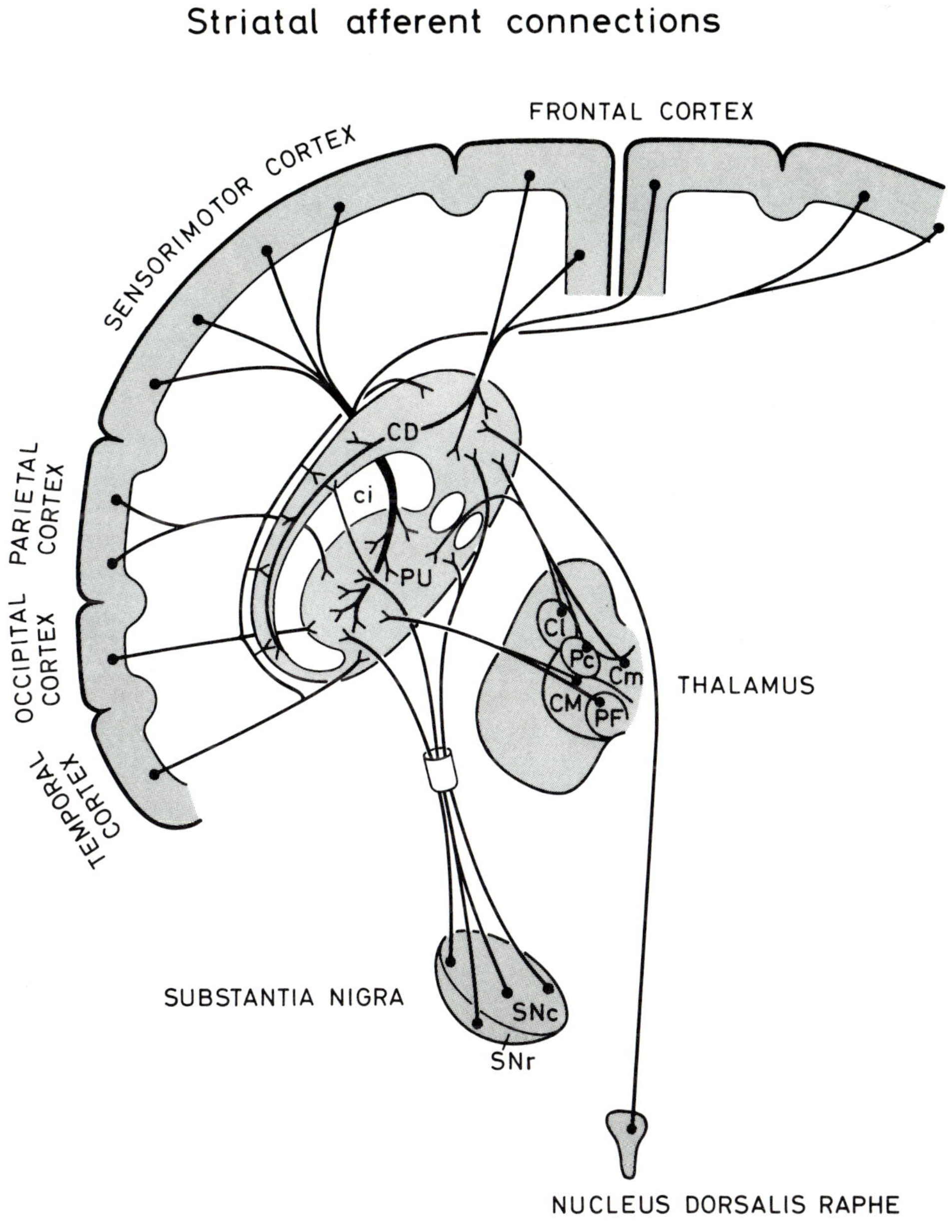

Fig. 1

Schematic diagram summarizing the organization of striatal input. CD, nucleus caudatus; ci, capsula interna; Cl, nucleus centralis lateralis; Cm, nucleus centralis medialis; CM, nucleus centrum medianum; Pc, nucleus paracentralis; PF, nucleus parafascicularis; PU, putamen; SNc, pars compacta of the substantia nigra; SNr, pars reticulata of the substantia nigra.

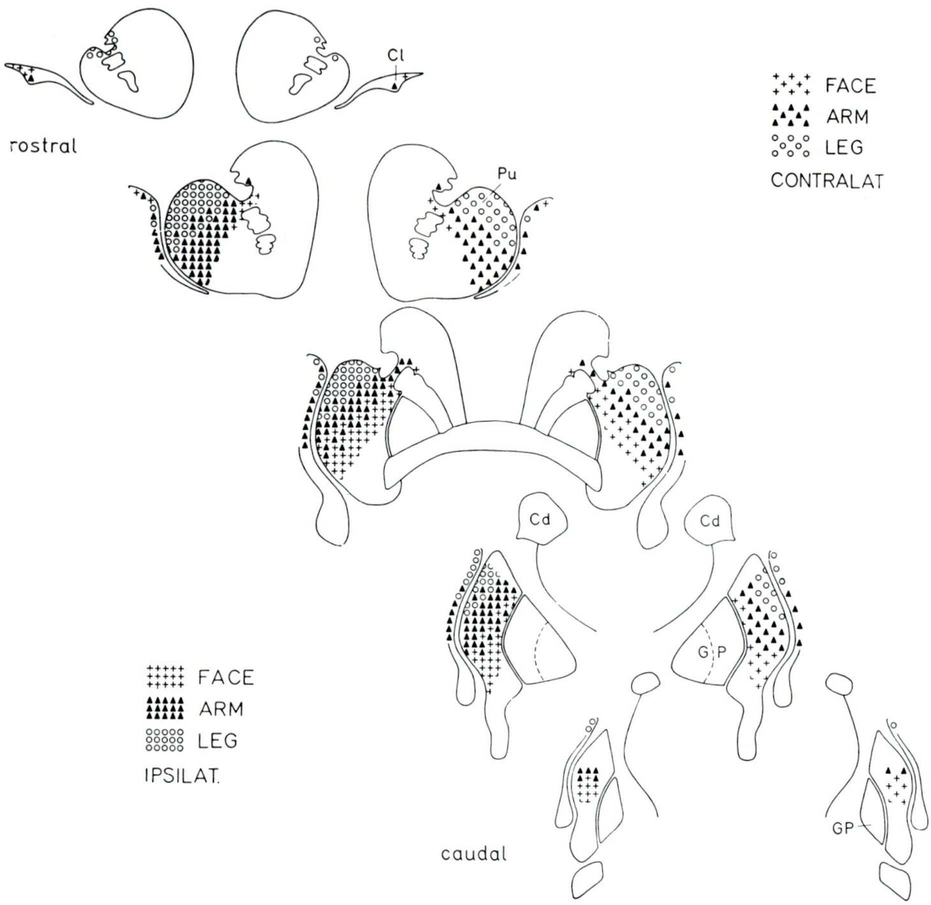

Fig. 2

Diagrammatic representation of the bilateral terminal field within the basal ganglia following differentially localized unilateral injections into the motor cortex. (From Künzle, 77.)

terminals are arranged within the striatum as a mosaic and despite of seemingly extensive overlap of terminals from different cortical areas each small patch of the striatum may be more specifically influenced by some area than by other. Grouping of corticostriatal terminals in the caudate nucleus has also been noted electronmicroscopically (66).

Cells of origin and mode of termination. The exact localization and types of cortical cells projecting to NS have been recently revealed by means of HRP tracing technique in the rat (52,133) and in the monkey (61). Such cells are localized to the cortical layer V, are uniform in size (mean diameters 14-16 μm) and correspond to small or medium size pyramidal cells (61). The cells of origin of the corticostriatal pathway appear to form a group that is distinct from the cells of origin of other subcortical projections, particularly the corticospinal. The present evidence (61,133) thus seems to indicate that the corticostriatal fibers represent an independent system and not collateral branches of axons coursing in the internal capsule and destined for other sites, as suggested by earlier studies (13,129).

The corticostriatal boutons are small with round synaptic vesicles and establish asymmetrical synapses predominantly with the dendritic spines of spiny striatal neurons (6,66,67). A putative transmitter substance in this pathway seems to be glutamate (Fonnum and Walaas, the present volume).

Thalamostriatal Fibers

The projection from the thalamus to NS has been shown in human neuropathological material (84,112,127), by experimental studies using retrograde cellular degeneration (25,31,32,106,107) and orthograde fiber degeneration methods (59,85,94) as well as retrograde transport of HRP (60,76,88,93,110) and anterograde transport of labeled proteins (62, 111). This projection originates in the intralaminar thalamic nuclei, i.e. nucleus centralis lateralis, centralis medialis, paracentralis, nucleus centrum medianum and nucleus parafascicularis. It is also possible that some other thalamic nuclei, such as the nucleus mediodorsalis, project to NS (68, 120,121).

The thalamostriatal projection is strictly ipsilateral and topographically organized in the anteroposterior direction (60). It is possible, however, that the distribution of thalamostriatal fibers within NS may vary in different species. Thus, in two recent AR studies, fibers from the centromedian nucleus in the monkey are reported to terminate exclusively in the putamen (62), whereas in the cat (111), they are said to distribute mainly to the caudate nucleus along its entire anteroposterior extent. It is interesting that both authors (62,111) report a patchy distribution of the thalamofugal fibers within the NS. At present it is still unknown whether the thalamostriatal fibers overlap or not with the patchily distributed corticostriatal fibers described in AR studies (45,77,78,79,80,134). Moreover, it is not known whether the thalamostriatal fibers are collaterals of the thalamocortical projection or whether they arise from a separate population of cells in the intralaminar nuclei (60).

Thalamostriatal boutons establish asymmetrical synapses mostly on spines but also on the shafts of dendrites (67). Physiological evidence indicates that some of them terminate on the same medium spiny cells as the corticostriatal boutons (72). The transmitter in this pathway remains unknown (Fonnum and Walaas, the present volume).

Nigrostriatal Fibers

The nigrostriatal pathway has been suggested from retrograde cell degeneration studies (30,38,43,55,90,102,109) but it could be demonstrated neither with Marchi method (69,87,109) nor with Nauta's technique (2,18,20,23,35). The first convincing evidence that the substantia nigra indeed projects to the striatum was provided by biochemical and fluorescent histochemical techniques (3,4,8,44,56,57,96, 105,122). Shortly after, the nigrostriatal projection was anatomically demonstrated in various species by means of Fink-Heimer and Wiitanen silver impregnation techniques(19,53,58,82,89,117,123),by retrograde transport of HRP (12,22,28,33,36, 71,76,88,93,110,113,119), and by anterograde transport of tritiated proteins (21, 28,29).

The nigrostriatal fibers arise mainly but not exclusively from cells in the pars compacta of the substantia nigra. Some neurons in the pars reticulata of the substantia nigra and paranigral cell groups in the ventral and ventrolateral tegmental areas (including the cat retrorubral nucleus) also contribute to this projection (12,36,71,110,113,119). The nigrostriatal fibers ascend along the dorsomedial border of the substantia nigra to the prerubral area and lateral hypothalamus, then enter the medial part of the internal capsule and run in a dorsorostral direction to reach the head of the caudate nucleus and the rostral part of the putamen. Fibers destined for the posterior parts of putamen and nucleus caudatus leave the main bundle in Forel's field, course laterally, dorsal to the subthalamic nucleus, and penetrate through the posterior part of the internal capsule and through the globus pallidus. This projection also appears to be topographically organized in anteroposterior, mediolateral and dorsoventral directions, matching reciprocal striatonigral connection (19,22,119,123). Preliminary results of a detailed study on the organization of the nigro-striato-nigral loop in the rat by means of autoradiographic and HRP techniques confirm a distinct mediolateral organization of the nigrostriatal fibers (29). The nigrostriatal connection is generally supposed to be ipsilateral; however, a few labeled cells have recently been described in the contralateral pars compacta following HRP injections in the cat nucleus caudatus (110).

Electronmicroscopically, the nigrostriatal terminals are characterized by the presence of small granular vesicles (57). Following large mesencephalic lesions (which probably interrupted also striatal afferents from the raphe), Kemp and Powell (66) found only very sparse degeneration in the caudate nucleus without any evidence of grouping. The degenerating terminals invariably established asymmetrical synapses, most commonly with dendritic spines. There exists physiological evidence that nigrostriatal fibers terminate on the same medium spiny neurons as thalamostriatal and corticostriatal fibers (72).

The nigrostriatal connection is generally referred to as dopaminergic. It should be emphasized, however, that not all nigral dopaminergic neurons project to the NS (34). On the other hand, there exists some evidence that a portion of the nigrostriatal neurons are non-catecholaminergic (37,39). HRP tracing of nigrostriatal neurons combined with immunohistochemistry could neither confirm nor exclude this possibility (81).

Raphe-striatal Fibers

Serotoninergic projection from the mesencephalic raphe nuclei to the striatum has been suggested from biochemical and histofluorescence studies. It has been shown that lesions of these nuclei or interruption of fibers presumably arising in these nuclei result in an extensive reduction of striatal serotonin and its synthetizing enzyme tryptophan hydroxylase (73,74,75,103,104). Furthermore, an increased release of serotonin from the striatum has been reported following raphe stimulation (54). The existence of a pathway from the nucleus dorsalis raphe to the striatum has been confirmed with morphological tracing techniques, both with HRP (24,88,93, 98,113) and with autoradiography (5,9,10).

Fibers arising from the nucleus dorsalis raphe course for a short distance rostrally through the fasciculus longitudinalis medialis, then dip down along each side of the midline through the brachium conjunctivum to reach the ventral tegmental area of Tsai. Further rostrally they course through the lateral hypothalamus. Fibers destined for the putamen run laterally through zona incerta, those for the caudate nucleus leave the main system at the diencephalo-telencephalic junction and ascend along medial border of the internal capsule towards fundus striati. They terminate throughout the entire extent of the striatum and especially profusely in its caudal parts (10).

Other Afferents

A substantial bilateral projection from the locus coeruleus to the striatum was recently reported in the rat by means of the HRP tracing technique (24), but fibers distributing to the striatum have not been seen when the AR technique was used (101). At present, this connection is doubtful.

EFFERENT CONNECTIONS

The nucleus caudatus and putamen give rise to a large number of extremely thin (average diameter = 0.6 μm) poorly myelinated fibers (1,42,125) which collect already within the nuclei into tiny bundles, Wilson's pencils (132). These bundles convey radially upon the globus pallidus and pass through and along this nucleus. Further caudally, the striatal fibers traverse the cerebral peduncle as the ventral component of Edinger's comb system and then enter the substantia nigra (40,41,42, 92). The globus pallidus and the substantia nigra are the only well documented targets of striatal efferents. Golgi and electronmicroscopic studies indicate that at least some of the striatopallidal fibers are collaterals of the striatonigral projection (40,42).

Cells of origin of the striatofugal fibers. According to the classical concept (126) only the large striatal neurons which represent at most 5% of the total striatal cell population, (see 47 for references) were considered to be the output cells. From HRP studies of striatopallidal and striatonigral connections it has been repeatedly demonstrated that a much larger number of striatal cells (30-50%) project outside the striatum and that these cells belong to the category of medium-sized striatal neurons (11,12,46,47,63,119). Similar results were obtained also by means of retrograde transport of herpes simplex virus (7), and they are further confirmed by recent Golgi studies (27,97, Rafols, personal communication, Pasik, the present volume). Following HRP injections in the substantia nigra labeled neurons within the cat's nucleus caudatus and putamen are not regularly dispersed but tend to be grouped in clusters (personal observation). It is tempting to speculate that this arrangement might be related to the patchy distribution of corticostriatal and/or thalamostriatal terminals seen in autoradiographic studies (45,61,62,77,78, 79,80,111,134).

Topographical organization of striatopallidal and striatonigral fibers. Many investigators studying striatal efferent systems in various species have noted their precise topographical organization (11,26,28,48,70,92,95,114,115,116,118,119,124, 128). However, the only systematic studies on the topographical organization of striatopallidal and striatonigral fibers were made by Szabo in the monkey (114,115, 116,118) and by Knook in the rat (70). The following description is based mainly on our unpublished observations in the cat made partly on silver impregnated sections following small electrolytic lesions in various parts of the striatum, partly on preparations treated according to the method of Graham and Karnovsky following HRP injections in various parts of the substantia nigra. The topographical principles of the organization of striatopallidal and striatonigral fibers in the cat are similar to those described by Szabo in the monkey (114,115,116,118) and to the striatonigral fibers in the rat (28). They are schematically depicted in Fig. 3.

Striatopallidal fibers distribute to both the globus pallidus and the nucleus entopeduncularis which in lower mammals represents the medial pallidal segment. Fibers from the caudate nucleus supply the dorsal regions of the globus pallidus and nucleus entopeduncularis, while those from the putamen course and terminate ventrally. There is a distinct mediolateral organization of caudatal efferents. Ventromedial parts of the head of the caudate nucleus project most medially, its dorsolateral parts most laterally to both parts of the pallidum (Fig. 3A). The anteroposterior

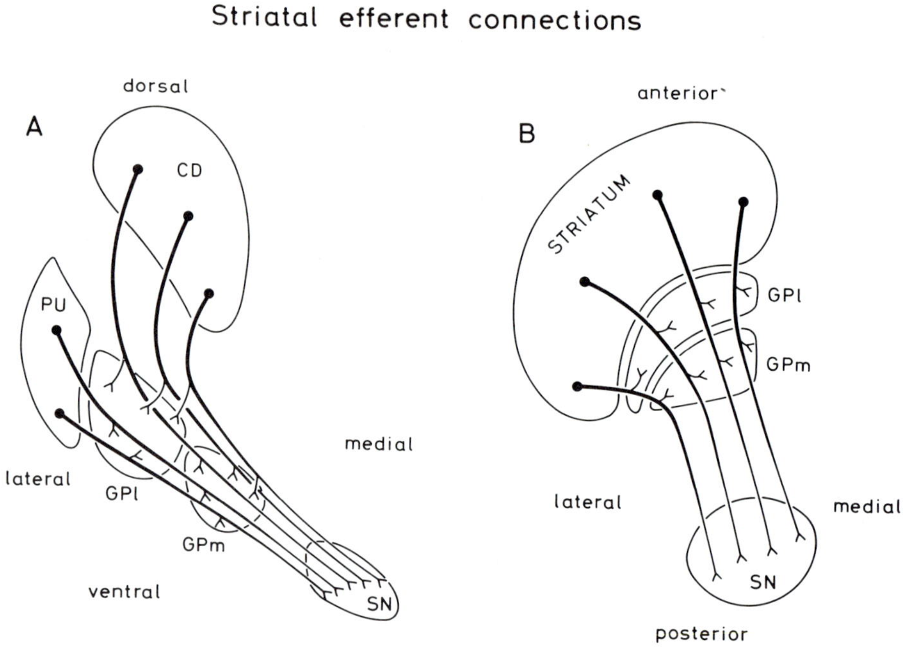

Fig. 3

Schematic representation of the topographic organization of striatopallidal and striatonigral projections. Dorsoventral and mediolateral organizations are shown in transverse plane (A), anteroposterior and mediolateral in horizontal plane (B). CD, nucleus caudatus; GPl, lateral pallidal segment; GPm, medial pallidal segment (= nucleus entopeduncularis); SN, substantia nigra.

organization of striatopallidal fibers is less obvious and it can be best understood from horizontal sections (Fig. 3B). It reflects not only the anteroposterior but also the mediolateral relationships between these nuclei. Thus the rostralmost pole of the globus pallidus and entopeduncular nucleus represent also the medialmost regions of these nuclei, and consequently are related to the ventromedial regions of the rostral striatum. The dorsolateral regions of the rostral striatum, on the other hand, are related to more posterior and lateral parts of the globus pallidus and nucleus entopeduncularis.

The fibers from the ventromedial and rostralmost part of the striatum are during their course through the posterior limb of the internal capsule and the cerebral peduncle gradually displaced dorsally by fibers originating in more caudal and dorsal striatal regions. The caudatal fibers thus enter the rostral pole of the substantia nigra in a reverse dorsoventral order.

Striatonigral fibers distribute to both parts of the substantia nigra in a distinct mediolateral topographical pattern. Fibers from the ventromedial part of the head of the caudate nucleus end most medially and those from the posterior putamen most laterally. The anteroposterior organization seen in silver impregnated sections appears doubtful in the light of recent autoradiographic studies in the cat (91) and rat (28). In the rat (28) there exists also a precise dorsoventral organization of the striatonigral fibers. Fibers from the ventralmost part of NS terminate most dorsally and those from the dorsalmost part of NS most ventrally within the substantia nigra.

The mode of termination of striatopallidal and striatonigral fibers. Fine structure of the globus pallidus and the substantia nigra is very similar (41,64,108). The most outstanding features of the neuropil are relatively smooth dendritic profiles almost completely covered by synaptic terminals and large masses of thin unmyelinated fibers surrounding the synaptic complexes. From experimental electronmicroscopic studies it is obvious that a large number of boutons in the globus pallidus (64) and in the substantia nigra (48,49,64) belong to the striatal afferents. Such boutons, of the 'en passage' type (41), contain pleomorphic synaptic vesicles and establish symmetrical synaptic contacts mainly on dendritis shafts.

Many striatonigral and striatopallidal fibers are GABAergic. In addition, substance P was suggested as a putative transmitter in the striatonigral pathway, and Leu-enkephalin in the striatopallidal connection (Fonnum and Walaas, the present volume).

Striato-nigro-striatal loop. It is still unclear whether the nigral dopaminergic neurons which project to the NS receive a substantial monosynaptic striatal input. Electronmicroscopic (48) and light microscopic AR studies (28,91) indicate that the striatonigral fibers terminate in the pars compacta as well as pars reticulata. Furthermore, dopaminergic nigral neurons projecting to the NS are located ventrally within the pars compacta and send long dendrites into the pars reticulata (34). These dendrites may well be contacted by striatonigral terminals. However, Hattori et al. (51) in an elegant AR electronmicroscopic study demonstrated that nigral dopaminergic cells are preferentially innervated from the globus pallidus.

REFERENCES

1. Adinolfi, A.M. and Pappas, G.D. The fine structure of the caudate nucleus of the cat. J. comp. Neurol. 133, 167-184 (1968).
2. Afifi, A. and Kaelberg, W.W. Efferent connections of the substantia nigra in the cat. Exp. Neurol. 11, 474-482 (1964).
3. Andén, N.-E., Carlsson, A., Dahlström, A., Fuxe, K., Hillarp, N.A. and Larsson, K. Demonstration and mapping out of nigro-neostriatal dopamine neurons. Life Sci. 3, 523-530 (1964).
4. Andén, N.E., Dahlström, A., Fuxe, K. and Larsson, K. Further evidence for the presence of nigro-neostriatal dopamine neurons in the rat. Amer. J. Anat. 116, 329-333 (1965).
5. Azmitia, E.C. and Segal, M. An autoradiographic analysis of the differential ascending projections of the dorsal and median raphe nuclei in the rat. J. comp. Neurol. 179, 641-668 (1978).
6. Bak, I.J., Choi, W.B., Hassler, R., Usunoff, K.G. and Wagner, A. Fine structural synaptic organization of the corpus striatum and substantia nigra in rat and cat. Adv. Neurol. 9, 25-41 (1975).
7. Bak, I.J., Markham, C.H., Cook, M.L. and Stevens, J.G. Ultrastructural and immunoperoxidase study of striatonigral neurons by means of retrograde axonal transport of herpes simplex virus. Brain Res. 143, 361-368 (1978).

8. Bédard, P., Larochelle, L., Parent, A. and Poirier, L.J. The nigrostriatal pathway - A correlative study based on neuroanatomical and neurochemical criteria in the cat and the monkey. Exp. Neurol. 25, 365-377 (1969).
9. Bobillier, P., Petitjean, F., Salvert, D., Ligier, M. and Seguin, S. Differential projections of the nucleus raphe dorsalis and nucleus raphe centralis as revealed by autoradiography. Brain Res. 85, 205-210 (1975).
10. Bobillier, P., Seguin, S., Petitjean, F., Salvert, D., Touret, M. and Jouvet, M. The raphe nuclei of the cat brain stem: A topographical atlas of their efferent projections as revealed by autoradiography. Brain Res. 113, 449-486 (1976).
11. Bunney, B.S. and Aghajanian, G.K. The precise localization of nigral afferents in the rat as determined by a retrograde tracing technique. Brain Res. 117, 423-435 (1976).
12. Butcher, L.L. and Giesler Jr., G.J. Nigro-neostriato-nigral relationships demonstrated by horseradish peroxidase (HRP) histochemistry. Neurosci. Abstr. 3, 35 (1977).
13. Cajal, R.S. Histologie du système nerveux de l'homme et des vertébrés. Vol. 1. Maloine, Paris (1909).
14. Carman, J.B., Cowan, W.M. and Powell, T.P.S. The organization of cortico-striate connexions in the rabbit. Brain 86, 525-562 (1963).
15. Carman, J.B., Cowan, W.M., Powell, T.P.S. and Webster, K.E. A bilateral cortico-striate projection. J. Neurol. Neurosurg. Psychiat. 28, 71-77 (1965).
16. Carpenter, M.B. Anatomy of the basal ganglia and related nuclei: A review. Adv. Neurol. 14, 7-48 (1976).
17. Carpenter, M.B. Anatomical organization of the corpus striatum and related nuclei. In: M.D. Yahr (Ed.), The Basal Ganglia, Raven Press, New York, 1-36 (1976).
18. Carpenter, M.B. and McMasters, R.E. Lesions of the substantia nigra in the rhesus monkey. Efferent fiber degeneration and behavioral observations. Amer. J. Anat. 114, 293-319 (1964).
19. Carpenter, M.B. and Peter, P. Nigrostriatal and nigrothalamic fibers in the rhesus monkey. J. comp. Neurol. 144, 93-116 (1972).
20. Carpenter, M.B. and Strominger, N.L. Efferent fibers of the subthalamic nucleus in the monkey. A comparison of the efferent projections of the subthalamic nucleus, substantia nigra and globus pallidus. Amer J. Anat. 121, 41-72 (1967).
21. Carpenter, M.B., Nakano, K. and Kim, R. Nigrothalamic projections in the monkey demonstrated by autoradiographic technics. J. comp. Neurol. 165, 401-416 (1976).
22. Carter, D.A. and Fibiger, H.C. Ascending projections of presumed dopamine-containing neurons in the ventral tegmentum of the rat as demonstrated by horseradish peroxidase. Neuroscience (Oxford) 2, 569-576 (1977).
23. Cole, M., Nauta, W.J.H. and Mehler, W.R. The ascending efferent projections of the substantia nigra. Trans. Amer. neurol. Ass. 89, 74-78 (1964).
24. Couch, J.R. and Goldstein, G.D. Projections of locus coeruleus and raphe nuclei to neostriatum. Neurosci. Abstr. 3, 36 (1977).
25. Cowan, W.M. and Powell, T.P.S. The projection of the midline and intralaminar nuclei of the thalamus of the rabbit. J. Neurol. Neurosurg. Psychiat. 18, 266-279 (1955).
26. Cowan, W.M. and Powell, T.P.S. Strio-pallidal projection in the monkey. J. Neurol. Neurosurg. Psychiat. 29, 426-439 (1966).
27. DiFiglia, M., Pasik, P. and Pasik, T. A Golgi study of neuronal types in the neostriatum of monkeys. Brain Res. 114, 245-256 (1976).
28. Domesick, V.B. The topographical organization of the striatonigral connection in the rat. Anat. Rec. 187, 567 (1977).
29. Domesick, V.B., Beckstead, R.M. and Nauta, W.J.H. Some ascending and descending projections of the substantia nigra and ventral tegmental area in the rat. Neurosci. Abstr. 2, 61 (1976).

30. Dresel, K. und Rothmann, H. Völliger Ausfall der Substantia nigra nach Extirpation von Grosshirn und Striatum. Z. ges. Neurol. Psychiat. 94, 781-789 (1925).
31. Droogleever-Fortuyn, J. On the configuration and the connections of the medioventral area and the midline cells in the thalamus of the rabbit. Folia psychiat. neerl. 53, 213-254 (1950).
32. Droogleever-Fortuyn, J. and Stefens, R. On the anatomical relations of the intralaminar and midline cells of the thalamus. Electroenceph. clin. Neurophysiol. 3, 393-400 (1951).
33. Fallon, J.H. and Moore, R.Y. Topography of mesencephalic dopamine cell projections to the telencephalon. Anat. Rec. 187, 764 (1977).
34. Fallon, J.H., Riley, J.N. and Moore, R.Y. Substantia nigra dopamine neurons: separate populations project to neostriatum and allocortex. Neurosci. Letters 7, 157-162 (1978).
35. Faull, R.L.M. and Carman, J.B. Ascending projections of the substantia nigra in the rat. J. comp. Neurol. 132, 73-92 (1968).
36. Faull, R.L.M. and Mehler, W.R. The cells of origin of nigrotectal, nigrothalamic and nigrostriatal projections in the rat. Neuroscience (Oxford) (1978) in press.
37. Feltz, P. and de Champlain, J. Persistence of caudate unitary responses to nigral stimulation after destruction and functional impairment of the striatal dopaminergic terminals. Brain Res. 43, 595-600 (1972).
38. Ferraro, A. Contributa sperimentale allo studio della substantia nigra normale a dei suoi rapporti con la corteccia cerebrale e con il corpo striato. Arch. gen. Neurol. Psichiat. 6, 26-117 (1925).
39. Fibiger, H.C., Pudritz, R.E., McGeer, P.L. and McGeer, E.G. Axonal transport in nigro-striatal and nigro-thalamic neurons: effects of medial forebrain bundle lesions and 6-hydroxy-dopamine. J. Neurochem. 19, 1697-1708 (1972).
40. Fox, C.A. and Rafols, J.A. The radial fibers in the globus pallidus. J. comp. Neurol. 159, 177-200 (1975).
41. Fox, C.A. and Rafols, J.A. The striatal efferents in the globus pallidus and in the substantia nigra. In: M.D. Yahr (Ed.), The Basal Ganglia, Raven Press, New York, 37-55 (1976).
42. Fox, C.A., Rafols, J.A. and Cowan, W.M. Computer measurements of axis cylinder diameters of radial fibers and comb bundle fibers. J. comp. Neurol. 159, 201-224 (1975).
43. Fox, C.A. and Schmitz, J.T. The substantia nigra and the entopeduncular nucleus in the cat. J. comp. Neurol. 80, 323-334 (1944).
44. Fuxe, K. and Andén, N.E. Studies on the central monoamine neurons with special reference to the nigro-neostriatal dopamine neuron system. In: E. Costa, L.J. Côté and M.D. Yahr (Eds.), Biochemistry and Pharmacology of the Basal Ganglia, Raven Press, New York, 123-129 (1966).
45. Goldman, P.S. and Nauta, W.J.H. An intricately patterned prefronto-caudate projection in the rhesus monkey. J. comp. Neurol. 171, 369-386 (1977).
46. Grofová, I. The identification of striatal and pallidal neurons projecting to substantia nigra. An experimental study by means of retrograde axonal transport of horseradish peroxidase. Brain Res. 91, 286-291 (1975).
47. Grofová, I. Types of striato-nigral neurons labeled by retrograde transport of horseradish peroxidase. Appl. Neurophysiol. (1978) in press.
48. Grofová, I. and Rinvik, E. An experimental electron microscopic study on the striatonigral projection in the cat. Exp. Brain Res. 11, 249-262 (1970).
49. Hajdu, F., Hassler, R. and Bak, I.J. Electron microscopic study of the substantia nigra and the strio-nigral projection in the rat. Z. Zellforsch. 146, 207-221 (1973).
50. Hassler, R. and Chung, J.W. The discrimination of nine different types of synaptic boutons in the fundus striati (nucleus accumbens septi). Cell Tiss. Res. 168, 489-505 (1976).

51. Hattori, T., Fibiger, H.C. and McGeer, P.L. Demonstration of a pallidonigral projection innervating dopaminergic neurons. J. comp. Neurol. 162, 487-504 (1975).
52. Hedreen, J.C. Corticostriatal cells identified by the peroxidase method. Neurosci. Letters 4, 1-7 (1977).
53. Hedreen, J.C. and Chalmers, J.P. Neuronal degeneration in rat brain induced by 6-hydroxy-dopamine; a histological and biochemical study. Brain Res. 47, 1-36 (1972).
54. Holman, R.B. and Vogt, M. Release of 5-hydroxytryptamine from caudate nucleus and septum. J. Physiol. (Lond.) 223, 243-254 (1972).
55. Holmes, G. The nervous system of the dog without a forebrain. J. Physiol. (Lond.) 27, 1-25 (1901).
56. Hornykiewicz, O. Metabolism of brain dopamine in human parkinsonism: Neurochemical and clinical aspects. In: E. Costa, L.J. Coté and M.D. Yahr (Eds.), Biochemistry and Pharmacology of the Basal Ganglia, Raven Press, New York, 171-185 (1966).
57. Hökfelt, T. and Ungerstedt, U. Electron and fluorescence microscopical studies on the nucleus caudatus putamen of the rat after unilateral lesions of ascending nigroneostriatal dopamine neurons. Acta physiol. scand. 76, 415-426 (1969).
58. Ibata, Y., Nojyo, Y., Matsuura, T. and Sano, Y. Nigro-neostriatal projection - A correlative study with Fink-Heimer impregnation, fluorescence histochemistry and electron microscopy. Z. Zellforsch. 138, 333-344 (1973).
59. Johnson, T.N. Fiber connections between the dorsal thalamus and corpus striatum in the cat. Exp. Neurol. 3, 556-569 (1961).
60. Jones, E.G. and Leavitt, R.Y. Retrograde axonal transport and the demonstration of non-specific projections to the cerebral cortex and striatum from thalamic intralaminar nuclei in the rat, cat and monkey. J. comp. Neurol. 154, 349-378 (1974).
61. Jones, E.G., Coulter, J.D., Burton, H. and Porter, R. Cells of origin and terminal distribution of corticostriatal fibers arising in the sensory-motor cortex of monkeys. J. comp. Neurol. 173, 53-80 (1977).
62. Kalil, K. Patch-like termination of thalamic fibers in the putamen of the rhesus monkey: an autoradiographic study. Brain Res. 140, 333-339 (1978).
63. Kanazawa, I., Marshall, G.R. and Kelly, J.S. Afferents to the rat substantia nigra studied with horseradish peroxidase with special reference to fibers from the subthalamic nucleus. Brain Res. 115, 485-491 (1976).
64. Kemp, J.M. The termination of strio-pallidal and strio-nigral fibers. Brain Res. 17, 125-128 (1970).
65. Kemp, J.M. and Powell, T.P.S. The cortico-striate projection in the monkey. Brain 93, 525-546 (1970).
66. Kemp, J.M. and Powell, T.P.S. The site of termination of afferent fibers in the caudate nucleus. Phil. Trans. B. 262, 403-412 (1971).
67. Kemp. J.M. and Powell, T.P.S. The termination of fibers from the cerebral cortex and thalamus upon dendritic spines in the caudate nucleus: A study with the Golgi method. Phil. Trans. B. 262, 429-438 (1971).
68. Khalifeh, R.R., Kaelberg, W.W. and Ingram, W.R. Some efferent connections of the nucleus medialis dorsalis. An experimental study in the cat. Amer. J. Anat. 116, 341-354 (1965).
69. Kimmel, D.L. Nigro-striatal fibers in the cat. Anat. Rec. 82, 425 (1942).
70. Knook, H.L. The Fibre Connections of the Forebrain, (Thesis), Van Gorcum, Assen (1965).
71. Kocsis, J.D. and Vandermaelen, C.P. The caudate projection of the retrorubral nucleus and its relationship to the substantia nigra in the cat. Anat. Rec. 187, 628-629 (1977).
72. Kocsis, J.D., Sugimori, M. and Kitai, S.T. Convergence of excitatory synaptic inputs to caudate spiny neurons. Brain Res. 124, 403-413 (1977).

73. Kostowski, W., Giacalone, E., Garattini, S. and Valzelli, L. Studies on behavioural and biochemical changes in rats after lesions of midbrain raphe. Europ. J. Pharmacol. 4, 371-376 (1968).
74. Kuhar, M.J., Aghajanian, G.K. and Roth, R.H. Tryptophan hydroxylase activity and synaptosomal uptake of serotonin in discrete brain regions after midbrain raphe lesions: correlations with serotonin levels and histochemical fluorescence. Brain Res. 44, 165-176 (1972).
75. Kuhar, M.J., Roth, R.H. and Aghajanian, G.K. Selective reduction of tryptophan hydroxylase activity in rat forebrain after midbrain raphe lesions. Brain Res. 35, 167-176 (1971).
76. Kuypers, H.G.J.M., Kievit, J. and Groen-Klevant, A.C. Retrograde axonal transport of horseradish peroxidase in rat's forebrain. Brain Res. 67, 211-218 (1974).
77. Künzle, H. Bilateral projections from precentral motor cortex to the putamen and other parts of the basal ganglia. An autoradiographic study in Macaca fascicularis. Brain Res. 88, 195-210 (1975).
78. Künzle, H. Projections from the primary somatosensory cortex to basal ganglia and thalamus in the monkey. Exp. Brain Res. 30, 481-492 (1977).
79. Künzle, H. An autoradiographic analysis of the efferent connections from premotor and adjacent prefrontal regions (areas 6 and 9) in Macaca fascicularis. Brain Behav. Evol. (1978) in press.
80. Künzle, H. and Akert, K. Efferent connections of cortical area 8 (frontal eye field) in Macaca fascicularis. A reinvestigation using the autoradiographic technique. J. comp. Neurol. 173, 147-164 (1977).
81. Ljungdahl, Å., Hökfelt, T., Goldstein, M. and Park, D. Retrograde peroxidase tracing of neurons combined with transmitter histochemistry. Brain Res. 84, 313-319 (1975).
82. Maler, L., Fibiger, H.C. and McGeer, P.L. Demonstration of the nigrostriatal projection by silver staining after nigral injections of 6-hydroxydopamine. Exp. Neurol. 40, 505-515 (1973).
83. Martin, G.F. and Hamel Jr., E.G. The striatum of the opossum, Didelphis virginiana. Description and experimental studies. J. comp. Neurol. 131, 491-516 (1967).
84. McLardy, T. Projection of the centromedian nucleus of the human thalamus. Brain 71, 290-303 (1948).
85. Mehler, W.R. Further notes on the centre median nucleus of Luys. In: D.P. Purpura and M.D. Yahr (Eds.), The Thalamus, Columbia University Press, New York, 109-127 (1966).
86. Mehler, W.R. and Nauta, W.J.H. Fiber connections within the extrapyramidal system. Confin neurol. (Basel) 35, 205-222 (1974).
87. Mettler, F.A. Extensive unilateral cerebral removals in the primate. Physiologic effects and resultant degeneration. J. comp. Neurol. 79, 185-243 (1943).
88. Miller, J.J., Richardson, T.L., Fibiger, H.C. and McLennan, H. Anatomical and electrophysiological identification of a projection from the mesencephalic raphe to the caudate-putamen in the rat. Brain Res. 97, 133-138 (1975).
89. Moore, R.Y., Bhatnagar, R. and Heller, A. Anatomical and chemical studies of a nigro-neostriatal projection in the cat. Brain Res. 30, 110-135 (1971).
90. Monakow, C.von Experimentelle und pathologisch-anatomische Untersuchungen über die Haubenregion, den Sehhügel und die Regio subthalamica, nebst Beiträgen zur Kenntnis früh erworbener Gross- und Kleinhirndefekte. Arch. Psychiat. Nervenkr. 27, 1-219 (1895).
91. Nauta, H.J.W. Efferent projections of the caudate nucleus, pallidal complex, and subthalamic nucleus in the cat. (Thesis), Case Western Reserve University, Cleveland, Ohio (1974).
92. Nauta, W.J.H. and Mehler, W.R. Projections of the lentiforme nucleus in the monkey. Brain Res. 1, 3-42 (1966).

93. Nauta, H.J.W., Pritz, M.B. and Lasek, R.J. Afferents to the rat caudoputamen studied with horseradish peroxidase. An evaluation of a retrograde neuroanatomical research method. Brain Res. 67, 219-238 (1974).
94. Nauta, W.J.H. and Whitlock, D.G. An anatomical analysis of the non-specific thalamic projection system. In: J.F. Delafresnaye (Ed.), Brain Mechanisms and Consciousness, Blackwells, Oxford, 81-104 (1954).
95. Niimi, K., Ikeda, T., Kawamura, S. and Inoshita, H. Efferent projections of the head of the caudate nucleus in the cat. Brain Res. 21, 327-343 (1970).
96. Parizek, J., Hassler, R. and Bak, I.J. Light and electron microscopic autoradiography of substantia nigra of rat after intraventricular administration of tritium labeled norepinephrine, dopamine, serotonin and the precursors. Z. Zellforsch. 115, 137-148 (1971).
97. Pasik, P., Pasik, T. and DiFiglia, M. Quantitative aspects of neuronal organization of the neostriatum of the macaque monkey. In: M.D. Yahr (Ed.), The Basal Ganglia, Raven Press, New York, 57-90 (1976).
98. Pasquier, D.A., Forbes, W.B., Kemper, T.L. and Morgane, P.J. Serotoninergic (dorsalis raphe nucleus) connections with the striatum. Anat. Rec. 187, 677 (1977).
99. Petras, J.M. Some fiber connections of the precentral and postcentral cortex with basal ganglia, thalamus and subthalamus. Trans. Amer. neurol. Ass. 91, 274-275 (1965).
100. Petras, J.M. Some efferent connections of the superior and inferior parietal lobules with the basal ganglia, diencephalon and midbrain in the rhesus monkey. Anat. Rec. 163, 243-244 (1969).
101. Pickel, V.M., Segal, M. and Bloom, F.E. A radioautographic study of the efferent pathways of the nucleus locus coeruleus. J. comp. Neurol. 155, 15-42 (1974).
102. Poirier, L.J. Experimental and histological study of midbrain dyskinesias. J. Neurophysiol. 23, 534-551 (1960).
103. Poirier, L.J., McGeer, E.G., Larochelle, L., McGeer, P.L., Bedard, P. and Boucher, R. The effect of brain stem lesions on tyrosine and tryptophan hydroxylase in various structures of the telencephalon of the cat. Brain Res. 14, 147-155 (1969).
104. Poirier, L.J., Singh, P., Boucher, R., Bouvier, G., Olivier, A. and Larochelle, P. Effect of brain lesions on striatal monoamines in the cat. Arch. Neurol. Psychiat. (Chic.) 17, 601-608 (1967).
105. Poirier, L.J. and Sourkes, T.L. Influence of the substantia nigra on the catecholamine content of the striatum. Brain 88, 181-192 (1965).
106. Powell, T.P.S. and Cowan, W.M. The connexions of the midline and intralaminar nuclei of the thalamus of the rat. J. Anat. (Lond.) 88, 307-319 (1954).
107. Powell, T.P.S. and Cowan, W.M. A study of thalamo-striate relations in the monkey. Brain 79, 364-390 (1956).
108. Rinvik, E. and Grofová, I. Observations on the fine structure of the substantia nigra in the cat. Exp. Brain Res. 11, 229-248 (1970).
109. Rosegay, H. An experimental investigation of the connections between the corpus striatum and substantia nigra in the cat. J. comp. Neurol. 80, 293-321 (1944).
110. Royce, G.J. Afferents to the caudate nucleus in the cat. Neurosci. Abstr. 3, 45 (1977).
111. Royce, G.J. Autoradiographic evidence for a discontinuous projection to the caudate nucleus from the centromedian nucleus in the cat. Brain Res. 146, 145-150 (1978).
112. Simma, K. Zur Projektion des Centrum medianum und Nucleus parafascicularis thalami beim Menschen. Mschr. Psychiat. Neurol. 122, 32-46 (1951).
113. Sotelo, C. and Riche, D. The smooth endoplasmic reticulum and the retrograde and fast orthograde transport of horseradish peroxidase in the nigro-striatonigral loop. Anat. Embryol. 146, 209-218 (1974).

114. Szabo, J. Topical distribution of the striatal efferents in the monkey. Exp. Neurol. 5, 21-36 (1962).
115. Szabo, J. The efferent projections of the putamen in the monkey. Exp. Neurol. 19, 463-476 (1967).
116. Szabo, J. Projections from the body of the caudate nucleus in the Rhesus monkey. Exp. Neurol. 27, 1-15 (1970).
117. Szabo, J. A silver impregnation study of nigrostriate projections in the cat. Anat. Rec. 169, 441 (1971).
118. Szabo, J. The course and distribution of the efferents from the tail of the caudate nucleus in the monkey. Exp. Neurol. 37, 562-572 (1972).
119. Szabo, J. Horseradish peroxidase study of the striatal and nigral efferents. Anat. Rec. 187, 726 (1977).
120. Tobias, T.J. Afferents to prefrontal cortex from the thalamic mediodorsal nucleus in the rhesus monkey. Brain Res. 83, 191-212 (1975).
121. Tobias, T.J. and Ebner, F.F. Thalamocortical projections from the mediodorsal nucleus in the virginia opossum. Brain Res. 52, 79-96 (1973).
122. Ungerstedt, U. Stereotaxic mapping of the monoamine pathways in the rat brain. Acta physiol. scand. (Suppl.) 367, 1-48 (1971).
123. Usunoff, K.G., Hassler, R., Romansky, K., Usunova, R.P. and Wagner, A. The nigrostriatal projection in the cat. Part I. Silver impregnation study. J. neurol. Sci. 28, 265-288 (1976).
124. Usunoff, K.G., Hassler, R., Wagner, A. and Bak, I.J. The efferent connections of the head of the caudate nucleus in the cat: an experimental morphological study with special reference to a projection to the raphe nuclei. Brain Res. 74, 143-148 (1974).
125. Verhaart, W.J.C. Fiber analysis of the basal ganglia. J. comp. Neurol. 93, 425-440 (1950).
126. Vogt, C. and Vogt, O. Zur Lehre der Erkrankungen des striären Systems. J. Psychol. Neurol. (Lpz.) 25, 627-648 (1920).
127. Vogt, C. and Vogt, O. Thalamusstudien I-III. J. Psychol. Neurol. (Lpz.) 50, 31-154 (1941).
128. Voneida, T.J. An experimental study of the course and destination of fibers arising in the head of the caudate nucleus in the cat and monkey. J. comp. Neurol. 115, 75-87 (1960).
129. Webster, K.E. Cortico-striate interrelations in the albino rat. J. Anat. (Lond.) 95, 532-544 (1961).
130. Webster, K.E. The cortico-striatal projection in the cat. J. Anat. (Lond.) 99, 329-337 (1965).
131. Whitlock, D.G. and Nauta, W.J.H. Subcortical projections from the temporal neocortex in macaca mulatta. J. comp. Neurol. 106, 183-212 (1956).
132. Wilson, S.A.K. An experimental research into the anatomy and physiology of the corpus striatum. Brain 36, 427-492 (1914).
133. Wise, S.P. and Jones, E.G. Cells of origin and terminal distribution of descending projections of the rat somatic sensory cortex. J. comp. Neurol. 175, 129-158 (1977).
134. Yeterian, E.H. and van Hoesen, G.W. Cortico-striate projections in the rhesus monkey: the organization of certain cortico-caudate connections. Brain Res. 139, 43-63 (1978).

Localization of Neurotransmitter Candidates in Neostriatum

Frode Fonnum and Ivar Walaas

Norwegian Defence Research Establishment, Division for Toxicology, P.O. Box 25, N-2007 Kjeller, Norway

This review will deal with the distribution of neurotransmitter candidates and receptors in the neostriatum (caudate-putamen), n. accumbens and pallidum both the external and internal segment which in rodents is called nucleus entopeduncularis. As a whole these regions will also be described by the term basal ganglia. For discussion on the morphological basis for this division the reader is referred to the chapters by Webster and Heimer. For discussion on morphological evidence for afferent and efferent fibers the reader is referred to the chapters by Pasik et al and Grofová.

It is a pleasant task to discuss the localization of neurotransmitters in striatum and its connections since there has been a tremendous development in the last few years, particularly with regard to the role of amino acids and peptides.

Four groups of transmitter candidates have been found in these regions: acetylcholine (ACh), monoamines, peptides and amino acids. The distribution of cholinergic parameters (Table 1) demonstrates that the activity of choline acetyltransferase (ChAT), and the endogenous level of ACh are high in neostriatum and n. accumbens and low in globus pallidus (10, 19, 20). Apart from n. interpeduncularis they are among the highest levels in brain. The ACh turnover in neostriatum and n. accumbens is rapid, which further indicates a considerable cholinergic activity in these regions (43). A high density of ACh receptors is also present in neostriatum. They are mostly muscarinic in nature, as measured by quinuclidinyl benzilate (QNB) binding (67, 79), but a sizeable population of proteins binding to bungarotoxin is also present (51, 69). It is at present not known to what extent this binding correspond to the nicotinic receptor.

The levels of catecholamine related parameters (Table 2) demonstrate a very high activity of dopamine (DA) syntesizing enzymes and high concentrations of DA in neostriatum and n. accumbens (20, 46, 59, 76), while globus pallidus has intermediate levels of these parameters. Noradrenaline (NA) and the specific NA marker dopamine-β-hydroxylase display negligible levels in neostriatum, but NA is present in globus pallidus and n. accumbens (57, 76). Adrenaline is not found in the basal ganglia (75), whereas serotonin (5HT) is present in all of them, with the highest concentration in the globus pallidus (58). The level of histamine is rather low in both neostriatum and globus pallidus (66), but it is higher than in hippocampus

TABLE 1 Distribution of cholinergic parameters in neostriatum, globus pallidus and nucleus accumbens of rat brain

	Units	Neo-striatum	Globus pallidus	N. accumbens Lat	N. accumbens Med	Max. level in brain/region	Ref
ChAT	μmole/h/g dry wt	83	21	41	84	311/n.inter-peduncularis	19
ACh	nmole/mg protein	0.60	0.16	0.60		1.27/n.inter-peduncularis	10
ACh-turn-over	h	0.10	0.16	0.12		-	43
Bungaro-toxin binding	pmoles/mg protein	0.16	-	-		0.69/inferior colliculus	51, 69
QNB binding	pmoles/g protein	478	-	-		478/neostria-tum	79

Abbreviations: ACh - acetylcholine
ChAT - acetyltransferase
QNB - quinuclidinyl benzilate

where histamine is thought to function as a neurotransmitter (3). Histaminergic fibers are therefore probably present in striatum. The distribution of receptors for DA, NA and 5HT in different parts of striatum seems generally to correlate with the levels of presynaptic markers (15, 17). However, a surprisingly high density of the β-adrenergic receptor (dihydroalprenolol binding) is found in human neostriatum, indicating that the human neostriatum may receive some NA fibers (16), or alternatively that the β-adrenergic receptor is unsuitable as a marker for NA synapses.

Several peptide transmitter candidates have been found in the basal ganglia (Table 3). Methionine (met)-enkephalin, whose Ca^{++} dependent release from globus pallidus has been demonstrated (35), is concentrated in globus pallidus, but it is also present in neostriatum and n. accumbens (33). In contrast, opiate receptors are, at least in human brain, more concentrated in the neostriatum (15) and n. accumbens than in the globus pallidus. The distribution of these receptors in the neostriatum and n. accumbens appears to be patchy (2). Substance P is more evenly distributed in the different parts of the basal ganglia, but it is much more concentrated in the substantia nigra. In rats only n. accumbens contains sizeable amounts of neurotensin (24), while in calf brain, high levels have also been reported in caudate and globus pallidus (23). This difference could also reflect differences in the specificity of the different antibodies used. Somatostatin, mostly found in the mediobasal hypothalamus, is also present in the n. accumbens (39). Vasoactive intestinal polypeptide and a gastrinlike peptide which both have been suggested as possible transmitter candidates (28, 56) have also been found in neostriatum by immunohistochemical technique, with especially high levels of the gastrinlike peptide.

TABLE 2 Distribution of monoaminergic parameters in neostriatum, globus pallidus and nucleus accumbens of rat brain

	Units	Neo-striatum	Globus pallidus	N. accumbens Lat	N. accumbens Med	Max. level in brain/region	Ref
AAD	μmol/h/g dry wt	14.2	6.5	10.4	16.6	28/ventral tegmental area	19
TyrH	nmole/h/mg dry wt	11	-	6		28/ventral tegmental area	59
TyrH[x]	nmole/h/g wet wt	112	36	-		141/substantia nigra	46
DBH	nmole/h/g wet wt	8	-	-		198/medial hypothalamus	57
DA	pg/μg prot	70.5	39.6	87.2		94.5/olfactory tubercle	76
NA	pg/μg prot	n.d.	5.3	9.3		88.9/n.dorsomedialis hypothalami	76
A	ng/mg prot	n.d.	n.d.	n.d.		2.5/n.periventricularis hypothalami	75
TryH	pmole/45 min/mg prot	57	-	-		268/hypothalamus	
5HT	ng/mg protein	11.0	19.0	14.7		26.6/n.raphe dorsalis	58
Histamine[x]	ng/mg protein	0.7	1.8	-		26/corpus mammillare	66
Spiroperidol binding	fmole/mg protein	270	101	-		270/neostriatum	17
Haloperidol[xx] binding	fmole/mg protein	35	<10	-		35/neostriatum	15
5HT binding[xx]	fmole/mg protein	71	87	-		152/frontal cortex	15 15
DHA binding[xx]	fmole/mg protein	105	104	-		261/medulla oblongata	15

[x]Monkey [xx]Human n.d. - not detectable

Abbreviations:
A - adrenaline
DA - dopamine
DHA - dihydroalprenolol
NA - noradrenaline
TryH - tyrosine hydroxylase
AAD - aromatic amino acid decarboxylase
DBH - dopamine-β-hydroxylase
5HT - 5 hydroxytryptamine (serotonin)
TryH - tryptophan hydroxylase

TABLE 3 Distribution of peptide transmitter candidates and related parameters in neostriatum, globus pallidus and nucleus accumbens of rat brain

	Units	Neo-striatum	Globus pallidus	N.accumbens	Max. level in brain/region	Ref
Enkephalin (met)	ng/mg protein	10	76	10	76/globus pallidus	33
Naloxone binding[x]	fmole/mg protein	35	19	-	93/frontal cortex	
Substance P	ng/g wet wt	335	448	363	2325/substantia nigra	36
Neurotensin	ng/mg protein	0.03	-	1.37	3.38/median eminence 1.69/preoptic region	39
Somatostatin	ng/mg protein	0.6	-	2.4	73.1/mediobasal hypothalamus 4.3/preoptic region	39
VIP	-	++	-	+	cortex + amygdala	24, 28
Gastrinlike peptide	-	+	-	-	cortex + striatum	56

[x]Human brain

Abbreviations: VIP - vasoactive intestinal polypeptide

Two amino acids are strong transmitter candidates in these regions, γ-aminobutyric acid (GABA) and glutamate. We have recently studied the distribution of glutamate decaboxylase (GAD), a marker for GABA terminals in the basal ganglia (18, 19, 20, 21), and found that the n. accumbens and globus pallidus have a considerable GAD activity, with the rostral neostriatum having only half that value (Table 4). However, these values were much lower than the activities found in substantia innominata and substantia nigra, pars reticulata, where major parts of the outflow from n. accumbens and neostriatum terminate, respectively (7, 78). In contrast, GABA binding to receptors in humans is less concentrated in globus pallidus than in neostriatum and in n. accumbens (41). Strychnine binding, which is a measure of glycine receptors, is negligible in neostriatum (80). Glycine is therefore not an important neurotransmitter candidate in this region.

Glutamate has recently emerged as a major excitatory transmitter candidate in several brain regions, among them neostriatum and n. accumbens. Thus, we have found that the high affinity synaptosomal uptake of this amino acid, which seems to be a suitable marker for such neurons (22, 70), is very active in the neostriatum and slightly less so in the n. accumbens (77). In contrast the globus pallidus seems to be almost devoid of putative glutamatergic terminals. In the neostriatum and n. accumbens, iontophoretic application of glutamate has been shown to excite neurons (44), and a Ca^{++} dependent release of the glutamate analogue D-aspartate from striatal slices has been observed (Fonnum, unpublished observation). Con-

TABLE 4 Distribution of amino acid-ergic parameters in neostriatum, globus pallidus and nucleus accumbens of rat brain

	Units	Neo-striatum	Globus pallidus	N. accumbens Lat	N. accumbens Med	Max. level in brain/region	Ref
GAD	µmole/h/g dry wt	90	198	163	220	1042/substantia innominata 614/substantia nigra	19
GABA-T[x]	µmol/h/g dry wt	114	52	-		294/inferior olive	15
GABA-binding[xx]	fmole/mg protein	67	17	58		328/cerebellar cortex	15
Strychnine[x] binding	cpm/mg protein	<80	-	-		2315/spinal cord	80
High affinity glutamate uptake	µmole/h/g	4.1	0.4	2.7		8.2/Thalamus	20 77

[x]Monkey [xx]Human

Abbreviations: GAD - glutamate decarboxylase GABA-T - GABA transaminase

siderable evidence therefore supports the transmitter role for glutamate in these regions.

The detailed topographical distribution of some of these transmitter markers in the neostriatum of rats and monkeys has also been studied (Table 5, Fig. 1, 2, 3). The cholinergic terminals are rather evenly distributed rostrocaudally in both species (60, 20) and this is parallelled by the muscarinic receptor (67). Enkephalin levels and opiate receptors are more concentrated rostrally in monkey (64), while substance P seems to be concentrated caudally in rat (60). The latter rostrocaudal difference was, however, not statistically significant. GABAergic neurons are evenly distributed in the rostro-caudal direction (20). Of the mono-amine-systems, the DA terminals and DA receptors are concentrated predominantly in the rostral neostriatum (4, 71), while 5HT is concentrated more in the caudal part (72)

TABLE 5 Topographical distribution of some parameters in monkey caudate nucleus (CN) and putamen

	Unit	CN Head	CN Body	CN Tail	Putamen	Ref
Met-Enkephalin	units/mg protein	5,28	3,04	1,72	0,96	64
Opiate receptor	femtomole/ mg protein	19.4	9.0	8.9	11.7	74
ChAT	nmole/h/mg protein	73	90	-	111	67
QNB binding	pmole/g	976	1061	-	1126	67

Abbreviations as in Table 1.

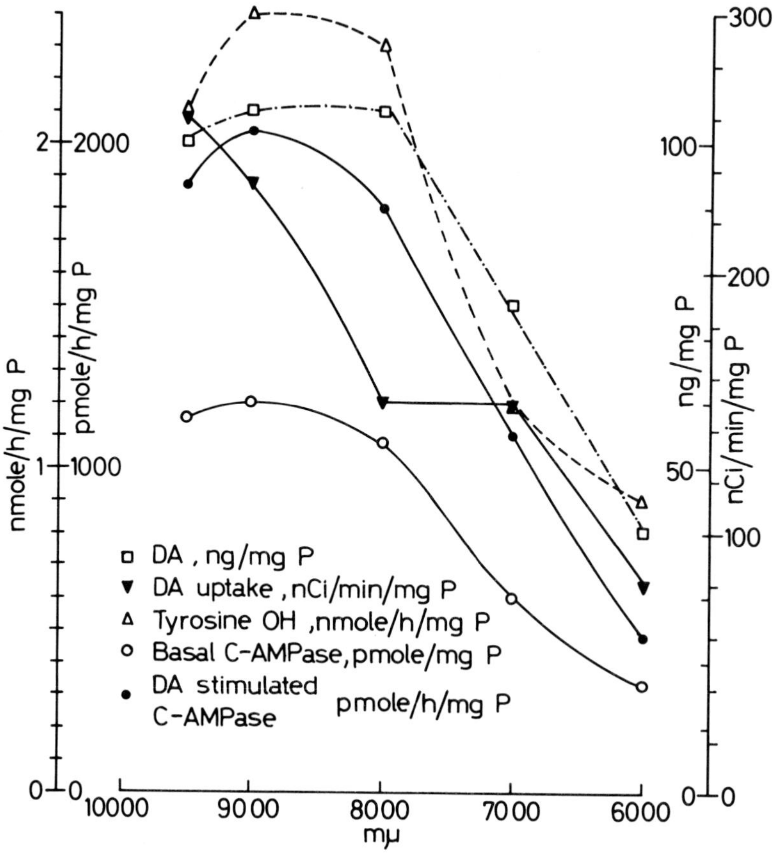

Fig. 1. Rostrocaudal distribution of dopamine-related parameters in rat neostriatum DA - dopamine, Tyrosine OH - tyrosine hydroxylase, C-AMPase - adenyalte cyclase. Data taken from ref 4, 60, 71.

To understand the anatomical and functional organization of the transmitter systems reviewed above, the neurons which contain the different transmitter candidates and their receptors must be identified. Such studies were started with the monoamine histofluorescence technique, which demonstrated the now well-known DA content of the nigrostriatal fibers (1). This tract which is responsible for all DA in striatum, is at present probably the best known of all fiber projections with identified transmitter in the mammalian central nervous system, and it will therefore not be discussed further here (for a recent review, see (5)). A similar DA pathway from mesencephalon to n. accumbens has been described (74). Histofluorescent techniques have also demonstrated a 5HT-containing projection from the raphe nuclei to the neostriatum (23), which by lesion techniques was shown (42) to originate solely from the dorsal raphe. This has later been confirmed (27) (Table 6). The origin of the 5HT in the globus pallidus and n. accumbens has not been studied in the same detail.

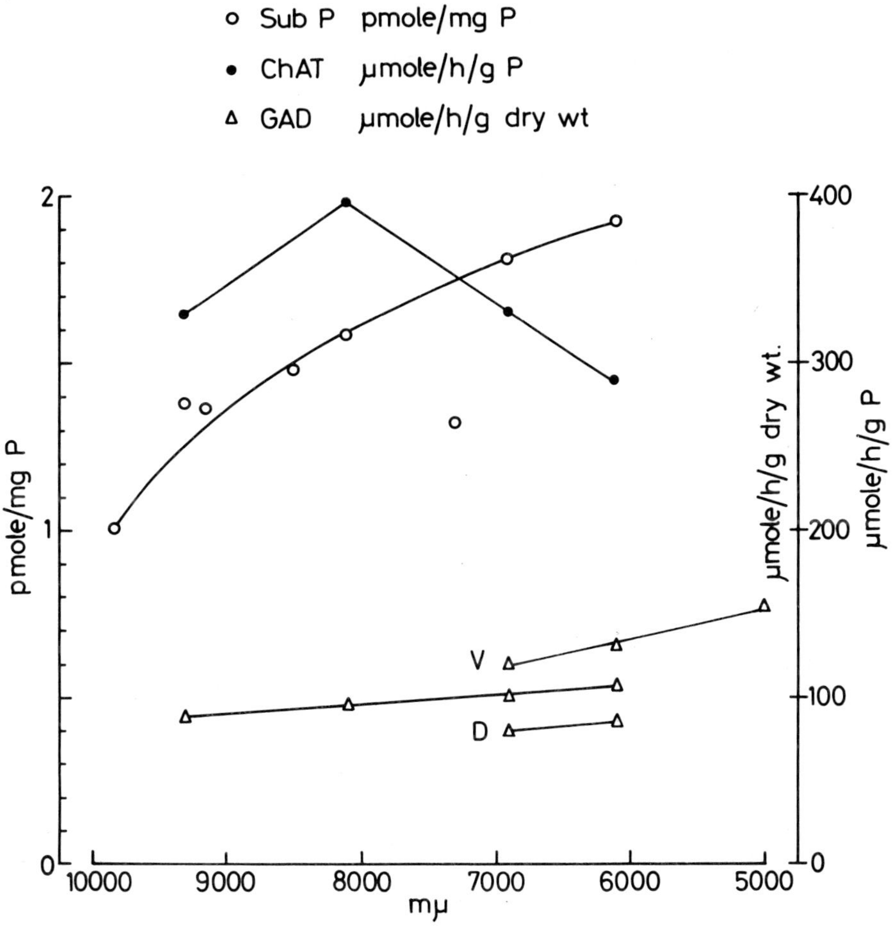

Fig. 2. Rostrocaudal distribution of substance P, choline acetyltransferase (ChAT) and glutamate decarboxylase (GAD) in rat neostriatum. Data compiled from ref 20 and 60.

TABLE 6 Origin of serotoninergic fibers to rat neo-striatum

Lesioned region	Per cent change in striatum	Parameter	Ref
Dorsal raphe	-54	5-HT	42
Dorsal raphe	-72	TryH	27
Rostral dorsal raphe	-66	5-HT	42
Mid-caudal dorsal raphe	-37	5-HT	42
Median raphe	+3	5-HT	42

Abbreviations as in Table 2.

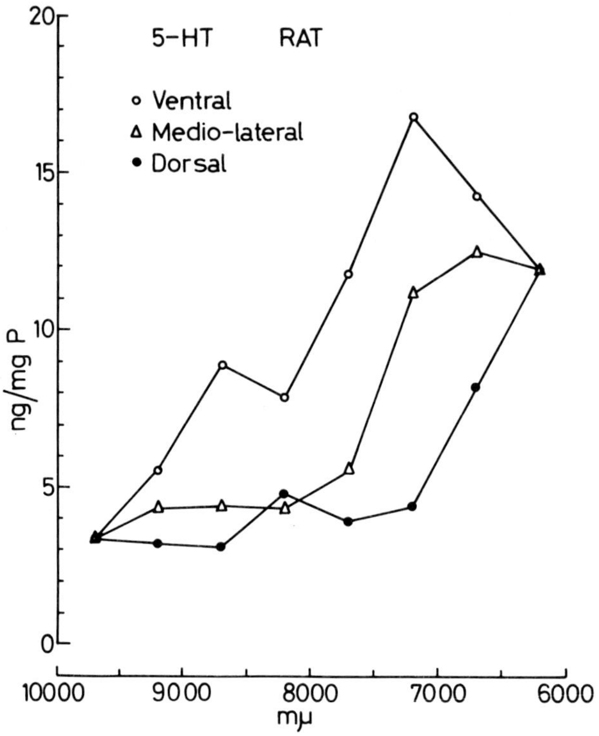

Fig. 3 Rostrocaudal distribution of serotonin (5HT) in rat neostriatum. Data compiled from ref 72

Lesion techniques have also been used on a third major afferent projection to neostriatum, the corticostriatal fibers. Ablation of frontal, but not caudal, neocortex was shown to selectively decrease the high affinity uptake of glutamate in the frontal neostriatum (Table 7), without affecting other neuronal markers (14). This finding was confirmed (50), and extended (38) by showing that frontal cortical lesion selectively decreased the level of endogenous glutamate, but left the other amino acids unchanged in neostriatum. Thus fibers in the cortico-neostriatal pathway, at least from the frontal cortex, probably use glutamate as transmitter. A dec decrease in glutamate uptake in the contralateral striatum showed that the glutamatergic pathway is partly crossed. Recently, a comparable allocortical-accumbens glutamatergic projection has also been found (77).

Conflicting results have been presented regarding the fourth major afferent projection, the thalamostriatal fibers. (65) have reported a decrease in ChAT activity in rostral neostriatum after thalamic lesions, a finding at variance with the conclusions of other groups (8, 45). We have also not been able to demonstrate a decrease of the ChAT activity in rostral neostriatum after hemitransections between thalamus and striatum. Thus, the presence of cholinergic fibers in this projection is still under debate.

In order to identify transmitter candidates in intrinsic neurons, previous workers have mostly used an indirect approach, e.g. they lesioned afferent fibers and assumed that the transmitter was present in local neurons if no decrease was found

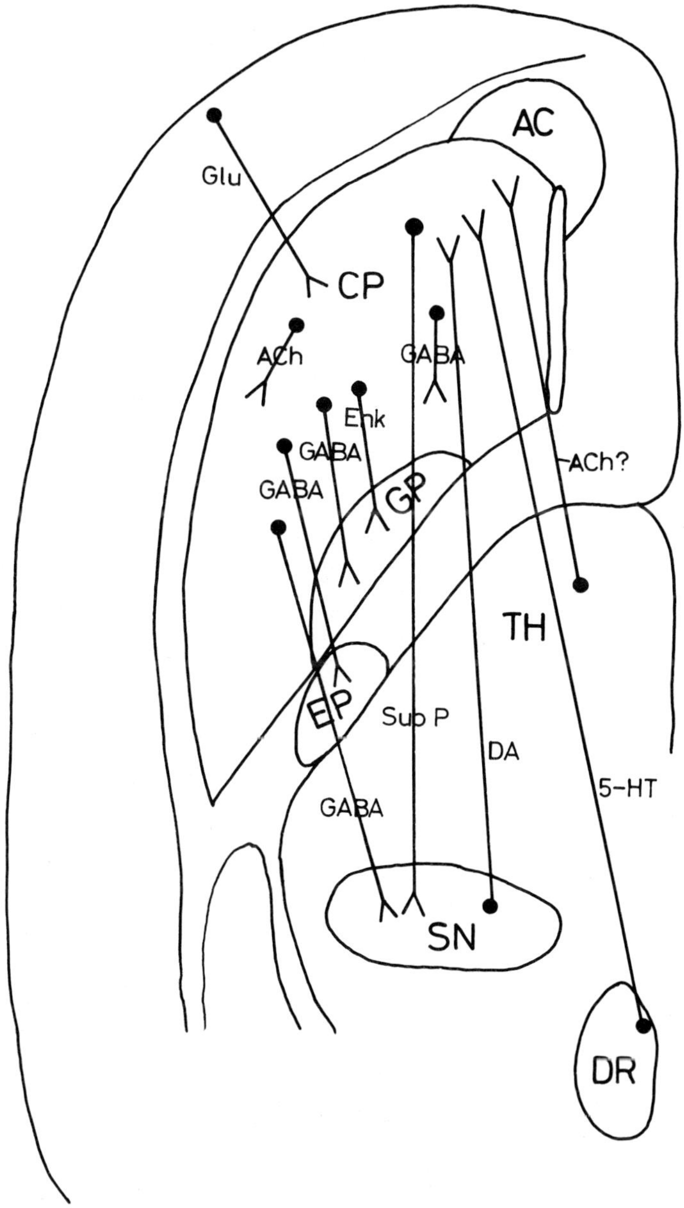

Fig. 4. Identity of neurotransmitters of connections of the rat neostriatum. CP - caudateputamen, GP - globus pallidus, EP - n.entopeduncularis, SN - substantia nigra, DR - dorsal raphe, Ac - n.accumbens.

TABLE 7 Origin of glutamergic fibers to rat neo-striatum

Lesioned region	Per cent change in striatum			
	Ipsilateral		Contralateral	
	Rostral	Caudal	Rostral	Caudal
Frontal cortex	-65	-52	-23	-13
Occipital cortex	+ 7	- 1	+12	+14
Hemicortex	-70	-75	-13	-20
Thalamus	+ 3	+10	+ 5	- 5

Data from ref 21

in the transmitter markers. By such means, ACh and GABA were suggested to be present in local neurons in the neostriatum (45, 47). Recent studies have shown that local application of kainic acid, a glutamate analogue, will destroy local cell bodies and dendrites but will leave afferent axons and terminals intact (32, 53). Application of kainic acid to striatum has shown that most - if not all - cholinergic and GABAergic terminals in the neostriatum are derived from local neurons. Afferent DA or 5HT fibers were not affected significantly number (Table 8). Similar studies in n. accumbens (77) demonstrated that this nucleus also contains ACh and GABA interneurons, and in addition some glutamatergic or aspartergic cells.

Recent technical development have made it possible to study the receptors for DA, 5HT, ACh, GABA and enkephalin (Table 9). Pharmacological techniques have indicated that the DA terminals contain both ACh (muscarinic and nicotinic) and GABA receptors (13, 29, 30). Axo-somatic terminals on the afferent fibers do not, however, occur frequently in neostriatum (Paskin et al, previous chapter). Kainic acid lesion shows that a major part of the 5HT receptors (61), all those DA receptors which are coupled to adenylate cyclase (48, 62), and almost half of the muscarinic ACh receptors are located on intrinsic cells. In contrast, all GABA receptors seem to be present on afferent fibers, some of them on the DA fibers (9), and almost half of the haloperidol-binding DA receptors are located on the corticostriatal glutamatergic fibers (63). Enkephalin receptors have also been found on DA-terminals (55).

The identity of neurotransmitters in the efferent fiber projections from the neostriatum have been investigated recently (Table 10, 11, 12). The striatopallidal fibers have recently been shown to consist of both GABAergic and enkephalinergic fibers (12, 20), the striato-entopeduncular projection has some GABAergic fibers (20), and the striatonigral projection has both GABAergic and substance P containing fibers (6, 18, 20, 37, 34, 52). The GABAergic projections to globus pallidus and entopeduncularis could in part be colaterals of the striato-nigral projection. It is now well established that the GABA and substance P containing fibers originate in different parts of neostriatum, e.g. substance P predominantly in the rostral part and GABA mainly in the caudal part (6, 25). A pallidonigral GABAergic projection has also been reported (31). Interestingly, DA receptors coupled to adenylate cyclase are found on the striatonigral terminals in the substantia nigra where they probably are activated by DA released from local dendrites (26, 40). Thus DA probably controls the activity of both the striatal cell bodies and their terminal projections by means of independent distinct populations of DA receptors. Whether these receptors are located on both GABA and substance P neurons is presently not known (54, 68). Detailed knowledge on the localization of the neurons

TABLE 8 The effect of kainic acid injections on transmitter parameters in rat neostriatum

	Per cent change	Tool	Ref
Acetylcholine fibers	-70	ChAT	62
GABA fibers	-68	GAD	62
Serotonin fibers	- 1	H.A. 5-HT uptake	61
Dopamine fibers	+47[x] -13	TyrH DA	68
Monoamine oxidase	-10	-	68

[x]V_{max} constant, Km for pteridine cofactor reduced.

Abbreviations as in Table 1, 2, 4.

TABLE 9 The localization of neurotransmitter receptors in rat neostriatum

Lesioned fibers	GABA binding	QNB binding	DA-cyclase	Haloperidol binding	5-HT binding
	Per cent change after lesion				
Intrinsic neurons (kainic acid)	+200[x]	-43	-85	-35	-63
Nigrostriatal fibers	- 22	-	-	-	-
Corticostriatal fibers	-	-	-15[xx]	-32	-
References	9	62	48, 63	63	61

[x]Number of binding sites constant, affinity for GABA increased.

[xx]Not significantly different from control.

Pharmacological studies also indicate the presence of ACh, DA and GABA presynaptic receptors on the nigrostriatal fibers.

which contain the other peptide transmitter candidates present in neostriatum (see above) is not yet available.

In conclusion (Fig. 4), the evidence presented in this review indicates that the neostriatum receives DA fibers from substantia nigra, which terminate on receptors located on neostriatal neurons. The afferents from the dorsal raphe are serotoninergic. Afferents from frontal cortex are glutamatergic, and judged from the sensitivity to kainic acid, may have receptors on cholinergic, GABAergic and substance P cells. Some of the afferents from thalamus to the rostral neostriatum may be cholinergic. Intrinsic cells in neostriatum, n. accumbens and globus pallidus are cholinergic or GABAergic. There are cholinergic receptors on the DA-terminals and on neostriatal neurons. GABAergic receptors may be localized on afferent terminals. The efferent fibers to globus pallidus contain GABA and enkephalin, fibers to

TABLE 10 Effect of neostriatal lesions on neurotransmitters in rat globus pallidus

Parameter	Per cent change	Ref
GAD	-69	20
AAD	0	20
Substance P	0	34
Enkephalin	(loss of fluor-escense)	12

Abbreviations as in Table 2, 4.

TABLE 11 Effect of striatal lesions on neurotransmitters in rat nucleus entopeduncularis (globus pallidus internal segment)

Parameter	Per cent change	Ref
GAD	-52	20
GAD[x]	-72	20

[x]Frontal tip of globus pallidus lesioned.
Abbreviations as in Table 4.

TABLE 12 Effect of striatal lesions on neurotransmitter parameters in rat substantia nigra

Parameter	Per cent change	Ref
GAD	-33	20
GAD[x]	-80	20
Substance P[x]	-70	34
Dopamine adenylate cyclase	-90	68

[x]Frontal tip of globus pallidus lesioned.
Abbreviations as in Table 4.

n. entopeduncularis contain GABA, and fibers to substantia nigra contain GABA and substance P. The major afferent pathway which at present is largely undefined neurochemically, is the thalamostriatal fibers. Although transmitters of the major afferent and efferent fibers in striatum have been identified, the existence of several new transmitter candidates such as peptides, make it possible that some of the projections may contain fibers with different transmitters. Future work will in addition have to concentrate on the localization of receptors and the study of the interplay between of the transmitter systems.

(1) Anden N.E., Carlsson A., Dahlstrøm., Fuxe K., Hillarp N.A. & Larsson K., Demonstration and mapping out of nigroneostriatal dopamine neurons, Life Sci. 3, 523-530 (1964).
(2) Atweh S.F. & Kuhar M.J., Autoradiographic localization of opiate receptors in rat brain. III. The telencephalon, Brain Res. 134, 393-405 (1977).
(3) Barbin G., Garbarg M., Schwartz J.-G. & Storm-Mathisen J., Histamine synthesizing afferents to the hippocampal region, J. Neurochem. 26, 259-263 (1976).
(4) Bockaert J., Premont J., Glowinski J., Thierry A.M. & Tassin P., Topographical distribution of dopaminergic innervation and dopaminergic receptors in the rat striatum. II. Distribution and characteristics of dopamine adenylate cyclase - interaction of D-LSD with dopaminergic receptors, Brain Res. 107, 303-315 (1976).
(5) Björklund A. & Lindvall O., The meso-telencephalic dopamine neuron system. In The continuing evolution of the limbic system concept (Livingston K. & Hornykiewicz O., eds), Plenum, New York, in press (1978).
(6) Brownstein M.J., Mroz E.A., Tappaz M.L. & Leeman S.E., On the origin of substance P and glutamic acid decarboxylase (GAD) in the substantia nigra, Brain Res. 135, 315-323 (1977).
(7) Bunney B.S. & Aghajanian G.K., The precise localization of nigral afferents in the rat as determined by a retrograde tracing technique, Brain Res. 117, 423-435 (1976).
(8) Butcher S.G. & Butcher L.L., Origin and modulation of acetylcholine activity in the neostriatum, Brain Res. 71, 167-171 (1974).
(9) Campochiaro P., Schwarcz & Coyle J.T., GABA receptor binding in rat striatum: Localization and effects of denervation, Brain Res. 136, 501-511 (1977).
(10) Cheney D.L., Racagni G. & Costa E., Distribution of acetylcholine and choline acetyltransferase in specific nuclei and tracts of rat brain. In Biology of Cholinergic Function (Goldberg A.M. & Hanin I., eds), pp 655-659, Raven Press, New York (1976).
(11) Coyle J.T. & Schwarcz R., Lesion of striatal neurons with kainic acid provides a model for Huntington's chorea, Nature (London) 263, 244-246 (1976).
(12) Cuello A.C. & Paxinos G., Evidence for a long Leu-enkephalin striopallidal pathway in rat brain, Nature (London) 271, 178-180 (1978).
(13) De Belleroche J. & Bradford H.F., Biochemical evidence for the presence of presynaptic receptors on dopaminergic nerve terminals, Brain Res. 142, 53-68 (1978).
(14) Divac I., Fonnum F. & Storm-Mathisen J., High affinity uptake of glutamate in terminals of corticostriatal axons, Nature (London) 266, 377-378 (1977).
(15) Enna S.J., Bennett J.P., Bylund D.B., Creese I., Burt D.R., Charness M.E., Yamamura H.I., Simantov R. & Snyder S.H., Neurotransmitter receptor binding: regional distribution in human brain, J. Neurochem. 28, 233-236 (1977).
(16) Farley I. & Hornykiewicz O., Noradrenaline distribution in subcortical areas of the human brain, Brain Res. 126, 53-62 (1977).
(17) Fields J.Z., Reisine T.D. & Yamamura H.J., Biochemical demonstration of dopaminergic receptors in rat and human brain using (^{3}H) spiroperidol, Brain Res. 136, 578-584 (1977).
(18) Fonnum F., Grofová I., Rinvik E., Storm-Mathisen J. & Walberg F., Origin and distribution of glutamate decarboxylase in substantia nigra of the cat, Brain Res. 71, 77-92 (1974).

(19) Fonnum F., Walaas I. & Iversen E., Localization of GABAergic, cholinergic and aminergic structures in the mesolimbic system, J. Neurochem. 29, 221-230 (1977).
(20) Fonnum F., Gottesfeld Z. & Grofová I., Distribution of glutamate decarboxylase, choline acetyltransferase and aromatic amino acid decarboxylase in the basal ganglia of normal and operated rats. Evidence for striatopallidal, striatoentopeduncular and striatonigral GABAergic fibers, Brain Res. 143, 125-138 (1978)
(21) Fonnum F., Storm-Mathisen J. & Divac I., Glutamate as neurotransmitter of the cortico-striatal and cortico-thalamic projections, to be published (1979).
(22) Fonnum, F. & Walaas I., The effect of intrahippocampal kainic acid injections and surgical lesions on neurotransmitters in hippocampus and septum (submitted for publication), (1978).
(23) Fuxe K. & Jonsson G., Further mapping of central 5-hydroxytryptamine neurons: studies with the neurotoxic dihydroxytryptamines. In Advances in Biochemical Psychopharmacology 10, pp 1-12, Raven Press, New York (1974).
(24) Fuxe K., Hökfelt T., Said S.I. & Mutt V., Vasoactive intestinal polypeptide and the nervous system: immunohistochemical evidence for localization in central and peripheral neurons, particularly intracortical neurons of the cerebral cortex, Neuroscience Letts. 5, 241-246 (1977).
(25) Gale K., Hong J.-S. & Guidotti A., Presence of substance P and GABA in separate striatonigral neurons, Brain Res. 136, 371-375 (1977).
(26) Geffen L, Jessell T., Cuello A. & Iversen L.L., Release of dopamine from dendrites in rat substantia nigra, Nature (London) 260, 258-260 (1976).
(27) Geyer M.A., Puerto A., Dawsey W.J., Knapp S., Bullard W.P. & Mandell A.J., Histologic and enzymatic studies of the mesolimbic and mesostriatal serotonergic pathways, Brain Res. 106, 241-256 (1976).
(28) Giachetti A., Said S.I., Reynolds R.C. & Koniges F.C., Vasoactive intestinal polypeptide in brain: Localization in and release from isolated nerve terminals, Proc. Natl. Acad. Sci. U.S.A. 74, 3424-3428 (1977).
(29) Giorguieff M.F., Le Floc'h M.L., Glowinski J. & Besson M.J., Involvement of cholinergic presynaptic receptors of nicotinic and muscarinic types in the control of the spontaneous release of dopamine from striatal dopaminergic terminals in the rat, J. Pharmacol. Exp. Ther. 200, 535-544 (1977).
(30) Giorguieff M.F., Kemel M.L., Glowinski J. & Bessom M.J., Stimulation of dopamine release by GABA in rat striatal slices, Brain Res. 139, 115-130 (1978).
(31) Hattori T., McGeer P.L., Fibiger H.C. & McGeer E.G., On the source of GABA-containing terminals in the substantia nigra. Electon microscopic autoradiographic and biochemical studies, Brain Res. 54, 103-114 (1973).
(32) Hattori T. & McGeer E., Fine structural changes in the rat striatum after local injections of kainic acid, Brain Res. 129, 174-180 (1977).
(33) Hong J.S., Yang H.-Y.T., Fratta W. & Costa E., Determination of methionine enkephalin in discrete regions of rat brain, Brain Res. 134, 383-386 (1977).
(34) Hong J.S., Yang H.-Y., Racagni G. & Costa E., Projections of substance P containing neurons from neostriatum to substantia nigra, Brain Res. 122, 541-544 (1977).

(35) Iversen L.L., Iversen S.D., Bloom F.E., Vargo T. & Guillemin R., Release of enkephalin from rat globus pallidus in vitro, Nature (London) 271, 679-681 (1978).
(36) Kanazawa I. & Jessel T., Postmortem changes and regional distribution of substance P in the rat and mouse nervous system, Brain Res. 117, 362-367 (1976).
(37) Kanazawa I., Emson P.C. & Cuello A.C., Evidence for the existence of substance P-containing fibers in striatonigral and pallidonigral pathways in rat brain, Brain Res. 119. 447-453 (1977).
(38) Kim J.-S., Hassler R., Haug P. & Paik K.-S., Effect of frontal cortex ablation on striatal glutamic acid level in rat, Brain Res. 132, 370-374 (1977).
(39) Kobayashi R.M., Brown M. & Vale W., Regional distribution of neurotensin and somatostatin in rat brain, Brain Res. 126, 584-588 (1977).
(40) Korf J., Zieleman M. & Westerink B., Dopamine release in substantia nigra ? Nature (London) 260, 257-258 (1976).
(41) Lloyd K.G., Shemen L. & Hornykiewicz O., Distribution of high affinity sodium-independent (^{3}H)-gamma aminobutyric acid ((^{3}H) GABA) binding in the human brain: alterations in Parkinson's disease, Brain Res. 127, 264-278 (1977).
(42) Lorens S.A. & Guldberg H.C., Regional 5-hydroxytryptamine following selective midbrain raphe lesions in the rat, Brain Res. 78, 45-46 (1974).
(43) Mao C.C., Cheney D.L., Mario E., Revuelta A. & Costa E., Turnover times of gamma-aminobutyric acid and acetylcholine in nucleus caudatus, nucleus accumbens, globus pallidus and substantia nigra: effects of repeated administration of haloperidol, Brain Res. 132, 375-379 (1977).
(44) McCarthy P.S., Walker R.J. & Woodruff G.N., On the depressant action of dopamine in rat caudate nucleus and nucleus accumbens, Brit. J. Pharmacol. 59, 469P-470P (1977).
(45) McGeer P.L., McGeer E.G., Fibiger H.C. & Wickson V., Neostriatal cholineacetylase and cholinesterase following selective brain lesions, Brain Res. 35, 308-314 (1971).
(46) McGeer E.G., McGeer P.L. & Wada J.A., Distribution of tyrosine hydroxylase in human and animal brain, J. Neurochem. 18, 1647-1658 (1971).
(47) McGeer P.L. & McGeer E.G., Evidence for glutamic acid decarboxylase - containing interneurons in the neostriatum, Brain Res. 91, 331-335 (1975).
(48) McGeer E.G., Innanen V.T. & McGeer P.L., Evidence on the cellular localization of adenyl cyclase in the neostriatum, Brain Res. 118, 356-358 (1976).
(49) McGeer E.G. & McGeer P.L., Duplication of biochemical changes of Huntington's chorea by intrastriatal injections of glutamic and kainic acids, Nature (London) 263, 517-519 (1976).
(50) McGeer P.L., McGeer E.G., Scherer U. & Singh K., A glutamatergic corticostriatal path ? Brain Res. 128, 369-373 (1977).
(51) Morley B.J., Lorden J.F., Brown G.B., Kemp G.E. & Bradley R.J., Regional distribution of nicotinic acetylcholine receptor in rat brain, Brain Res. 134, 161-166 (1977).
(52) Mroz E.A., Brownstein M.J. & Leeman S.E., Evidence for substance P in the striato-nigral tract, Brain Res. 125, 305-311 (1977).
(53) Olney J.W., Rhee V. & Ho O.L., Kainic acid: a powerful neurotoxic analogue of glutamate, Brain Res. 77, 507-512 (1974).
(54) Phillipson O.T., Emson P.C., Horn A.S. & Jessel T., Evidence concerning the anatomical location of the dopamine stimulated

adenylate cyclase in the substantia nigra, Brain Res. 136, 45-58 (1977).

(55) Pollard H., Llorens-Cortes C. & Schwartz J.-C., Enkephalin receptors on dopaminergic neurons in rat striatum, Nature (London) 268, 745-747 (1977).

(56) Robberecht P., Deschodt-Lanckman M. & Vanderhaeghen J.J., Demonstration of biological activity of brain gastrinlike peptidic material in the human: its relationship with the COOH-terminal octapeptide of choliecystokinin, Proc. Natl. Acad. Sci. U.S.A. 75, 524-528 (1978).

(57) Ross R.A. & Reis D.J., Effects of lesions of locus coeruleus on regional distribution of dopamine-β-hydroxylase activity in rat brain, Brain Res. 73, 161-166 (1974).

(58) Saavedra J.M., Brownstein M. & Palkovits M., Serotonin distribution in the limbic system of the rat, Brain Res. 79, 437-441 (1974).

(59) Saito M., Hirano M., Uchimura H., Nakahara T. & Ito M., Tyrosine hydroxylase activity in the catecholamine nerve terminals and cell bodies of the rat brain, J. Neurochem. 29, 161-165 (1977).

(60) Scally M.C., Ulus I.H., Wurtman R.J. & Pettibone D.J., Regional distribution of neurotransmitter-synthesizing enzymes and substance P within the rat corpus striatum, Brain Res. 143, 556-560 (1978).

(61) Schwarcz R., Bennett J.P. jr. & Coyle J.T., Loss of striatal serotonin synaptic receptor binding induced by kainic acid lesions: Correlations with Huntington's Disease, J. Neurochem. 28, 867-869 (1977).

(62) Schwarcz R. & Coyle J.T., Striatal lesions with kainic acid: neurochemical characteristics, Brain Res. 127, 235-249 (1977).

(63) Schwarcz R., Creese I., Coyle J.T. & Snyder S.H., Dopamine receptors localised on cerebral cortical afferents to rat corpus striatum, Nature (London), 271, 766-768 (1978).

(64) Simantov R., Kuhar M.J., Pasternak G.W. & Snyder S.H., The regional distribution of a morphine-like factor enkephalin in monkey brain, Brain Res. 106, 189-197 (1976).

(65) Simke J.P. & Saelens J.K., Evidence for a cholinergic fiber tract connecting the thalamus with the head of the striatum of the rat, Brain Res. 126, 487-495 (1977).

(66) Snyder S.H. & Taylor K.M., Histamine in the brain: a Neurotransmitter ? In Perspectives in Neuropharmacology (Snyder S.H., ed.), pp 43-73, Oxford University Press, New York (1972).

(67) Snyder S.H., Chang K.J., Kuhar M.J. & Yamamura H.I., Biochemical identification of the mammalian muscarinic cholinergic receptor, Federation Proc. 34, 1915-1921 (1975).

(68) Spano P.F., Trabucchi M. & di Chiara G., Localization of nigral dopamine-sensitive adenylate cyclase on neurons originating from the corpus striatum, Science 196, 1343-1345 (1977).

(69) Speth R.C., Chen F.M., Lindström J.M., Kobayashi R.M. & Yamamura H.I., Nicotinic cholinergic receptors in rat grain identified by (^{125}I) Naja naja siamensis α-toxin binding, Brain Res. 131, 350-355 (1977).

(70) Storm-Mathisen J., Glutamic acid and excitatory nerve endings: reduction of glutamic acid uptake after axotomy, Brain Res. 120, 379-386 (1977).

(71) Tassin J.P., Cheramy A., Blanc G., Thierry A.M. & Glowinski J., Topographic distribution of dopaminergic innervation and of dopaminergic receptors in the rat striatum. I. Microestimation of (^{3}H)dopamine uptake and dopamine content in micro-

discs, Brain Res. 107, 291-301 (1976).
(72) Ternaux J.P., Héry F., Bourgoin S., Adrien J., Glowinski J. & Hamon M., The topographical distribution of serotoninergic terminals in the neostriatum of the rat and the caudate nucleus of the cat, Brain Res. 121, 311-376 (1977).
(73) Uhl G.R. & Snyder S.H., Regional and subcellular distributions of brain neurotensin, Life Sci. 19,1827-1832 (1976).
(74) Ungerstedt U., Stereotaxic mapping of the monoamine pathways in the rat brain, Acta physiol. scand. suppl. 367, 1-48 (1971).
(75) van der Gugten J., Palkovits M., Wijnen H.L.J.M. & Versteeg D.H.G., Regional distribution of adrenaline in rat brain, Brain Res. 107, 171-175 (1976).
(76) Versteeg D.H.G., van der Gugten J., de Jong W. & Palkovits M., Regional concentrations of noradrenaline and dopamine in rat brain, Brain Res. 113, 563-574 (1976).
(77) Walaas I. & Fonnum F., The effects of surgical and chemical lesions on neurotransmitter candidates in nucleus accumbens, (submitted for publication 1978).
(78) Williams D.J., Crossman A.R. & Slater P., The efferent projections of the nucleus accumbens in the rat, Brain Res. 130, 217-227 (1977).
(79) Yamamura H.I. & Snyder S.H., Muscarinic cholinergic Binding in Rat Brain, Proc. Natl. Acad. Sci. U.S.A. 71, 1725-1729 (1974).
(80) Young A.B. & Snyder S.H., Strychnine binding associated with glycine receptors of the central nervous system, Procl. Natl. Acad. Sci. U.S.A. 70, 2832-2836 (1973).

A Survey of Electrophysiology and Pharmacology of Neostriatal Input-Output Relations

J. Feger, J. M. Deniau, J. de Champlain and P. Feltz

Laboratoire de physiologie des centres nerveux
Université Pierre et Marie Curie, 4 place Jussieu, 75230 Paris Cedex 05

The aim of this review is to establish a correlation between morphological and neurochemical data on the organization of the major input-output relationships of the neostriatum (see reviews of PASIK, FONNUM and GROFOVA in this volume). In this respect, the striatal connections with cortex, thalamus, basal ganglia and brain stem will be examined. In a first part, the electrophysiological tracing of afferent and efferent pathways of the neostriatum will be described. Most of these studies rely on the method of antidromic activation. This electrophysiological method is based on the possibility to activate a cell retrogradely by axonal stimulation and provides information about the neurons sending their axon near the sites which are stimulated. Some pitfalls of this methodology will be discussed.

The second part of the review will be concerned with the nature of the synaptic events within the Neostriatum. The possibility of convergence of various neostriatal afferents on individual cells will be discussed. These cells will be tentatively divided in the ouptut neurons and interneurons on the basis of antidromic activation. On the other hand, the survey of the literature summarizes the inhibitory and excitatory effects of electrical stimulation of the neostriatal output reaching the pallidum and substantia nigra.

In a final discussion, concerning the electrophysiology and pharmacology of intrinsic neostriatal connections, the function of the nigro-striatal dopaminergic input will be examined. Understanding of synaptic actions of dopamine on striatal neurons which receive nigral inputs depends on the available informations about intrinsic synaptic organization of neostriatum. One must for instance distinguish effects of activation of interneurons from the effects attributable to the branching of both input and outputs fibres. This applies to the study of other neostriatal putative transmitters and/or neuro-modulators.

I - ELECTROPHYSIOLOGICAL TRACING OF STRIATAL INPUTS AND OUTPUTS

A - Inputs

1 - Cortical afferents

Cortico-striate neurons were recently identified electrophysiologically by KITAI et al (101). It appeared clear that : - the corticostriatal pathway could be activated antidromically ; - these identified cortical neurons were not antidromically activated after pyramidal stimulation and did not respond postsynaptically to thalamic stimulation at the level of VL Nucleus (Fig 1). The conclusion is that the cortico-striate neurons are different from those giving rise to the cortico-bulbar or corticospinal (pyramidal) tracts. Histological observations after peroxidase injection in the striatum of the cat indicate that cortico-striate neurons are medium size pyramidal cells located predominantly in layer III of the anterior sigmoid gyrus in the pericruciate cortex. There is however a discrepancy with other anatomical data (80,94) on the precise determination of the layers of origin of cortico-striatal neurons. This question is of fundamental interest for correlating cortical electrophysiological studies.

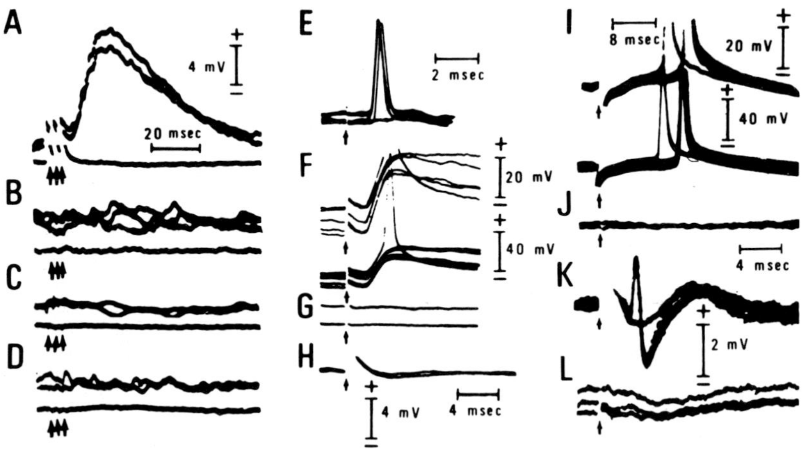

Figure 1 : Electrophysiological characterization of cat cortical neurons projecting to Striatum : A - D : intra- and extracellular (bottom trace) record of Cd neuron ; note large EPSPs to Cortical stimulation (A) and no clear short latency response to Pyramidal (B), V_L Thalamic (C) and basilar Pons (D). E - H : response pattern of a cortical neuron which is antidromically activated after Pyramidal stimulation (E), monosynaptically excited after VL stimulation (F,G) and unaffected by Striatal stimulation (H). I-J and K-L : these two cortical neurons are antidromically activated after Striatal stimulation and give no response after Pyramidal stimulation (J,L). From KITAI et al (101).

Other electrophysiological studies (3,4,18,24,88,100,101,116,140,154,164) report results obtained from orthodromic activation of cortico-striate neurons and extra- or intracellular records in the striatum. They provide electrophysiological confirmation of an excitatory relation between cortex and striatum. Moreover one of these studies (140) demonstrates that the caudate nucleus receives collaterals of cortico-pontine or cortico-olivary neurons with cell bodies in area 6 and thus provides an explanation for the occurence of evoked activity observed in the cerebellar cortex after caudate nucleus stimulation.

2 - Thalamo-striate afferents

Antidromic activation of units localized in centrum medianum was observed after stimulation in the caudate nucleus (Fig 2) (106). In another more recent study (148), it was observed that the same thalamic neuron can be antidromically activated by both pericruciate and putamen stimulation. The authors concluded that thalamic neuron bifurcate, sending one branch to putamen and another to cortex.

Orthodromic stimulation of thalamo-striatal pathway provides an excitatory input to striatal neurons, observed in both extra and intracellular records (3,4,24,105,126,187).

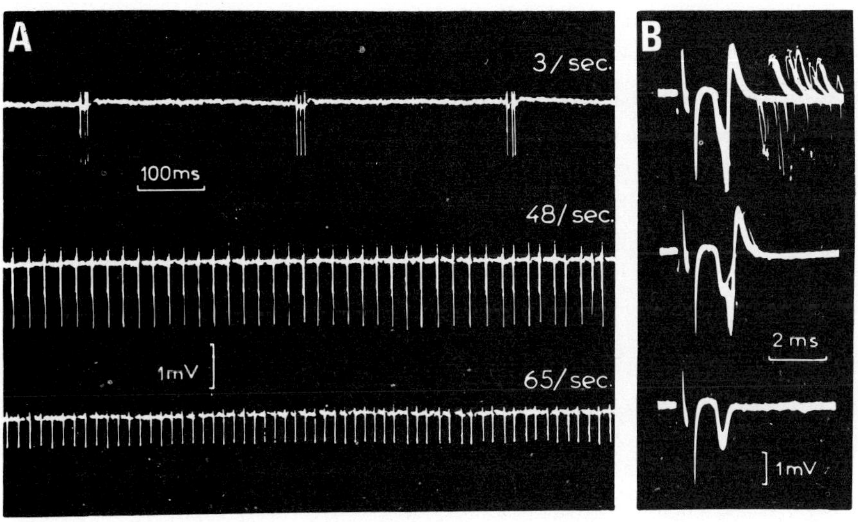

Figure 2 : Characterization of thalamic neurons projecting to Striatum: Antidromic invasion of a unit in cat Centrum Medianum. Thalamic stimulation induces fast antidromic response followed by burst discharge (A,B,first row). Postsynaptic bursting response disappears with fast rates of stimulation. Note stability of remaining antidromic response (B, second sweep, dissociation of IS-SD segment of spike and final failure of SD as stimulus rate was increased. From KRAUTHAMER et al (106).

3 - Nigral afferents

There are many electrophysiological experiments devoted to nigro-striatal projections. These afferents are generally considered to be excitatory and only a few authors report depression of active cells in contrast to excitation of inactive or silent striatal cells (24,30,31,51,53,57,62,64,72,88,89,100,102, 103,104,105,116,124,139,149,187).

The existence of two functionnally different nigrostriatal pathways was put forward by FRIGYESI and PURPURA (62), and was strongly suggested by FELTZ and DE CHAMPLAIN (57). The last authors observed persistance of striatal responses to nigral stimulation in animals where the striatal dopamine level was completely reduced by a combination of 6-hydroxydopamine, inhibition of monoamine bio-synthesis and monoamine depletion (Fig 4). FIBIGER el al (60) also infer the existence of two nigro-striatal projections from studies on anterograde transport of labelled amino-acids (see also 81,118). Another type of evidence is provided by the observation that some nigral neurons, labelled after retrograde transport of peroxidase following striatal injection, did not react to the cytochemical test for tyrosine hydroxylase (117) (Fig 3 and 4).

A recent electrophysiological study in the rat confirms the presence of two different nigrostriatal pathways (Fig 3) (38) : - two groups of nigral cells are antidromically activated after striatal stimulation. They differ with respect to both their localization in substantia nigra (pars compacta or reticulata) and to their electrophysiological characteristics (spontaneous firing rate ; conduction velocity 0,33 to 1 m/sec, or 1,9 to 10 m/sec respectively). Moreover electrophysiological studies have established that only the neurons localized in the pars reticulata of substantia nigra send branched axons reaching striatum, thalamic nuclei (VL and VM) and also superior colliculus ; the nigro-thalamic and nigro-collicular projections are bilateral. These nigral neurons exert an inhibitory control on thalamic (37) and collicular neurons (39). This result means that individual nigral neurons can inhibit simultaneously thalamic, collicular and perhaps striatal cells.

4 - Raphe afferents

The occurence of serotoninergic afferents from raphe nuclei was suspected on the basis of biochemical and cytochemical studies (7,20,22) and also from lesion or stimulation experiments (85,108). The projection was confirmed by anatomical studies with retrograde transport labelling cells in dorsal and medial raphe nuclei (see report of GROFOVA in this volume). Electrical stimulation of raphe nuclei inhibits caudate cells (131). This inhibition may however also depends on synaptic activity within the substantia nigra which also receives serotoninergic innervation (20, see references in 43 and 131).

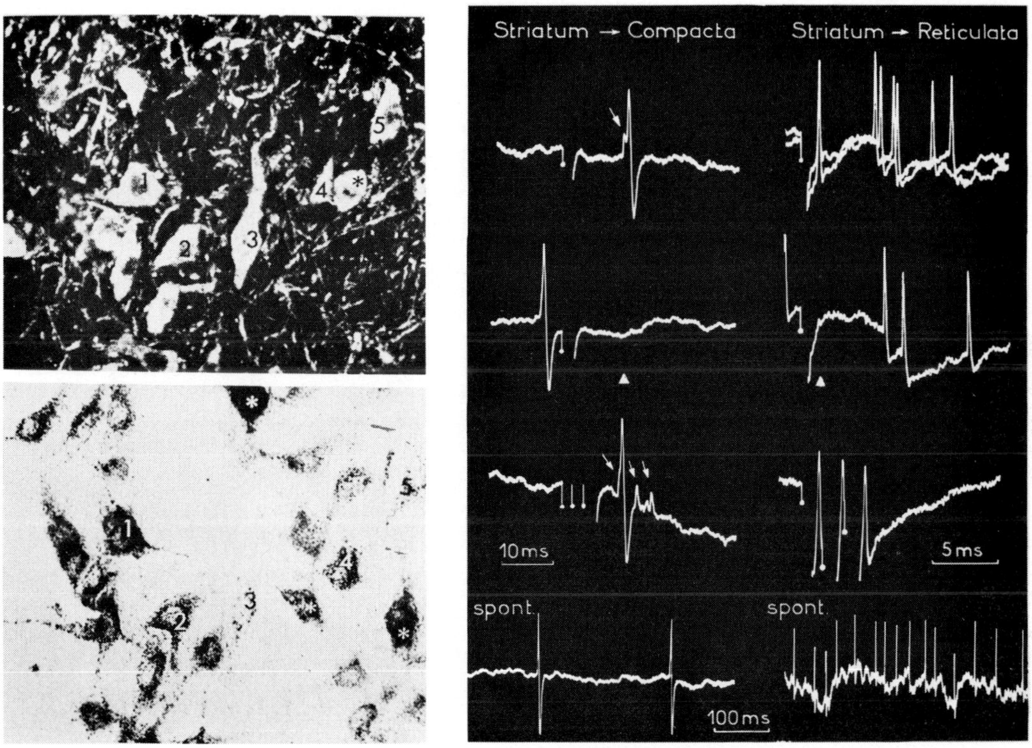

Figure 3 : Histological and electrophysiological demonstration of the existence of two different types of nigral neurons projecting to the striatum. A and B are adjacent sections of the substantia nigra. A shows tyrosine hydroxylase (TH) reacting cells. B shows cells labelled retrogradely by horseradish peroxidase (HRP) injected into the head of the caudate nucleus. Many cells contain both TH and HRP (numbered cells) but some cells contain only TH (black asterisk, A) while others are only HRP positive (white asterisks in B). From LJUNGDAHL et al (117). C : neuron in pars compacta is antidromically invaded, following striatal stimulation (dots under traces), with a latency of 12 msec. Second trace shows collision with a spontaneous spike. Third trace shows spike fractionation with high stimulation rates. Bottom trace shows slow spontaneous firing rate ($<$ 10Hz). D : neuron in pars reticulata also responds antidromically to caudate stimulation but with a shorter latency (1.2 msec). The whole spike follows high stimulation rates (third trace). The lowest trace shows the high (50 Hz) spontaneous firing rate. From DENIAU et al (38).

B - Outputs

Striatal efferents project only in a descending direction, to external segment of Globus Pallidus, to its internal segment (or entopeduncular nucleus), to substantia nigra and to some raphe nuclei. There is no anatomical evidence for a projection to the subthalamic nucleus as suggested from electrophysiological data (140).

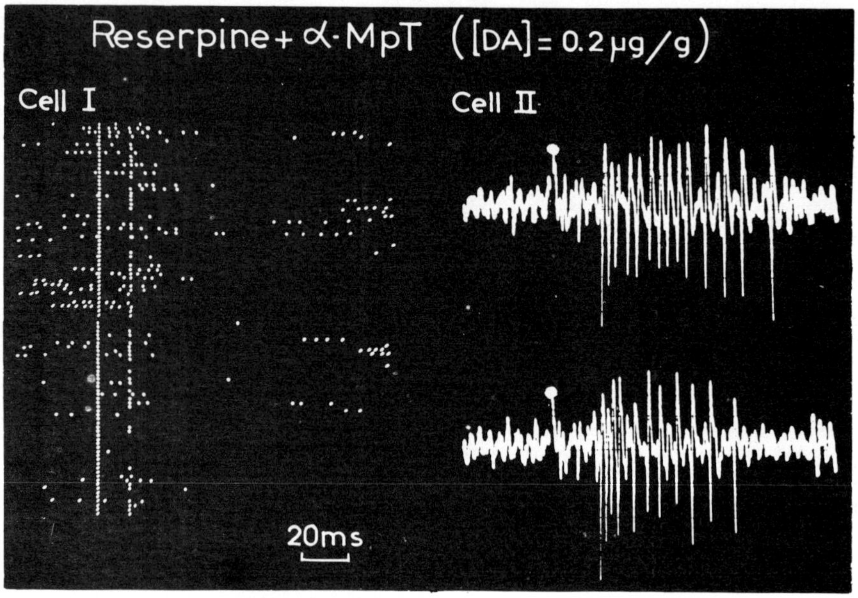

Figure 4 : The persistance on striatal cells of nigral evoked excitation and inhibition after dopaminergic denervation (see references in FELTZ, 1978) was confirmed in experiments on cats in which inactivation of the DA system was caused with reserpine and alpha-methyl-p-tyrosine. Number refers to DA concentration remaining in the whole caudate nucleus ; control DA concentration was 11,6 ± 2,3 microg. per g. (13 cats).
- Cell 1 : raster display shows that after the stimulus artefact (unbroken vertical line) the cell was usually excited, with a fixed latency of 13 msec (incomplete vertical line). Negative results of collision tests showed this not to be antidromic activation. The excitation is followed by inhibition. Monosynaptic excitatory impulses may reach this caudate cell along a nigrostriatal input which is not dopaminergic (pathway with question mark in SN ; see scheme of Fig 11). Inhibition may be either due to disynaptic interneuronal activity (see cell II) or due to collaterals of GABAergic striato-nigral projection cells stimulated antidromically (see branching axon of triangular cell in schema of Fig 11).
- Cell II : bursting discharges evoked by nigral stimulation (dot above traces). The cell was found in the vicinity of cell A in the same experiment. This cell could play the role of an inhibitory interneurone. Note the diversity of firing patterns which remains after DA containing afferents have presumably been inactivated. Similar results were obtained after the total destruction of dopaminergic terminals with 6-hydroxydopamine.

1 - Striato-pallidal pathway

There is agreement in the literature that striatal stimulation induces an inhibition of pallidal cells (IPSPs in the case of intracellular studies) sometimes preceded by an initial facilitation (EPSP) (110,126). A study performed on the monkey confirms the occurence of a topographical organization of the projections arising from the caudate nucleus or putamen. Pallidal cells in the external or internal segment, localized in the dorso-medial part, are inhibited only after caudate stimulation whereas those localized in the ventro-lateral part respond only to the putamen (140) (Fig 5).

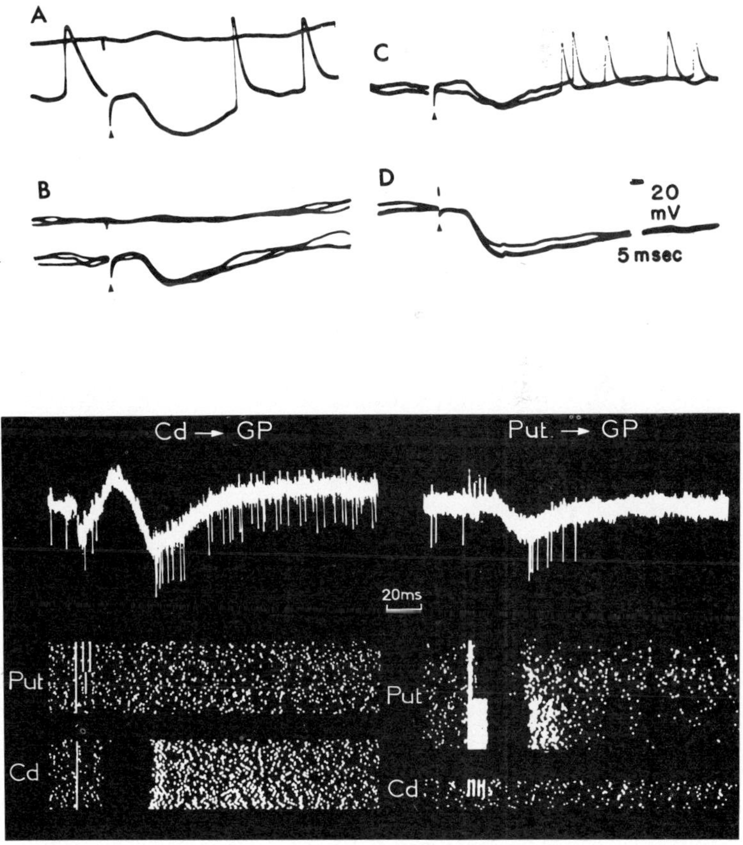

Figure 5 : Intracellular responses observed in three different (A,C,B and D) entopeduncular neurons in cat after caudate stimulation (arrows under traces) stimulation. Note the small variability of caudate evoked IPSP latency indicating that inhibitory potential may be direct. Inhibition of spontaneous activity can be observed during the IPSP. From MALLIANI and PURPURA (126). Extracellular responses observed in Globus Pallidus (GP) in awake Monkey after caudate (Cd) or Putamen (Put) stimulation. One cell was localized in dorso-medial part, another in ventro-lateral part. Arrest of spontaneous activity is followed by rebound excitation. The raster displays demonstrate that even with a varied number of stimuli (vertical lines drawn in) the synaptic activation was specific to only one of the two stimulation sites. From OHYE et al (140).

2 - Striato-nigral pathway

In the case of the substantia nigra, there is a general consensus concerning the IPSP or inhibition observed after striatal stimulation (32,42,50, 52,62,63,73,125,183). This inhibition involves both the cells in pars compacta and in pars reticulata. The latter neurons were identified as nigro-striatal, nigro-thalamic or nigro-collicular (35,36). According to KITAI et al (99) there is no reciprocal relation between striato-nigral and nigro-striatal neurons. Some evidences indicate that both entopeduncular nucleus and substantia nigra receive collaterals of the same neostriatal neurons (184,185).

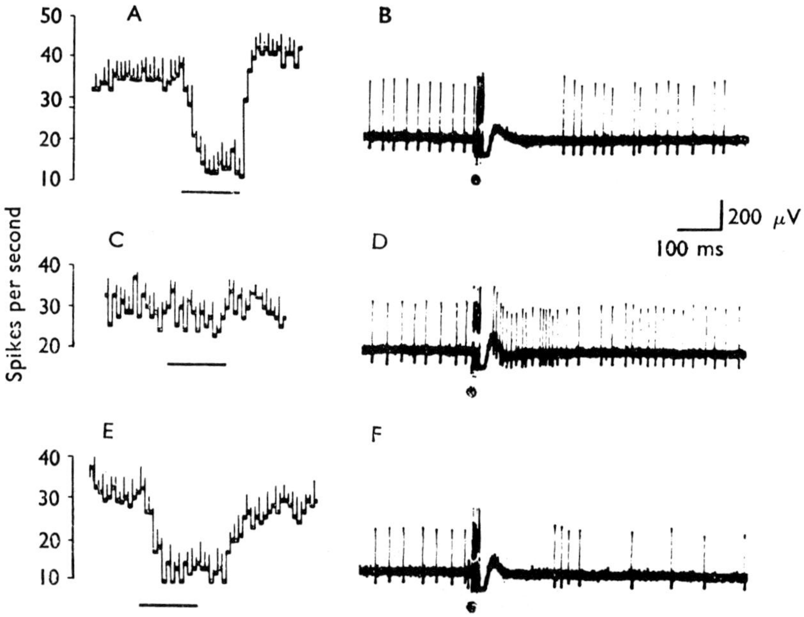

Figure 6 : Demonstration of a GABA-mediated inhibitory pathway from striatum to substantia nigra in the rat. A,C,E : ratemeter records of spontaneous cell firing showing,in control A, the depressant effect of iontophoretic GABA (40 nA, 15 sec, shown by line under trace). B,D,F : inhibition of spontaneous firing by caudate stimulation (dot under records). B is control. C and D demonstrate antagonism, by iontophoretic picrotoxin, of the GABA responses and of caudate inhibition. E and F were recorded 10 mins after the picrotoxin ejection was discontinued. From GROSSMAN et al (32).

The inhibition of nigral neurons observed after striatal stimulation is mimicked by iontophoretic application of GABA (32,52) and is antagonized by systemic injections of picrotoxin (32,146) or bicuculline (42,43) (Fig 6). Some authors (42,50) have noted that inhibition can be preceded by excitation (spikes rising from EPSPs). Possibly these inhibitory responses and sequences of excitation-inhibition take place on different groups of cells (50). One can also speculate about distinct excitatory and inhibitory pathways. A projection by peptide-containing striatal cells (33,45,65,86,95)could also be responsible. But the iontophoretic application of substance P causes only slowly-rising excitatory effects which contrast with the observed potent and short latency excitatory effects after electrical stimulation of the striatum. Thus, substance P projecting neurons may rather "modulate" excitability (34,175).

3 - Striato-subthalamic pathway

Although there is no anatomical evidence for a direct projection from the striatum to the subthalamic nucleus, electrophysiological data suggest that there is a striato-subthalamic connection in rat and monkey. Thus striatal stimulation evoked a biphasic response in neurons accurately localized in the subthalamic nucleus : an initial excitation followed by a depression of spontaneous activity. In the rat and monkey the initial response can in turn be divided into two components. It could thus be concluded that the first component of the excitatory response is caused by stimulation of cortico-subthalamic fibers since it disappeared in chronically decorticated animals. The second part of the excitation cannot be explained by the activation of a direct striato-subthalamic pathway since recent anatomical results denied its existence. One explanation is that the stimulation of the inhibitory striato-pallidal pathway could relieve the subthalamic cells from the inhibition exerted by the Pallidum.

Classically, the subthalamic nucleus projects only to the globus pallidus. Electrophysiological evidence suggests that there is also an excitatory monosynaptic pathway ending in the substantia nigra (78). This finding confirms a recent anatomical study based on the use of retrograde transport of peroxidase (94).

4 - Striato-raphe pathway

After caudate lesions in the cat degenerating fibers have been observed in the raphe nuclei as well as in the entopeduncular nucleus and in the rostral part of the substantia nigra (174).

GENERAL REMARKS

In studies of synaptic relations of distant brain formations it is very important to identify interneurons and output neurons and to determine the targets of the latter ones. This can be achieved by antidromic activation. It can be expected, for example, that antidromic activation of nigral cells would be observed in experiments with the striatal stimulation. This has however rarely been reported and it is not easy to explain why such antidromic responses are missing. Several possible explanations can be put forward. Firstly the reduction of axon diameter at each branch-points can prevent antidromic conduction. Both nigro-striatal pathway (dopaminergic and non dopaminergic) are branched : to separate sites within the striatum as well as between the striatum and others structures. Secondly the cat striatum is too large for all striato-nigral fibers to be activated simultaneously, so that the positionning of striatal stimulating electrodes becomes important. This problem is augmented by the topographical organization of the nigro-striatal projection. Both problems can however be overcome by performing experiments in rats in which a greater proportion of caudate efferent fibres can be activated. Thirdly, in the case of nigro-striatal fibers with a slow speed of conduction, mainly belonging to the dopaminergic neurons, antidromic activation of nigral somata could be prevented by the hyperpolarization resulting from the accompanying stimulation of the inhibitory striato-nigral pathway. Finally this lack of antidromic activation of nigral cells could also be due to stabilization of the membrane, occuring as a side effect of anaesthetics such as barbiturate (5).

Another general point of interest is that neostriatal inputs arising from the brain stem, thalamus and some areas of cortex project to other brain structures. In the case of the nigro-striatal dopaminergic input there is already evidence for an extensive branching within the striatum and perhaps also within the substantia nigra. In the latter structure the dopamine could be released from dendrites. The organization of the serotonergic system supplying both substantia nigra and striatum provides another example of wide divergence of information arising from one group of neurones. A well defined organization may however emarge from further studies, and the apparent lack of specificity of neuronal activity may be resolved if future neurochemical experiments are performed with the increased resolutions which now appears to be both necessary and feasible.

It is clear from these considerations that electrical activation of the substantia nigra has numerous effects on the neostriatal neuronal network. Intended stimulation of a specific pathway produces confounded effects unless selective lesions of groups of cells (i.e. DA neurons) or nerve endings (i.e. gabaminergic striato-nigral pathway) have been made. Thus, for instance, the characteristics of a branched pathway projecting from the nigra to both thalamus and neostriatum could be examined without interference from the dopaminergic nigro-striatal system. Similarly in studies of nigro-striatal inhibitory relations, antidromic activation of the striato-nigral gabaergic system could be avoided by e.g. use of GABA antagonists or selective lesions of fibers bundles. Similar problems apply to other neostriatal inputs and outputs.

II - SYNAPTIC EVENTS WITHIN THE STRIATUM

The understanding of neuronal processing in the neostriatum requires to consider the three following aspects of input-output relationships :

- the postsynaptic effects of the main striatal inputs : cortical, thalamic and nigral. The neurochemical correlates of these different inputs are described elsewhere (FONNUM, this volume).
- the nature of the neurons receiving these inputs : interneurons or output neurons.
- the occurence of convergence of the different inputs onto striatal neurons.

There is agreement that stimulation of the cortical, thalamic and nigral inputs to the striatum causes, in each case, a monosynaptic EPSP. There is, however, slight disagreement about the latency of these EPSPs which do not always generate a spike. Some authors have differenciated three categories of synaptic responses to these inputs : namely an EPSP-IPSP sequence, a pure EPSP and, more rarely, a pure IPSP (24,64,88,89,99,100,101,102,104,105). Output neostriatal neurons identified by antidromic activation from entopeduncular nucleus or substantia nigra, exhibit large EPSP and small IPSP, when either cortical, thalamic or nigral afferents are activated (64). This may be related to the relatively hyperpolarized state of neostriatal cells. Interneurons showed an EPSP-IPSP sequence with a larger IPSP.

Recently, nigral stimulation has been found to produce a synaptic response in the striatum with a dual (i.e. two successive EPSP C_1 and C_2 (Fig 7). The mean latency of the first components is 3.17 ms and that of the second component C_2 is 16,7 ms (105). The slow potential is also two to three times larger than the fast potential. Striatal neurons recorded for this study (105) also receive cortical and thalamic inputs which induce a depolarization similar to the C_2 response : occasionally thalamic stimulation evoked a C_1-C_2 like sequence. Experiments with paired conditionning cortical or thalamic stimulation and test nigral stimulation demonstrate the persistance of C_1 only, and the reduction of all other EPSPs. The authors presented a diagrammatic representation of the synaptic organization in neostriatum : the nigral input, providing the C_1 component, is localized proximally on the soma or initial spine-free part of the dendrites. The others excitatory inputs are localized more distally, at the level of dendritic spines.

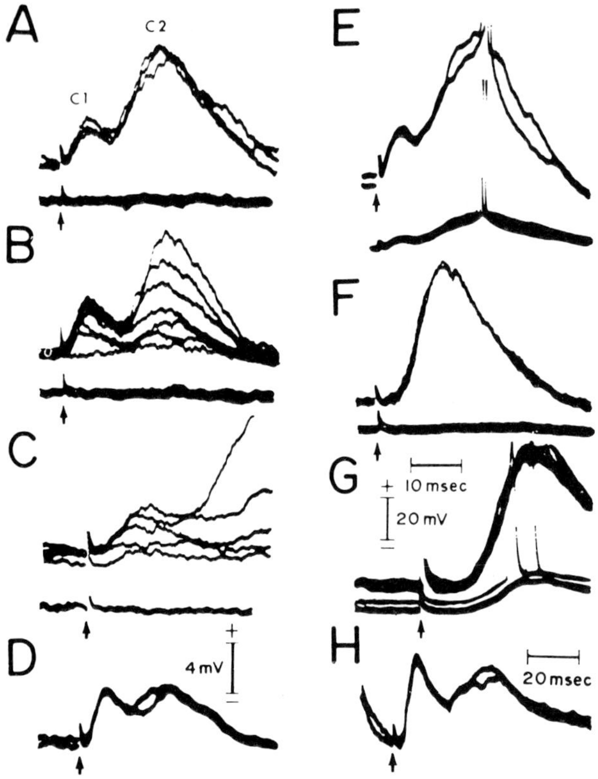

Figure 7 : Intracellular records of caudate neurons to show EPSP evoked from various sites. A,B,C : records from one cell showing a two component EPSP induced from nigral stimulation (arrow). Note in B the graded nature with variations in stimulus intensity and in C the constant latency (shown at faster sweep speed). Lower traces of each pair are extracellular controls. E,F,G : EPSPs recorded from nigra (E), cortex (F) and thalamus CMP (G). Note spikes in E and G. D,H : records from another caudate neuron. D show control nigral evoked dual EPSP and H shows the effect of 1-2nA hyperpolarization. From KOCSIS and KITAI (105).

This report (105) is in agreement with the majority of others indicating convergence of cortical, thalamic and nigral input on most striatal cells (24,64, 88,100,101,103,104,105). The type of the target neostriatal cell is determined either by histological examination after intracellular injection of peroxydase or by means of electrophysiological characteristics (antidromic activation after nigral or entopeduncular stimulation (104). It can be concluded that convergence occurs on the common medium-size spiny neurons which make up 96 % of caudate neurons (17) ; see also PASIK et al, this volume). It is these spiny cells, and not the"giant neurons" which are labelled after retrograde transport of peroxidase injected in the substantia nigra (see 75 and report of GROFOVA in this volume). This aspect of striatal organization is discussed in relation to neurochemical correlates in the following part (see also FONNUM, this volume). It can thus be concluded that convergence occurs on interneurons as well as on output neurons, antidromically activated by entopeduncular or nigral stimulation (104).

III - ELECTROPHYSIOLOGY AND PHARMACOLOGY OF INTRINSIC NEOSTRIATAL CONNECTIONS

In this section our aim is not to review exhaustively the current literature on the neurophysiology and pharmacology of the striatum ; rather it is to draw attention to the complexity of the intrastriatal neuronal network. Concepts of this network have been modified recently by the realisation that nigral stimulation has some effects not resulting directly from activation of dopaminergic synapses. Evidence is presented on the topographical and functional segregation both of synaptic and receptor sites. The difficulties in reliably correlating pharmacological, neurochemical and electrophysiological studies will also be emphasized.

Diversity of dopaminergic mechanisms

The problem of unequivocally establishing the synaptic actions of dopamine on striatal neurons has recently been reviewed by one of us (54). It seems possible that two classes of dopamine receptors exist : either as a multireceptor system on one cell (as in invertebrates, refs 9,13,49,66) or as different and specific receptors separated either on different elements on one cell or on different groups of cells. Indeed both cellular and regional differences in receptor distribution have recently become more obvious from neurochemical experiments (21,76,120,121,129,144,150,155,160,161,171, see below). Such a situation would provide a ready explanation for the divergent reports of various types of excitation and inhibition by iontophoretic dopamine (for reviews see 55,160,182).

Thus firstly, dopamine produces fast onset depolarizations mimicking fast EPSPs (101,102). An experimentally critical parameter may be the duration of iontophoretic application which could alter the balance of receptor types activated (for instance compare 102 and 103, with 17). In the case of fast depolarizations, there is a potentiation by dopamine of nigro-striatal EPSPs (102). This observation has been taken to support the dopaminergic nature of this EPSP, But in fact it would be most unusual for an exogenously administered transmitter to potentiate in this way the action of the same transmitter released synaptically when they both act on the same synaptic receptor to cause a typical increase in membrane conductance (see 10 and 54).

Figure 8 (I): Responses of a striato-nigral projection cell to nigral stimulation (A to G) and to iontophoretic ejections of acetylcholine (Ach) and dopamine (DA). Photographic records show antidromic activation and collision with a glutamate evoked spike (B, lower trace). In F and G it can be seen that antidromic activation is followed by a depression of glutamate induced firing. The ratemeter records show the lack of effect of high currents of Na^+ ions, and powerful depressant effects of Ach (several ejecting currents) and DA (compare Ach effect with that shown in Fig 9. This cell could correspond to a triangle cell in the scheme of Fig 11).

Figure 8 (II): Striatal cell activated monosynaptically by nigral stimulation via a non-dopaminergic synapse (excitation persisting after chemical destruction of dopaminergic terminals). Ratemater records show cell firing to regular pulses of iontophoretic glutamate. Both GABA and serotonin (5HT) depress this firing. Photographic records show in A the monosynaptic response to nigral stimulation (dot over trace). In B and C, stimulation of raphe dorsalis (arrow) does not inhibit nigral evoked monosynaptic excitation (tested at various intervals in C) although electrophoretic 5HT depressed cell firing. Moreover the possibility of effects of nigral stimulation being mediated by activation of intranigral collaterals of the raphe projection (dotted line in Fig 11) is excluded firstly because raphe stimulation cannot mimic the excitatory effects of nigral stimulation, and secondly the effects of nigral stimulation persist in cats treated with reserpine and para-chlorophenylalanine (PCPA), which would inactivate serotoninergic pathways.

Secondly, dopamine can cause a slowly rising atypical depolarization associated with a slight fall in membrane conductance (17). This response shows similarities to one of the types of acetylcholine responses ; for instance the membrane may not react via the normal opening of ionic channels. There are however differences, for the slow acetylcholine depolarization increases membrane excitability (Fig 9) whereas dopamine reduces excitability. The prolonged time course of the effects of both acetylcholine and dopamine (i.e. several seconds) could be related to certain long-lasting effects of repetitive nigral stimulation (51,88, see also Fig 6 in ref 54).

Thirdly, dopamine can produce genuine long-lasting hyperpolarizing effects with a slow onset. These effects are similar to those of cyclic AMP ; indeed in neurochemical experiments cyclic AMP appears to be a valuable correlate of the "inhibitory" effects of dopamine (for reviews see 54,91,96,160 ; but note contradictory data in 47,114). Unfortunately it is hard to rule out the possibility that these hyperpolarizing effects are indirect, since spread of iontophoretic dopamine to presynaptic and/or interneuronal sites can hardly be prevented.

Fourthly, there is strong evidence for presynaptic dopamine receptors which mediate rather simply a self-regulation of catecholamine release (see ref in 40) in line with the general concept of autoreceptors which have been clearly

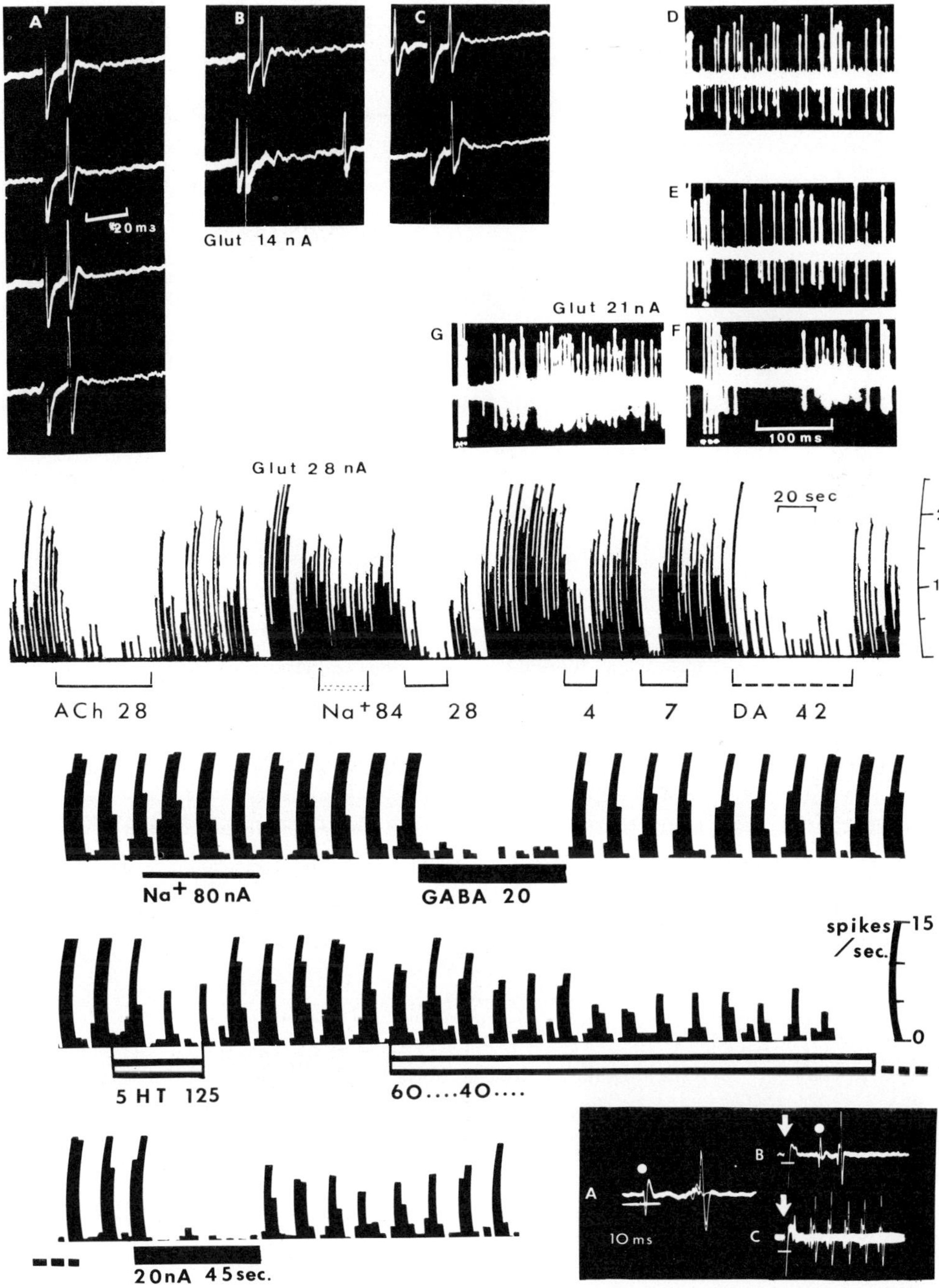

Figure 8 : Pharmacological testing of the sensitivity to putative transmitters on two caudate cells identified with respect to nigral stimulation. Fig 8(I) refers to projection cells shown by triangle (▲) in scheme of Fig 11 ; Fig 8(II) describes characteristics of striatal interneurons also shown in concluding scheme (black circles). Legend follows on the top of facing page.

described in the periphery (see reviews 108,165). Moreover dopamine can (as in the substantia nigra, see 144,151,161) facilitate the release of other transmitters. Thus, in the nigra and probably in the caudate as well GABA-ergic nerve endings are likely to be stimulated by a direct action of exogenous dopamine possibly on specific presynaptic receptors. Specificity means acting via some specific adenylate cyclase found on strio-nigral terminals (see 91,96). In this way short-acting GABA-ergic inhibition could be turned on regardless of neostriatal interneuronal activity or firing. Since the mechanism of dopamine-produced inhibition is not known, a role of GABA must be considered. Whether or not GABA plays a role in sole of these inhibitions could be tested in many ways : e.g. -by the use of variety of neuroleptics and ergot alcaloïds as well as dopamine agonists (54,139,149,160,187), by attempts to block dopamine-produced activation of adenylate cyclase (54,160), and by use of agonists and antagonists of GABA transmission (16,55-56). The use of such combined approach may yield conclusive data. However, owing to demonstrated lack of specificity of many pharmacological tools (e.g. neuroleptics), the intrastriatal synaptic mechanisms are not yet fully resolved on electrophysiological grounds (54). As a matter of fact, attempts to understand transmitter actions give yet rise to rather general questions about the interdependance between the various receptors for different putative transmitter, as activated either by exogenous compounds or by synaptic activation.

Finally, it should be pointed out that a neuroglial receptor for dopamine may exist (82) and that there is neurochemical evidence relating dopamine to the regulation of basic metabolism (see e.g. 6,136,180). It is thus clear that, in general, a wide variety of functional patterns of dopamine should be examined, particularly from a physiological viewpoint.

GABAergic versus dopaminergic inhibitions

It has recently been shown that exogenous dopamine often potentiates a possibly dopaminergic synaptic activation of some as yet unidentified cells. If these cells are inhibitory regardless of whether they exert inhibitory action as interneurones or via collaterals of projecting cells, then a simple assumption would be that dopamine afferents could effect a transneuronal and disynaptic inhibition. So far there is evidence for dopaminergic EPSPs only on those cells on which there is a convergence of cortical and dual-nigral (103,105) EPSPs : these cells are thus the only candidates for mediating an indirect inhibition by dopamine. In addition, genuilely direct dopaminergic inhibitions may reach various types of caudate neurons (see reviews 54,160, and Fig4,8)but such inhibitions should be discriminated from the more pronounced inhibitory influence of

GABAergic synapses (14,16,54,56,149,162). It seems probable that many intrastriatal IPSPs are mediated by GABA, because in this case, the most stringent criteria for identifying synaptic transmitters have been met (demonstration of changes in membrane potential and conductance together with a study with transmitter antagonists ; see Fig 5-9 and BERNARDI et al : 14,16).

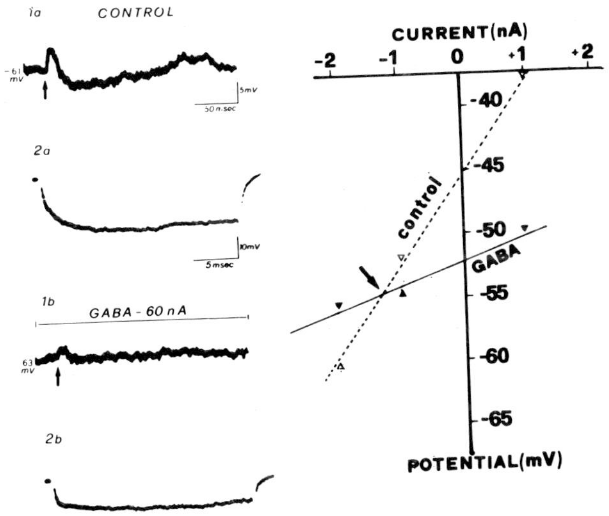

Figure 9 : Action of GABA on rat caudate neurons recorded intracellularly, On the left side : effect of GABA on EPSP-IPSP sequence and on membrane resistance of a caudate neuron intracellularly recorded (spike amplitude : 45 mV). Arrow, cortical stimulation. Pulse, 25 msec, 2 nA inward current. 1a-2a, control ; 1b-2b, same responses during 60 nA GABA application. On the right side : current-voltage plot obtained from another caudate cell. 80 nA of GABA were applied. Slopes were drawn by eye. Pulses of 20 msec were used for resistance measurement. From BERNARDI et al (14).

Both interneurones and projecting cells are subject to rate-limiting control mechanisms mechanisms via postsynaptic receptors for GABA (55,56). These inhibitions probably function independently of dopamine inhibitions effects regardless of whether these latter act via excitation of GABAergic inhibitory interneurones (54,57,59,160,187).Some target cells of the dopaminergic input have indeed been shown to inhibit both one another and other striatal neurons (53) possibly by releasing GABA (circles in Fig 11 and see ref 55 and 56). Such neurons are in addition activated monosynaptically from the substantia nigra by means of non-dopaminergic afferents (Fig 11) which probably release an amino-acid rather than acetylcholine (54,58, and see ref 163) ; they do not release serotonin or any type of catecholamine (ref 54 and Fig 3 and 7).

These various systems do not, however, preclude other roles of dopaminergic terminals in this network. In our previous review (54) we particularly emphasized the way in which denervation supersensitivity can be used as a criterion for assessing on electrophysiological grounds the divergence of dopaminergic nerve terminals onto different groups of striatal neurons (Fig 10 ; see also tables and plate 1 in ref 54).

Interdependance between dopamine and other transmitters

Consideration of the topographical and cellular distribution of receptors for dopamine as well as other transmitters may help to clarify receptor interactions. It is however desirable to avoid postulationg excessive numbers of interneuronal types when describing transmitter functions.

As an example of topographical variations in distribution of cell receptors one can select the dopaminergic and serotonergic systems. Striatal serotonergic receptors are likely to be activated following nigral stimulation of collateral branches or passing fibers of the raphe-striatal pathway (20,44,131,142,176). Many biochemical data suggest that dopamine receptor concentration is lower in those striatal areas where there is a heavy serotonergic innervation ; the inverse is also the case (21,22,26,171,172). Such differences have unfortunately not yet been explored physiologically. Another example of specific distribution pattern of membrane receptors comes from the suggestion that the intrinsic striatal cholinergic neurons are the preferential target for dopaminergic afferents (72,74,76,77,120,121,155,156,164,173). Perhaps also those cells which store GABA require to be further classified with respect differential distribution on them of pre- and postsynaptic dopaminergic mechanisms (8,12,120,121,128,129,150). From such considerations it would seem that the organization of the dopamine input exhibits a more selective pattern than is commonly the case with monoaminergic systems (see basic ref in 2 and 92).

From structural studies in mammalian brain and from valuable estimates of transmitter mechanisms in invertebrates (see reviews 10,66), it is clear that much of the specificity of the activity in neuronal network is related to the fact that there is a multiplicity of basic receptor mechanisms ; that is, there are multiple actions of one transmitter and complex interactions between different transmitters. Such a complexity poses problems in the interpretation of neurophysiological experiments, since physiological and even pharmacological responses may be altered in many ways depending on the state of activation of the various receptor complexes.

It is worth noting that the study of the transmitter actions of dopamine in ganglia of molluscs and other neuronal networks in the periphery has clearly demonstrated the cays in which a single cell and a single transmitter can have several functions (see 13,46,47,67,112,114,115,116,134,135,186). This of fundamental interest whenever a transmitter system with branching axons requires to be analyzed. However both structural and even appropriate electrophysiological studies have yielded insufficient information on the possibility of multiple functions of a single transmitter released, by one cell, on different target receptors. The possibility that such a diversity exists within the cholinergic system of the caudate nucleus should now be investigated. The difficulty with the dopamine system is that there are only indirect ways to reveal the relation between somatic and dendritic receptor sites. Somewhat easier systems are those which can be identified on physiological and/or histochemical grounds, for instance, the GABAergic projection cells (see 23,65,155 and FONNUM, present volume) and substance "P" containing cells (see 23,33,65,84,95). But another difficulties in the fact that the distribution of the numerous postsynaptic receptors parallels that of recently discovered presynaptic receptor sites, in particular those on dopamine afferents (40,69,145,167,178,179).

Further concepts may still emerge from future studies on the complex functionnal relations of the nigro-striatal dopaminergic input to its intrastriatal target-cells (see ref in 8). It should be possible to characterize these target cells on electrophysiological grounds (see reviews 54,103) and to correlate the findings with neurochemical data (see reviews 71,128). In that case the concepts derived from observations of these dopaminergic mechanisms can then be applied to describing the relations between the wide variety of neurotransmitters and receptors operating throughout the caudate nucleus and putamen. In addition, it is worth knowing which transmitters are likely to be released by the major target neurons innervated by dopaminergic terminals because some of these transmitters can regulate directly incoming dopaminergic activity. From these facts it may be possible to predict at least some properties of the intrinsic neostriatal pathways.

Post synaptic versus presynaptic sites of interactions

The electrophysiological basis of neuronal integration in the striatum has recently been reviewed by KITAI and KOCSIS (103). These authors have drawn attention to the ways in which excitations and inhibitions interact differently depending on the geometrical location of the inputs on the soma-dendritic profile of one class of neurones. That is, axospinous excitatory currents on the distal parts of dendrites (C2 EPSP, see Fig 4) can be shunted by inhibitory inputs in the absence of an IPSP as recorded from the soma. By contrast, the excitatory currents driving other EPSPs (C1 in Fig 7) appear immune to inhibitory shunting effects and probably are formed on the proximal part of the dendritic tree. Such geometrical aspects of neuronal integration as well as other evidence indicating interactions of several transmitters in the neostriatum suggest that interactions of a higher order may occur in the neostriatal neuronal network. The variety of functional patterns could be far greater than the number of transmitters involved.

Thus, release of only one transmitter onto highly specific receptors could result in qualitatively different responses depending firstly on the exact location of the synapses and secondly on the activation of any possible combination of at least three ionophores (mediating either a conductance increase or a decrease to NA^+, Cl^- or K^+). In addition, two or more classes of receptors for a single transmitter may coexist in various possible geometric patterns, as was suggested earlier for dopamine. In view of the different receptors already described in other brain sites for both acetylcholine and serotonin, it is quite likely that these, and perhaps other, transmitters act in the striatum on a variety of receptor types. In addition to all these mechanisms, there could be a differential channelling of information into two branches of a single axon as has been demonstrated in lobsters (75,143). That is, if one cell releasing one transmitter activates one class of receptor-ionophore complex which is distributed on distinct target cells, there is nevertheless a possibility of observing two kinds of postsynaptic patterns simply because the presynaptic axonal branches conducting afferent impulses may display high firing rates along one branch and a lower rate along collaterals. Furthermore, synaptic efficacy can be modified by the level of activity in the presynaptic terminal, as for example at the giant synapse of the squid (48). Lastly, the readiness with which many putative neurotransmitters cause the release of another compound or generate typical "presynaptic" inhibition suggests that specific postsynaptic actions could perhaps be entirely masked. Thus there is evidence that dopamine, serotonin, acetylcholine, GABA and even glutamate, administered into the striatum, are active at pre- as well as postsynaptic sites.

There are thus considerable problems in interpreting conventional electrophysiological and micro-iontophoretic experiments in terms of precise transmitter identification in a particular pathway. Similar problems exist in interpreting related neurochemical data because of possible interactions of dopaminergic and glutamatergic inputs from substantia nigra and the cortex (41,98,163) as well as local pools of cholinergic interneurons (although there may also be a thalamic cholinergic pathway ; see 79,121) and of GABAergic and peptidergic projection cells. We wish to highlight the fact that the interaction of transmitters and their modulators and antagonists may occur in a complex fashion at both pre- and postsynaptic sites. It is thus important to incorporate in experiments means by which "indirect" (i.e. pre- or disynaptic) effects can be ruled out (see review by FELTZ, 55).For instance, selective destruction of dopaminergic terminals results in a situation acceptable firstly for characterizing the inhibitory effect of acetylcholine on striatal projection neurons (ref 54, 57, 59 and Fig 8) ; secondly for assessing the differences between the direct postsynaptic inhibitory effects of dopamine (54,57,59,160) and the indirect -disynaptic- inhibitions mediated by GABAergic interneurons (14,16,54,55,56,162 and Fig 4,11) ; and thirdly for checking the evidence that serotonin exerts a direct inhibitory effect -that is without an intermediaire release of dopamine- ; (see refs 54,131,142,149 and Fig 8). Nonetheless more definitive conclusions on these and similar questions must await investigations in which indirect actions are more rigorously excluded by using high concentrations of Mg^{++} or Co^{++} (as for example has been done in the study of the complex effects of glutamate in the spinal cord of the lamprey ; 130). Such experiments could test more specifically than any pharmacological antagonist the possibility that dopamine may depress neuronal firing indirectly by excitation of inhibitory interneurons as suggested by RICHARDSON et al (149). Another determinant in the exact identification of transmitter effects is the selectivity of the antagonist used. It is worth noting that whilst the ionic mechanisms following receptor activation are ignored, the action of antagonists can not be strictly specified. This is, for instance, important for analysis of dopamine effects which may be direct or mediated by release of other transmitters such as GABA and acetylcholine from intrinsic neurons and glutamate and 5-HT from extra-striatal afferents. Knowledge of ionic events triggered by these transmitters and use of appropriate antagonists may help resolving such problems. Much of this was made obvious in electrophysiological studies of neuronal networks in molluscs : in this situation antagonists appear to be almost equipotent against different transmitters whenever these transmitters act on the same ionophore, even when the transmitter receptors (pre- or postsynaptic) are different (see review 10,66 and refs 27,49,181). Yet another experimental difficulty results from the process of desentization : for instance there appears little doubt that many of the conflicting results about the function of the dopaminergic input are due in part to the fact that the excitatory action of dopamine is much more susceptible to desensitization than is its inhibitory action. Such problems have been appreciated with dopamine receptors in molluscs (9,13,170) : as a results it is clear that the magnitude and duration of currents used to apply dopamine by iontophoresis are experimentally critical parameters, expecially when extracellular recordings are employed. It is important to bear this in mind when comparing the iontophoretic studies of KITAI et al (102) and BERNARDI et al (17). Acute in vivo experiments suffer also from the problem posed by the presence of anaesthetics ; thus for example barbiturates can depress cation selective conductance systems activated by transmitters whereas anionic conductances changes may be preferentially potentiated (1,11,138). The interpretation of the effects of dopamine observed in acute experiments in vivo therefore remains at the stage of a working hypothesis. Nevertheless further intracellular studies both in vivo (see 17,102,103) and in vitro (see 132) could yield more definite conclusions about the actions of dopamine, acetylcholine (15,132) and serotonin (see 44,131, see also 2). Such studies, taken in conjonction with structural data, should allow reconsideration of the many circuits which are commonly drawn to explain the interdependance of the different transmitters in the caudate nucleus.

Local regulations of transmitters release

Specific stimulation of the nigro-striatal dopaminergic pathway distally as compared to proximally in the striatum has yet not been achieved. Nevertheless, progressively more neurochemical studies have achieved regulated alterations of the local release of dopamine by topical administration of, for example, acetylcholine and both excitatory and inhibitory amino-acids (28,68,70,164,167, 177,178). Apparently in these experiments most putative transmitters overcome the self-inhibitory effect of catecholamines overflowing into the synaptics clefts. The amino acids and acetylcholine predominantly facitilate the release of dopamine induces by activation of nicotinic and muscarinic mechanisms (68,69,177,178).

In the striatum, unlike in the peripheral nervous system nicotinic and muscarinic systems are not mutually antagonistic. There are nonetheless considerable similarities in the pharmocology of striatal dopamine release and peripheral noradrenaline libaration. Such similarities are evident in the self-inhibitory effect of the catecholamine to be released (see ref in 40 and 109) as well as in possible presynaptic facilitatory effects of acetylcholine, excitatory amino-acids and perhaps also GABA (see reviews, 109,165,166 and ref 70,153). The effects of some of these transmitters are likely, however, to be unrelated to their occurence in neurons. Indeed, the role of neuroglial cells in transmitter interactions still cannot be ignored, firstly because it is likely that striatal glia possess receptors for dopamine (see 82,135,168,169) and secondly because in some situations, glia can exhibit slowly depolarizations in response to GABA (87,107). Such indirect effects must be ruled out before further conclusions can be reached as to whether or not the actions of those transmitters which can regulate dopamine release arise from presynaptic receptors. Possibly some concepts as yet unexplored may also be of interest, for instance the possible presence of presynaptic targets for cyclic nucleotides (158,179) or of presynaptic receptors for enkephalines (145). Some clues to show how all these transmitters act in the neostriatal network may be obtained from tests with tetrodotoxin (TTX) which can be used to provide a gross selection versus transsynaptic effects. For example, the release of dopamine induces by exogenous GABA is clearly sensitive to TTX whereas the release by exogenous acetylcholine is markedly resistant (68,69,70). This result must,however, be interpreted with caution, for the presence of presynaptic acetylcholine receptors in no way proves the presence of a cholinergic synapse on dopaminergic terminals ; one cannot therefore safely conclude that GABA is acting indirectly by inhibiting a cholinergic neuron. Nor can one safely assume that a direct presynaptic effect of exogenous GABA is not TTX-sensitive, since there is no available evidence on this question. If,however, the data are not simply a pharmacological artefact, then they demonstrate that the neuronal loop can be short circuited at presynaptic levels.

Some views on the intrinsic organisation of the neostriatum

Physiological evidence indicating that the dopaminergic input exerts a direct postsynaptic inhibition primarly on neostriatal cholinergic interneurons (77,120,156,164) has supplemented information based on neurochemical outlines of intrastriatal synapses. As originally suggested by TRABUCCHI et al (173), the cholinergic neurons with bifurcating axons may project both to each other and also to intrastriatal GABAergic projection cells ; the former connection has been suggested to be excitatory. The reciprocal excitatory connection between such interneurons or some others (squares and circles in Fig 11) may perhaps be those associated with the fast rising nicotinic cholinergic depolarization which have recently been described in an vitro preparation (132,133). Such depolarizations are however uncommonly seen in vivo (19,54,83,161,187) and differ firstly from the slowly rising muscarinic responses,(Fig 10) and secondly from some other truly inhibitory actions of acetylcholine on striato-nigral projection cells defined by antidromic activation (Fig 8). It is worth while pointing out some other neurochemical data on both acetylcholine and GABA. In keeping with an early hypothesis that dopamine acts inhibitory on cholinergic interneurons which in turn inhibit GABAergic projection cells (see ref 127,173), a disynaptic system controls perhaps all GABAergic activities, via the striato nigral

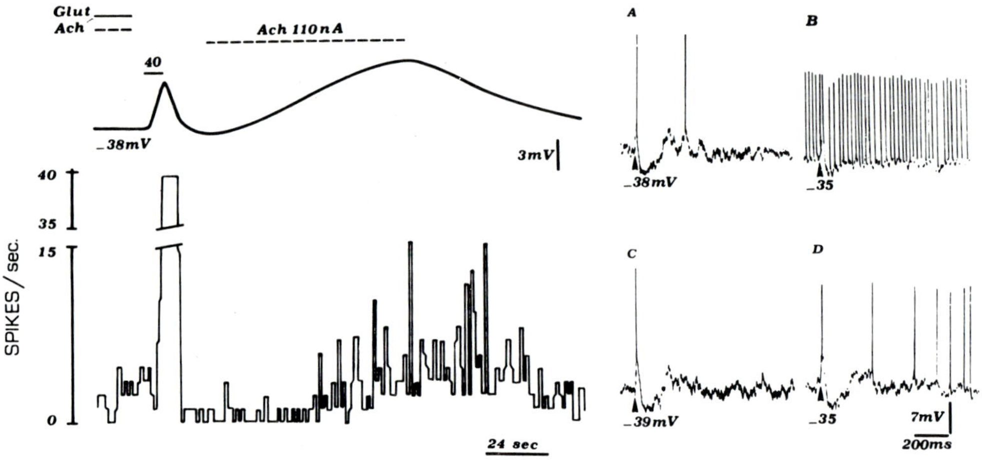

Figure 10 : Comparison of effects of electrophoretic acetylcholine (Ach) and L-glutamate on a caudate neuron. Upper left : cell membrane potential, initially - 38 mV, showing rapid depolarization to L-glut and slow depolarization to high ejection current of Ach. Lower left : ratemeter record of L-glut and Ach-evoked firing. Time base as for membrane potential record. A-D on the right are intracellular records showing an IPSP evoked by entopeduncular stimulation (arrowheads). A-C : controls. B : L-glut ejection depolarizes the cell from -38 to -35 mV, reduces the amplitude of the IPSP and evoked spikes on the IPSP. D : Ach ejection causes a smaller depolarization but a similar effect on the IPSP and a lesser increase in firing rate. From BERNARDI et al (15).

projections and their intrastriatal collaterals. That is, biochemical data show that switching of dopaminergic inhibition on cholinergic systems by pharmacological means causes enhanced release of interneural acetylcholine and hence cholinergic inhibition of GABAergic projection cells. According to more recent studies of the turnover rates of intrastriatal GABA it has been suggested that the two pools of GABAergic projection systems, towards the pallidum and the substantia nigra (see 61), behave differently with respect to their regulatory mechanisms (127, see also ref 122 about GABAergic interneuronal pools). There are similarities between both pools since they receive excitatory connections mediated possibly by glutamate (127) and arising from cortical (see ref in 98,163) but perhaps also nigral sources (see 54,57). The electrophysiological support of these views on the GABAergic striato-nigral pool may be the finding that cortico-striatal EPSPs occur, as noted above, on those striato-nigral projection cells which receive dual nigro-striatal EPSPs whenever nigral stimuli promote antidromic responses

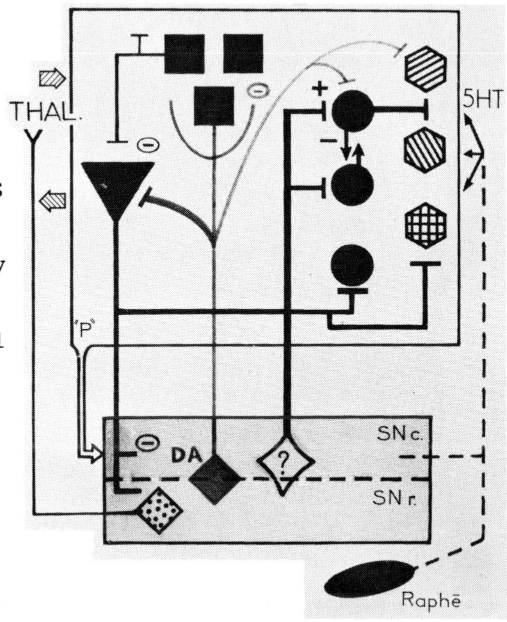

Figure 11 : Neuronal relationships between substantia nigra (SNc,SNr,lower frame) and striatum (upper frame) based on micro-electropharmacological data (54). Antidromically identified striatal neurons (GABA : black triangle ; or possibly substance "P") project to SN. Acetylcholine containing neurons (Ach ; black squares) receive inhibitory input (-) from dopamine -containing nigral cells (black diamond,DA); evidence for this inhibitory effect of DA afferents relies on biochemical data. Hypothetical non-DA neurons monosynaptically excites (excit.,+) striatal cells (black circles) believed to be inhibitory (inh.) on heterogenous population of other striatal cells (see ref 53,20,55-hexagonal symbols). Some of these cells could belong to the population of cholinergic interneurons ; they could also be GABAergic or peptidergic projection cells but have been found to be unresponsive to antidromic stimulation. On all cells, observed enhancements of slow DA-inhibition in 6-hydroxydopamine denervated cats (see plate and table 1 in ref 54) suggest existence of DA-terminations on all groups of striatal neurons identified on electrophysiogical grounds (black triangle and circles ; hexagone). Some of these DA-afferents may in addition promote DA-mediated EPSP, since fast depolarizing effects of DA have been observed on some cells (see ref 103). Pharmacological evidence indicates that some of the cells activated from the nigra by a non DA-synapses (+ on circles) may lead to subsequent GABAergic inhibition (reciprocally between circles, and also on hexagons). Relationships with thalamus (Thal) and raphe is also shown (left and right) ; the 5HT striatal input originating in raphe dorsalis may be inhibitory (−). The peptidergic ("P") output of the striatum is shown by arrow in left corner of upper frame ; this link may be excitatory (+) and may also have instriatal collateral branches similar to those shown for the inhibitory (-) projection cell (triangle).

(see KITAI et KOCSIS, 103,105 and sections I and II in this review). Those nigro-striatal monosynaptic excitations which are definitly not dopaminergic (see 54) should also be taken into account ; they too may be related to both striato-nigral and striato-pallidal pools of striatal GABAergic cells with respect to excitatory inputs, there is an apparent diversity about inhibition so far unsolved electrophysiologically. One difference between the two GABAergic projections is that, on the basis of neurochemical experiments, the metabolism of GABA in the striato-pallidal system is possibly related directly to inhibitory dopaminergic receptors, whereas an intermediate cholinergic step should be interposed between the dopaminergic afferents and the GABAergic striato-nigral projection cells (see 127,129). The predominant inhibitory effect of acetylcholine on projection cells, as we have shown in electrophysiological studies of striato-nigral neurons (see Fig 8) may well support parts of these neurochemical hypotheses but there remains another question : how to establish whether or not the depressant effect of exogenous dopamine on these same projection cells results from a direct activation of a type of postsynaptic dopaminergic receptor which "shares", to an unspecified degree, the role of cholinergic inputs. From the results of electrophysiological and pharmacological experiments it seems that interneuronal release of GABA or acetylcholine is not involved in this dopaminergic inhibition; nor does the serotonergic input play any significant role in mediating indirectly the depressant effects of dopamine which in fact exhibit many characteristics of post-synaptic responses (54,57,59,153). The synaptic activation by nigral afferents of such inhibitory postsynaptic receptors for both acetylcholine (disynaptically) and dopamine (monosynaptically) on striato-nigral projection cells remains subject to clarification. It can be suggested that many of the short acting inhibitions are mediated primarily by GABAergic collaterals of the relevant group of caudate projection neurons depending on the sites where the stimuli are applied -e.g. substantia nigra, pallidum cortex and also within the striatum (16,55,56,103,162).

CONCLUSION

Basic electrophysiological studies are increasingly becoming of outmust importance in clarifying the wealth of pharmacological and neurochemical data because the resolution of intracellular recording is so much greater. Thus even analysis of the differential interactions of dendritic inhibition with dendritic versus somatic EPSPs may be feasible. Interpretation of pharmacological studies on receptor localization depends on an understanding of how-morphologically and physiologically-excitatory inputs impinge on the soma-dendritic complex of the numerous caudate spiny neurons (KITAI, 103,105).Once this is established, the characteristics of the EPSPs can then be used to determine the relevant excitatory transmitters : -i.e. amino-acids, predominantly, nicotinicic acetylcholine or dopamine, as well as the most significant compounds having inhibitory functions -i.e. GABA, glycine, muscarinic acetylcholine and perhaps also serotonin and dopamine. Knowledge of the principal EPSPs and IPSPs should also provide the best key for testing any neuromodulator function of muscarinic acetylcholine, dopamine-dependant cyclase, serotonin and peptides. Again, it is physiological information on the synaptic organization of a specific type of caudate neuron which sould ultimatly lead to an analysis of the complex interactions between various neurotransmitters at both pre- and postsynaptic levels (interactions already described on biochemical grounds). This does not preclude designing combined electrophysiological and pharmacological studies, particularly in conjunction with neurotoxins (6-hydroxydopamine has already been used in such studies, e.g. 54,56,57,149,159, 160). Local injection of kainic acid could , for example, be used to study peptidergic projection cells. Widespread injection into the striatum of this amino-acid (which appears to be toxic to cell bodies but not terminals) has already been shown to decrease in parallel striatal levels of dopamine-dependent adenylate cyclase, choline acetyltransferase and glutamic acid decarboxylase ; tyosine hydroxylase, which one would expect to be localised in dopaminergic afferent terminals was unchanged. It is possible that more localised injections of kainate might cause differential effects on these or other enzymes, and that it would thereby be possible to correlate topographical variations in dopaminergic terminals with differences in the distribution of striatal cell types.

Acknowledgements : Our thanks are due to Dr M. HEADLEY, MRC fellow, for revision of this manuscript. Financial support from C.N.R.S. (ERA 237) and I.N.S.E.R.M. (ATP 76/71, CLR 76/I/205/6).

1. Adams, P.R., J. Physiol.(Lond) 260, 531-552 (1976).
2. Aghajanian, G.K., Cedarbaum, J.M. & Wang, R.Y., Brain Res. 136 (3), 570-577 (1977).
3. Albe-Fessard, D., Oswaldo-Cruz, E. & Rocha-Miranda, C, Electroenceph. Clin. Neurophysiol. 12, 405-420 (1960).
4. Albe-Fessard, D., Rocha-Miranda, C. & Oswaldo-Cruz, E., Electroenceph. Clin. Neurophysiol. 12, 649-661 (1960).
5. Albe-Fessard, D., Levante, A. & Rokyta, R., Inter. J. Neuroscience 1, 327-338 (1971).
6. Anchors, J.M. & Garcia-Rill, E., Brain Res. 133 (1), 183-189 (1977).
7. Anden, N.E., Corrodi, M., Dahlström, A., Fuxe, K. & Hökfelt, R., Life Science, 5, 561-568 (1966).
8. Ariano, M.A. and Adinolfi, A.M., Expl. Neurology 57, 426-433 (1977).
9. Ascher, P., J. Physiol.(Lond) 225,173-209 (1972).
10. Ascher,P. & Kehoe, J.S., Handbook of Psychopharmacology, vol 4, 265-310, Plevium Press, New York (1975).
11. Barker, J.L., Brain Res. 92, 35-55 (1975).
12. Bartholini, G. & Stadler, H., Neuropharmacology 16, 343-347 (1977).
13. Berry, M.S. & Cottrell, G.A., J. Physiol.(Lond) 244, 589-612 (1975).
14. Bernardi, G., Marciani, C., Morocutti, C. & Giacomini, P., Brain Res. 92, 511-515 (1975).
15. Bernardi, G., Floris, V., Marciani, M.G., Morocutti, C. & Stanzione, P., Brain Res. 114, 134-138 (1976).
16. Bernardi, G., Marciani, M.G., Morocutti, C. & Giacomini, P; Brain Res. 102, 379,384 (1976).
17. Bernardi, G., Marciani, M.G., Morocutti, C. & Stanzione, P.,In:Iontophoresis & transmitter mechanisms in the mammalian central nervous system. R.W. Ryall & J.S. Kelly (Eds), Elsevier North Holland Biomedical press, Amsterdam-New York, 44-46 (1978).
18. Blake, D.J., Zarzecki, P. and Somjen , G.G., J. of Neurobiology 7, 143-156 (1976).
19. Bloom,F.E., Costa, E. & Salmoiraghi, G.C., J. Pharmacol. Exp. Ther. 150, 244-252 (1965).
20. Bobillier, P., Petitjean, F., Salvert, D., Ligier, M. & Seguin, S., Brain Res. 85, 205-210 (1975).
21. Bockaert, J., Premont, J., Glowinski, J., Thierry, A.M. & Tassin, J.M., Brain Res. 107, 303-315 (1976).
22. Broch, O.J. & Marsden, C.A., Brain Res. 38, 425-428 (1972).
23. Brownstein, M.J., Mroz, E.A., Tappaz, M.L. & Leeman, S.E., Brain Res. 135 (2) 315-323 (1977).
24. Buchwald, N.A., Price, D.D., Vernon, L. & Hull, C.D., Exp. Neurol. 38, 311-323 (1973).
25. Bunney, B.S. & Aghajanian,G.K., In : The basal ganglia, M.D. Yahr (Eds), Research publications of the Assoc. For Res. in Nervous and Mental Disease (ARMD), vol. 55, Raven Press, New York, 249-267.
26. Calas, A., Besson, J.M., Gauchy, C., Alsonso, G. & Glowinski, J., Brain Res. 118, 1-13 (1976).
27. Carpenter, D.O., Swann, J.W. & Yarowsky, P.J., J. Neurobiology 8, 119-132 (1977).
28. Cheramy, A., Nieoullon, A. & Glowinski, J., Naunyn. Schmiedebergs Arch. Pharmacol. 297 (1), 31-37 (1977).
29. Cheramy, A., Nieoullon A. & Glowinski, J., J. Pharmacol. Exp. Ther. 203 (2), 283-293 (1977).
30. Connor. J.D., Science 160, 899-900 (1968).
31. Connor, J.D., J. Physiol.(Lond) 208, 691-703 (1970).
32. Crossman, A.R., Walker, R.J. & Woodruff, G.N., Brit. J. Pharmacol. 49, 696-698 (1973).
33. Cuello, A.C. & Kanazawa, I., J. Comp. Neurol. 178, 129-156 (1978).

34. Davies, J. & Dray, A., Brain Res. 107, 623-627 (1976).
35. Deniau, J.M., Feger, J. & Le Guyader, C., Brain Res. 156, 104-152 (1976).
36. Deniau, J.M., Hammond-Le Guyader, C., Feger, J. & Mc Kenzie, J.S., Neuroscience Letters 5, 45-50 (1977).
37. Deniau, J.M., Lackner, D. & Feger J., Brain Res. 145, 27-35 (1978).
38. Deniau, J.M., Hammond, C., Rizk, A. & Feger, J., Exp. Brain Res. 32 (1978).
39. Deniau, J.M., Chevalier, G., Feger, J., Neuroscience Letters, submitted for publication.
40. Di Chiara, G., Porceduff, M.L., Spano, P.F. & Gessa, G.L., Brain Res. 130 (2), 374-382 (1977).
41. Divac, I; Fonnum, F. & Storm-Mathisen, J., Nature 266 (5600), 377-378 (1977).
42. Dray, A. & Gonye, T.J., J. Physiol. 246, 88-89P (1975).
43. Dray, A., Gonye, T.J., & Oakley N.R., J. Physiol. 259, 825-849 (1976).
44. Dray, A., Gonye, T.J., Oakley, N.R. & Tanner, T., Brain Res. 113 (1), 45-57 (1976).
45. Duffy, M.J., Mulhall, D. & Ponell, D, J. Neurochem. 25, 305-307 (1975).
46. Dun, N. & Nishi, S., J. Physiol.(Lond) 239, 156-164 (1974).
47. Dun, N.J., Kaibarak, K. & Karczmar, A.G., Science 197, 778-780 (1977).
48. Erulkar, S.D. & Weight, F.F., J. Physiol. 266, 209-218 (1977).
49. Faber, D.S. & Klee, M.R., Brain Res. 65, 109-126 (1974).
50. Feger, J. & Ohye, C., Brain Res. 89, 155-159 (1975).
51. Feltz, P. & Mc Kenzie, J.S., Brain Res. 13, 612-616 (1969).
52. Feltz, P., Canad. J. Physiol. Pharmacol. 48, 1113-1115 (1971).
53. Feltz, P. & Albe-Fessard, D., Electroenceph. Clin. Neurophysiol. 33, 179-193 (1972).
54. Feltz, P., In : Iontophoresis and transmitter mechanisms in the mammalian central nervous system, R.W. Ryall & J.S. Kelly (Eds), Elsevier North-Holland Biomedical Press, Amsterdam-New York, 25-38 (1978).
55. Feltz, P., In : Symp. on Basal Ganglia-cellular and Functional aspects-, R. Hassler, N.A. Buchwald & S.T. Kitai (Eds), J. Applied Neurophysiol., Karger (Basel) in the press, 1977-78.
56. Feltz, P., De Champlain, J. & Dessama, J.M., In : Neuropsychopharmacology, J.R. Boissier, H. Hippius and P. Pichot (Eds) - "Pharmacology and Biochemistry of nigro-striatal system". Excerpt. Med. Int. Congress Series 359, 453-459 (1975).
57. Feltz, P. & De Champlain, J., Brain Res. 43, 595-600 (1972).
58. Feltz, P. & De Champlain J., Brain Res. 43, 601-605 (1972).
59. Feltz, P. & De Champlain, J., In :"Frontiers in catecholamine Research", E. Usdin & S.H. Snyder (Eds) , Pergamon Press, 950-956 (1973).
60. Fibiger, H.C., Pudritz, R.E., Mc Geer, P.L. & Mc Geer, E.G., J. Neurochem. 19, 1697-1708 (1972).
61. Fonnum, F., Gottesfeld, Z. & Grofova, I., Brain Res. 143, 125-138 (1978).
62. Frigyesi, T.L., & Purpura, P.D., Brain Res. 6, 440-456 (1967).
63. Frigyesi, T.L. & Szabo, J., Expl. Neurol. 49, 123-139 (1975).
64. Fuller, D.R.G., Hull, C.D. & Buchwald , N.A., Brain Res. 96, 337-341 (1975).
65. Gale, K., Hong, J.S., Guidotti, A., Brain Res. 136 (2), 371-375 (1977).
66. Gerschenfeld, H.M., Physiol. Rev. 53, 1 - 119 (1973).
67. Ginsborg, B.L., House, C.R. & Silinsky, E.M., J. Physiol.(Lond) 236, 723-731 (1974).
68. Giorguieff, M.F., Le Floc'h, M.L., Glowinski, J. & Besson, J.M., Brain Res. 106, 117-132 (1976).
69. Giorguieff, M.F., Kemel, M.L., Glowinski, J. & Besson, J.M., J. Pharmacol. Exp. Ther. 200 (3), 535-544 (1977).
70. Giorguieff, M.F., Kemel, M.L., Glowinski, J. & Besson, J.M., Brain Res. 139, 115-130 (1978).
71. Glowinski, J., Neurosciences, 4th study program, in press (40 p, NIT Press (1977).
72. Gonzalez Vegas, J.A., Brain Res. 80, 219-228 (1974).

73. Goswell, M.J. & Sedgwick, E.M., J. Physiol.(Lond) 218,84P-85P (1971).
74. Grofova, I. , Brain Res. 91, 286-291 (1975).
75. Grossman, A.R., Spira, M.E. & Parnas, I., Brain Res. 64, 379-386 (1973).
76. Guyenet, P., Agid, Y., Javoy, F., Beaujouan, J.C., Rossier, J. & Glowinski, J., Brain Res. 84, 227-244 (1975).
77. Guyenet, P., Euvrard, C., Javoy, F., Herbet, A. & Glowinski, J., Brain Res. 136 (3), 487-500 (1977).
78. Hammond, C., Deniau, J.M., Rizk, A. & Feger, J., Brain Res. 150, (1978).
79. Hassler, R., Chung, J.W., Rinne, U. & Wagner, A., Exp. Brain Res. 31,67-80 (1978).
80. Hedreen, J.C., Neuroscience Letters 4, 1-7 (1977).
81. Hedreen, J.C. & Chalmers, J.P., Brain Res. 47, 1-36 (1972).
82. Henn, F.A., Anderson, D.J. & Sellström, A., Nature 266, 637-638 (1977).
83. Herz, A. & Zieglgänsberger, W., Int. J. Neuropharmacol.7, 221-230 (1968).
84. Hökfelt, T., Kellerth, J.O., Nilsson, G. & Pernow, B., Science 190, 889-890 (1975).
85. Holman, R.B. & Vogt, M., J. Physiol.(Lond) 223,243-254 (1972).
86. Hong, J.S., Yang, H.Y.T., Racagni, G. & Costa, E., Brain Res. 122, 541-544 (1977).
87. Hösli, L., Andres, P.F. & Hösli, E., Neurosciences Letters 6, 79-83 (1977).
88. Hull, C.D., Bernardi, G. & Buchwald, N.A., Brain Res. 22, 163-179 (1970).
89. Hull, C.D., Bernardi, G., Price, D.D. & Buchwald, N.A., Exper. Neurol. 38, 324-336 (1973).
90. Hull, C.D., Levine, M.S., Buchwald, N.A., Heller, A. & Browning, R.A., Brain Res. 73, 241-261 (1974).
91. Iversen, L.K., J. Neurochem. 29, 5-12 (1977).
92. Jones, B.E., & Moore, R.Y., Brain Res. 127, 23-53 (1977).
93. Jones, E.G., Coulter, J.D., Burton, M. & Porter, R., J. Comp. Neurol. 173, 53-80 (1978).
94. Kanazana, I., Marshall, G.R. & Kelly, J.S., Brain Res. 115, 485-491 (1976).
95. Kanazawa, I., Emson, P.C. & Cuello, A.C., Brain Res. 119, 447-453 (1977).
96. Kebabian, J.W., Adv. Cyclic Nucleotide Res. 8, 421-508 (1977).
97. Kemp, J. & Powell, T.P.S., Phil. Trans. B., 262, 383-401 (1971).
98. Kim, J.S., Hassler, R., Haug, P. & Paik, K.S., Brain Res. 132(2), 370-374 (1977).
99. Kitai, S.T., Wagner, A., Precht, W. & Ohno, T., Brain Res. 85, 44-48 (1975).
100. Kitai, S.T., Kocsis, J.D., Preston, R.J. & Sugimori, M., Brain Res. 109, 601-606 (1976).
101. Kitai, S.T., Kocsis, J.D. & Wood, J., Brain Res. 118, 137-141 (1976).
102. Kitai, S.T., Sugimori, M. & Kocsis, J.D., Exp. Brain Res. 24, 351-363 (1976).
103. Kitai, S.T. & Kocsis, J.D., In : Iontophoresis and transmitter mechanisms in the mammalian central nervous system. R.W. Ryall and J.S. Kelly (Eds), Elsevier North-Holland Biomedical Press, Amsterdam-New York, 17-34 (1978).
104. Kocsis, J.D., Sugimori, M. & Kitai, S.T., Brain Res. 124, 403-413 (1977).
105. Kocsis, J.D. & Kitai, S.T., Brain Res. 138, 271-283 (1977).
106. Krauthamer, G., Feltz, P. & Albe-Fessard, D. , J. Neurophysiol. 30, 81-97 (1967).
107. Krnjevic, K. & Schwartz, S., Exp. Brain Res. 3, 306-319 (1966).
108. Kumar, M.M., Roth, R.M. & Aghajanian, G.R., Brain Res. 35, 167-176 (1971).
109. Langer S.Z, Br. J. Pharmacol. 60, 481-497 (1977).
110. Levine, M.S., Hull, C.D., & Buchwald, N.A., Exp. Neurol. 44, 448-460 (1974).
111. Levine, M.S., Hull, C.D., Buchwald, N.A., Garcia-Rill, E., Heller, A. & Erinoff, L., Brain Res. 131, 215-225 (1977).
112. Libet, B. & Owman, G.H., J. Physiol.(Lond) 237, 365-662 (1974).
113. Libet, B; & Tosaka, T., Proc. Nat. Acad. Sci. (USA) 67, 667-673 (1970).
114. Libet, B., Tanaka, T. & Tosaka, T., Life Sci. 20(11), 1863-1870 (1977).
115. Libet, B., Kobayashi, H. & Tanaka, T. , Nature (Lond) 258, 155-157 (1975).
116. Liles, S.L., Electroenceph. Clin. Neurophysiol. 35, 277-285 (1973).

117. Ljungdahl,A. Hökfelt, T., Goldstein, M. & Park, D., Brain Res. 84, 313-319 (1975).
118. Mc Geer, E.G., Fibiger, H.C., Mc Geer, P.L. & Brooke, S., Brain Res. 52, 289-300 (1973).
119. Mc Geer, E.G., Searl, K. & Fibiger, H.C., J. Neurochem. 24, 283-288 (1975).
120. Mc Geer, E.G., Mc Geer, P.L., Greewall, D.S. & Singh, V.K., J. Pharmacol. 6, 143-152 (1975).
121. Mc Geer, E.G., Innanen, V.T. & Mc Geer, P.L., Brain Res. 118, 356-358 (1976).
122. Mc Geer, P.L. & Mc Geer, E.G., Brain Res. 91, 331-335 (1975).
123. Mc Lennan, H. & York, D.H., J. Physiol. (Lond) 187, 163-175 (1966).
124. Mc Lennan, H. & York, D.H., J. Physiol. (Lond) 189, 393-402 (1967).
125. Mc Nair, J.L., Sutin, J. & Tsubokawa, T., Exp. Neurol. 37, 395-411 (1972).
126. Malliani, A. & Purpura, D.P., Brain Res. 6, 341-354 (1967).
127. Mao, C.C., Marco, E., Revuelta, A., Bertilsson, L. & Costa, E., Biol. Psychiatr. 12 (3), 359-371 (1977).
128. Mao, C.C., Cheney, D.L., Marco, E., Revuelta, I.A. & Costa, E., Brain Res. 132, 375-379 (1977).
129. Marco, E., Mao, C.C., Cheney, D.L., Reveulta, A. & Costa, E., Nature (Lond) 264, 263-266 (1976).
130. Martin, A.R., Wickelgreen, W.O. & Beranek, R., J. Physiol.(Lond) 207, 653-665
131. Miller, J.J., Richardson, T.L., Fibiger, H.C. & Mc Lennan, H., Brain Res. 97, 133-138 (1975).
132. Misgeld, U., In : Symp. on basal ganglia -cellular and functional aspects-, R. Hassler, N.A. Buchwald and S.T. Kitai (Eds), J. Applied Neurophysiol., Karger (Basel) in the press (1977-78).
133. Misgeld, U., In : Symp. on basal ganglia -cellular and functional aspects-, R. Hassler, N.A. Buchwald and S.T. Kita (Eds), J. Applied Neurophysiol., Karger (Basel) in the press (1977-78).
134. Myers, P.R., Livengood, D.R. & Shain, W., J. Cell. Physiol. 91 (1), 103-118 (1977).
135. Nelson, P.G., Physiol. Rev. 55, 1-161 (1975).
136. Nicklas, W.J. & Berl, S., Brain Res. 61, 343-356 (1973).
137. Nicklas, W.J., Berl, S. & Clarke, D.D., J. Neurochem. 23, 149-157 (1974).
138. Nicoll, R.A., Science (USA) 199, 451-452 (1978).
139. Norcross, K. & Spehlmann, R., Neurosci. Letters 6, 323-328 (1977).
140. Ohye, C., Le Guyader, C. & Feger, J., Brain Res. 111, 241-252 (1976).
141. Oka, K. & Kinnai, K., Exp. Brain Res. 31, 31-42 (1978).
142. Olpe, H.R. & Koella, W.P., Brain Res. 122, 357-360 (1977).
143. Parnas, I., J. Neurophysiol. 35, 930-914 (1972).
144. Phillipson, O.T., Emson, P.C., Horn, A.S., & Jessel, T., Brain Res. 136 (1) 45-58 (1977).
145. Pollard, H., Lloren-Cortes, C. & Schwartz, J.C., Nature 268 (5622), 745-747 (1977).
146. Precht, N. & Yoshida, M., Brain Res. 32, 229-233 (1971).
147. Purpura, D.P. & Malliani, A. , Brain Res. 6, 325-340 (1967).
148. Rasminski, M., Mauro, A.J. & Albe-Fessard, D., Brain Res.61, 69-77 (1973).
149. Richardson, T.L., Miller, J.J. & Mc Lennan, H., Brain Res.127(2), 219-234 (1977).
150. Racagni, G., Cheney, D.H., Trabbuchi, M. & Costa, E., J. Pharmacol. Exp. Ther. 196, 323-332 (1976).
151. Reubi, J.C., Iversen, L.L. & Jessel, T.M., Nature 268, 5621, 652-654 (1977).
152. Rocha-Miranda, C.E., Electroenceph. Clin. Neurophysiol. 19, 237-247 (1965).
153. Sangiah, S., Borowitz, J.L. & Yim, G.E.W., European J. Pharmacol. 27, 130-135 (1974).
154. Shultz, W. & Ungerstedt, U., Brain Res. 142, 357-362 (1978).
155. Schwarcz, R. & Coyle, J.T., Brain Res. 127 (2),235-249 (1977).
156. Sethy, U.H. & Van Woert, M.H., Nature (Lond) 251, 529-530 (1974).

157. Shimahara, T. & Tauc, L., J. Physiol.(Lond) 247 , 321-341 (1975).
158. Shimahara, T. & Tauc, L., Brain Res. 127, 168-172 (1977).
159. Siggins, G.R.,Neurosciences 4th study Program, MIT Press, (1977).
160. Siggins, G.R., Hoffer, B.J., Bloom, F.E. & Ungerstedt, U., In : The basal ganglia, M.D. Yahr (Eds), Research publications of the association for Res. in nervous and mental disease (ARMD), vol. 55, Raven Press, New York, 227-245 (1977).
161. Spano, P.F., Trabucchi, M. & Di-Chiarag, Science 196, 4296, 1343-1345 (1977).
162. Spehlmann, R., Norcross, K. & Grimmer, E.J., Experientia (Basel) 33, 623-625 (1977).
163. Spencer, J.J., Brain Res. 102, 91-101 (1976).
164. Stadler, H.,Lloyd, K.G., Gadea-Ciria, M. & Bartholini, G., Brain Res. 55 476-480 (1973).
165. Starke, K., Ergebn. Physiol., Rev. Physiol. Biochem. Pharmacol. 77, 1-124 (1977).
166. Starke, K., Taube, H.D. & Borowski, E., Biomechical Pharmacol. 26, 259-268 (1977).
167. Stoff, J.C. & Mulder, A.H., Eur. J. Pharmacol. 46 (2), 177-180 (1977).
168. Suria, A., Neuropharmacology 15, 11-19 (1976).
169. Suria, A. & Costa, E., Brain Res. 87, 102-106 (1975).
170. Swann, J.W. & Carpenter, D.O., Nature (Lond) 258, 751-754 (1975).
171. Tassin, J.P., Cheramy, A., Blanc, G., Thierry, A.M. & Glowinski J., Brain Res. 107, 291-301 (1976).
172. Ternaux, J.P., Hery, F., Bourgoin, S., Adrien, J., Glowinski, J. & Hamon, M., Brain Res. 121, 311-326 (1977).
173. Trabucchi, M., Cheney, D.L., Racagni, G. & Costa, E., Brain Res. 85, 130-134 (1975).
174. Usunoff, K.F., Hassler, R., Wagner, A. & Bak, I.J., Brain Res.74, 143-148, (1974).
175. Walker, R.J., Kemp, J.A., Yajima, M., Kitagawa, K. & Woodcruff, G.N., Experientia 32, 214-215 (1976).
176. Wang, R.Y. & Aghajanian,G.K., Brain Res. 132, 186-193 (1977).
177. Westfall, T.C., Neuropharmacology 13, 693-700 (1974).
178. Westfall, T.C., Besson, M.J., Giorguieff, M.F. & Glowinski, J., Naunyn Schmiedeberg's, Arch. Pharmacol. 292, 279-287 (1976).
179. Westfall, T.C., Kitai, D. & Wahl, G., J. Pharmacol., Exp. Ther. 199, 149-157 (1976).
180. Wilkening, D. & Makman, M.H., J. Neurochem. 28 (5), 1001-1007 (1977).
181. Yarowsky, P.J. & Carpenter, D.O., J. Neurophysiol. 41 (2), 20p (1978).
182. York, D.H., In : Handbook of psychopharmacology", L.L. Iversen, S.D. Iversen and S.H. Snyder (Eds), vol. 6, 23-61, Plenum Press, New York (1975).
183. Yoshida, M. & Precht, N., Brain Res. 32, 225-228 (1971).
184. Yoshida, M., Rabin, A. & Anderson, M., Brain Res. 30, 235-239 (1971).
185. Yoshida, M., Rabin, A. & Anderson, M., Exp. Brain Res. 15, 333-347 (1972).
186. Zapata, P., J. Physiol.(Lond) 244, 235-251 (1975).
187. Zarzecki, P., Blake, D.J. & Somjen, G.C., Brain Res. 115, 257-272 (1976).

Ontogeny and Phylogeny

Introduction

I. Divac and R. G. E. Öberg

The following section gives historical perspectives on the NS by treating its origin and development in phylogeny and ontogeny. These topics have rarely been discussed in the past few decades, reflecting a rather limited interest for the evolution and development of the brain in the neuroscientific community at large. This period of relative neglect seems surprising in retrospect: recent years have seen a surge in interest for species comparisons and developmental studies, and with the help of new research methods many fresh and important insights have been gained.

In the first chapter, Webster presents a thorough discussion of the comparative anatomy of the NS and emphasises many of the problems of comparative neurology. The difficulties notwithstanding, the hodological and biochemical findings of the past decade convincingly demonstrate that reptiles and birds are not "neostriatal preparations" as sometimes thought and taught, but that only parts of the telencephalon in these orders how correspondence to the NS in mammals. This (partial) correspondence gives impetus for future functional comparisons between the "ventral" striatum in reptiles, the "paleostriatum augmentatum" in birds, and the "neostriatum" in mammals. One interesting difference between these structures is, however, pointed out by Smart and Sturrock: NS in mammals develops from an embryonic generative zone which is missing in reptiles. Thus, strictly speaking, the NS may have no homologue in reptiles.

In the third chapter, Heimer and Van Hoesen present findings in support of the notion that nucleus accumbens and the olfactory tubercle (together constituting the ventral striatum) do not differ from NS in architecture and biochemical composition. Whereas NS is innervated from the neocortex, however, the ventral stiatum received projections from the cortex which is assoicated with the "limbic system". It appears, then, that the allocortical areas are related to the ventral striatum in a manner analogous to the relation between the neocortex and NS.

One approach to understanding the functions of NS is to compare the behaviour of decorticate animals with that of animals which are decorticate as well as destriate. This line of research was discussed by Dr. Steele-Russel during the meeting. Unfortunately, Dr. Steele-Russel has been unable to contribute to this volume.

Some Aspects of the Comparative Study of the Corpus Striatum

K. E. Webster

Department of Anatomy, University of London King's College, Strand, London WC2R 2LS, England

ABSTRACT

Large parts of what were originally considered to be the corpora striata (neostriatum and palaeostriatum) of non-mammals are now known to establish afferent-efferent relations similar in many ways to those of the mammalian neocortex. No matter how these observations are finally interpreted, it is clear that non-mammals can no longer be considered to be naturally occurring 'striatal' preparations. They do, however, possess neuron populations similar to those of the mammalian corpus striatum, although in many cases establishing more marked relationships with the brainstem than with other parts of the telencephalon.

The view that the behaviour of non-mammals could tell us directly much about the function of the mammalian corpus striatum was, until very recently, widely disseminated, if with no very great certainty as to its provenance or conviction of its validity. This view assumed first, that the non-mammalian forebrain lacks neocortex; secondly, that the mammalian corpus striatum and cortex are essentially independent, the 'highest' representatives of two separate channels for the outflow of 'motor information'; and thirdly that the nervous system evolved as an hierarchy, so that 'phylogenetically ancient' structures (i.e. non-cortical structures such as the corpus striatum) retain their structure and function even when the latter is somehow made subservient to more recently developed mechanisms (in this case the cerebral cortex)(120). Experimental evidence has necessitated drastic modification of the first two ideas (81,111,119). The third is discussed by Steele Russell (this volume). The forebrain of non-mammals has often been supposed to lack 'neocortex' because it is difficult to identify by simple examination of topography and cytoarchitecture any structure of the relative size, appearance and complexity of the neocortex of mammals. The hemisphere appears to be dominated by two components: structures concerned with olfaction, and others that in general terms appear to be corpus striatum - albeit a corpus striatum often of relatively enormous size and a cytoarchitectural complexity quite unknown in mammals. In contrast, the sole immediate claimant to the title 'neocortex' is small and poorly elaborated. This parcellation has been challenged in the past, more especially in relation to birds - the forebrain region known as the Wulst (hyperstriatum - v.i.) has been regarded by several authors as incorporating or representing, at least in part, a considerable cortical element (20,30). Elliot Smith (31, p.151) went so far as to remark that 'Birds...........have quite left the path leading to the formation of a neopallium and evolved a mechanism peculiar to their own class, which does much the same kind of work as the neopallium'. Before recent reinterpretations of the non-mammalian telencephalon (59,81), this comment (that hindsight at least must deem perceptive) went ignored. At best, it was often assumed that no more than a trivial 'primordium' of neocortex is present in the dorsal part of the hemisphere (DC, Figs. 1 and 2).

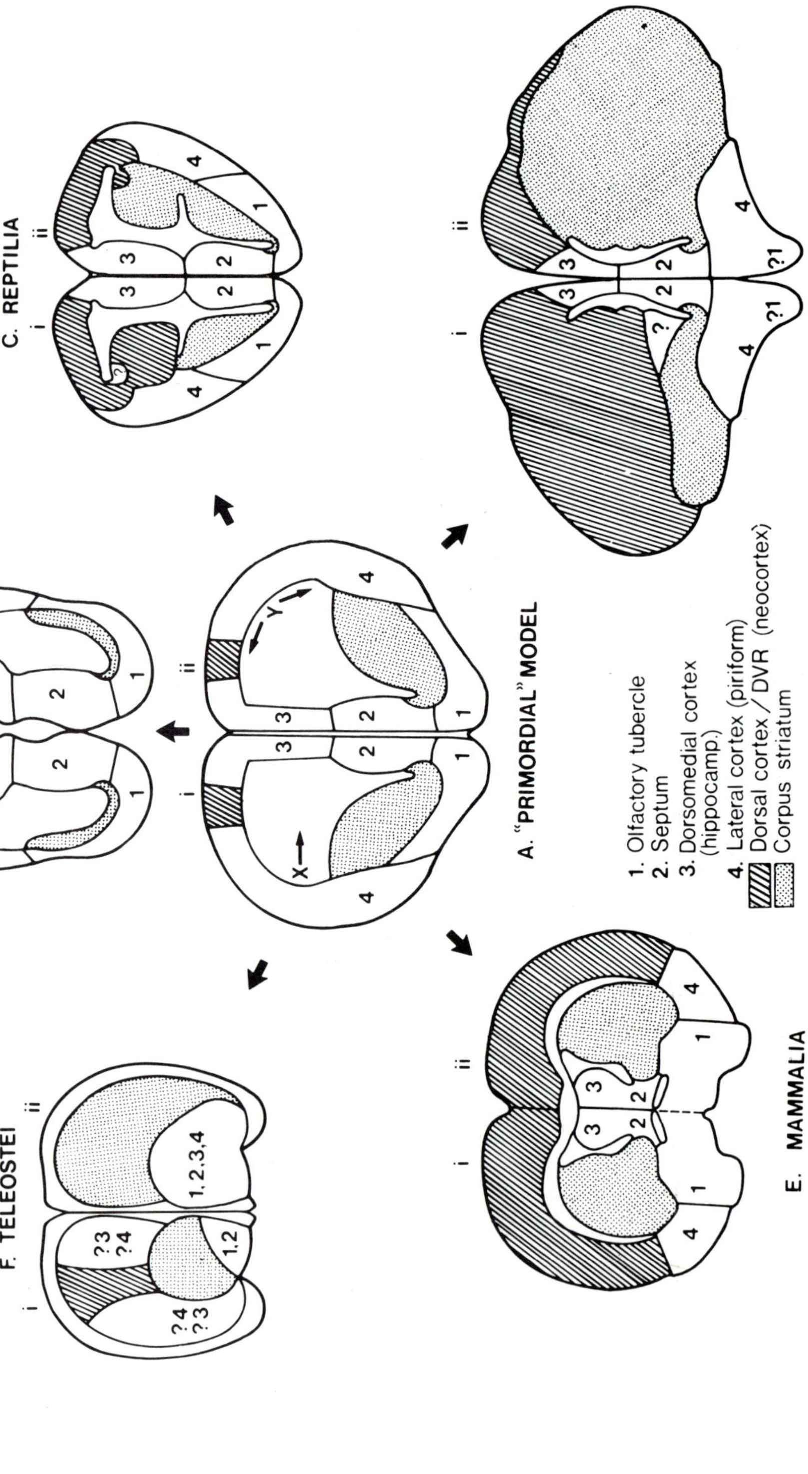

Fig. 1. Subdivisions in the vertebrate telencephalon: old (ii) and new (i) models.

The relatively larger lateral cortex (LC) was (and still is) generally thought to correspond to the mammalian olfactory (piriform) cortex (PiC), and the medial wall of the hemisphere to hippocampus above (DMC) and septum (Se) below (5,43,81,87,88,96) - see Figs. 1 and 2A,B,C. The basal ganglia of mammals were supposed to contain three basic subdivisions - the 'archistriatum' (amygdala); the 'palaeostriatum' (pallidum); and the 'neostriatum' (striatum or caudate-putamen). Each of these components was said to be individually identifiable as an equivalent nucleus (or, if need be, a collection of nuclei) in the lateral part of the telencephalon in question, where, in many non-mammals, such structures collectively make up the greater part of the hemisphere (Fig. 1Bii,Fii).

In fact, the configuration of the vertebrate forebrain shows an amazing variety to which this (and perhaps any) general hypothesis scarcely does justice. It is certainly impossible to give more than a general account in the space available, and I propose to limit the discussion largely to an examination of reptiles and birds, with an occasional general glance at the anamniote forebrain. It must be borne in mind that to speak of 'birds' or 'reptiles', rather than 'Columba' or 'Testudo' etc., is no less a convenience than to speak collectively of 'mammals'. It inevitably involves oversimplification which may in itself ultimately be confusing. It is becoming clear, for example, that there are considerable differences in the details of the organisation of the forebrain among various reptiles. In spite of these reservations, I hope to show that the general descriptive hypothesis outlined above can now be discarded. A new general thesis can be substituted, which, for all its defects, is a 'better fit' than the one it replaces.

The general topographical and cytoarchitectural arrangement of the reptilian telencephalon is illustrated at its most simple in Fig. 2B (see 3,5,36,87,96 and 103 for reviews). Judged on positional evidence, the structures labelled 'dorsal ventricular ridge' (DVR) and 'ventral striatum'* (Vst) appear to correspond to the mammalian corpus striatum - that is, to the 'neostriatum' and 'palaeostriatum' (c.f. Figs. 1Cii and 1E). The upper of these two subdivisions (which are not so clearly separated in all reptilian species - for example, the sulcus between the two is less well marked in Squamata and absent from Crocodylia) has been referred to as the 'dorsal striatum' (4), but the terms 'hypopallium' (32) and more especially 'dorsal ventricular ridge' (DVR) (44,45) have the advantage of neutrality. Indeed, Elliot Smith's insistence that the DVR is a pallial derivative might again be regarded as prescient, although the intuition was marred by his (and Johnston's) insistence that the DVR is best compared with the mammalian 'neostriatum' - i.e. the caudate-putamen (C-P). The ventral striatum they identified as 'palaeostriatum' - i.e. the pallidum (Pall) (see also 11,24). The DVR is continuous with the dorsal cortex (DC, Fig. 2B) above, and the continuity, originally referred to as the 'primordium neopallii' (21), is now known as the pallial thickening (Pall. Th., Fig. 2B). In some forms (e.g. Crocodylia) the dorsal cortex and DVR are apparently fused into one structure, and in all living reptiles there is a tendency for further nuclear subdivisions to appear in the ridge. In this hypothesis it is therefore often necessary to equate the cytoarchitecturally uniform caudate-putamen with a whole group of cytoarchitecturally distinct nuclei. The posterior (caudal) region of the DVR (the 'archistriatum') also shows much interspecies variation - in Crocodylia it is relatively simple; in Squamata it is elaborate, containing a well demarcated nucleus (nucleus sphaericus) which receives the afferents from the accessory olfactory bulb (45). At least part of the reptilian 'archistriatum' appears, therefore, to be comparable with the mammalian amygdala.

*This term is not synonymous with that used by Heimer (this volume) to describe parts of the mammalian olfactory tubercle (l in Fig. 1, and tub in Fig. 2), and for which equivalent areas in non-mammals are at present unknown.

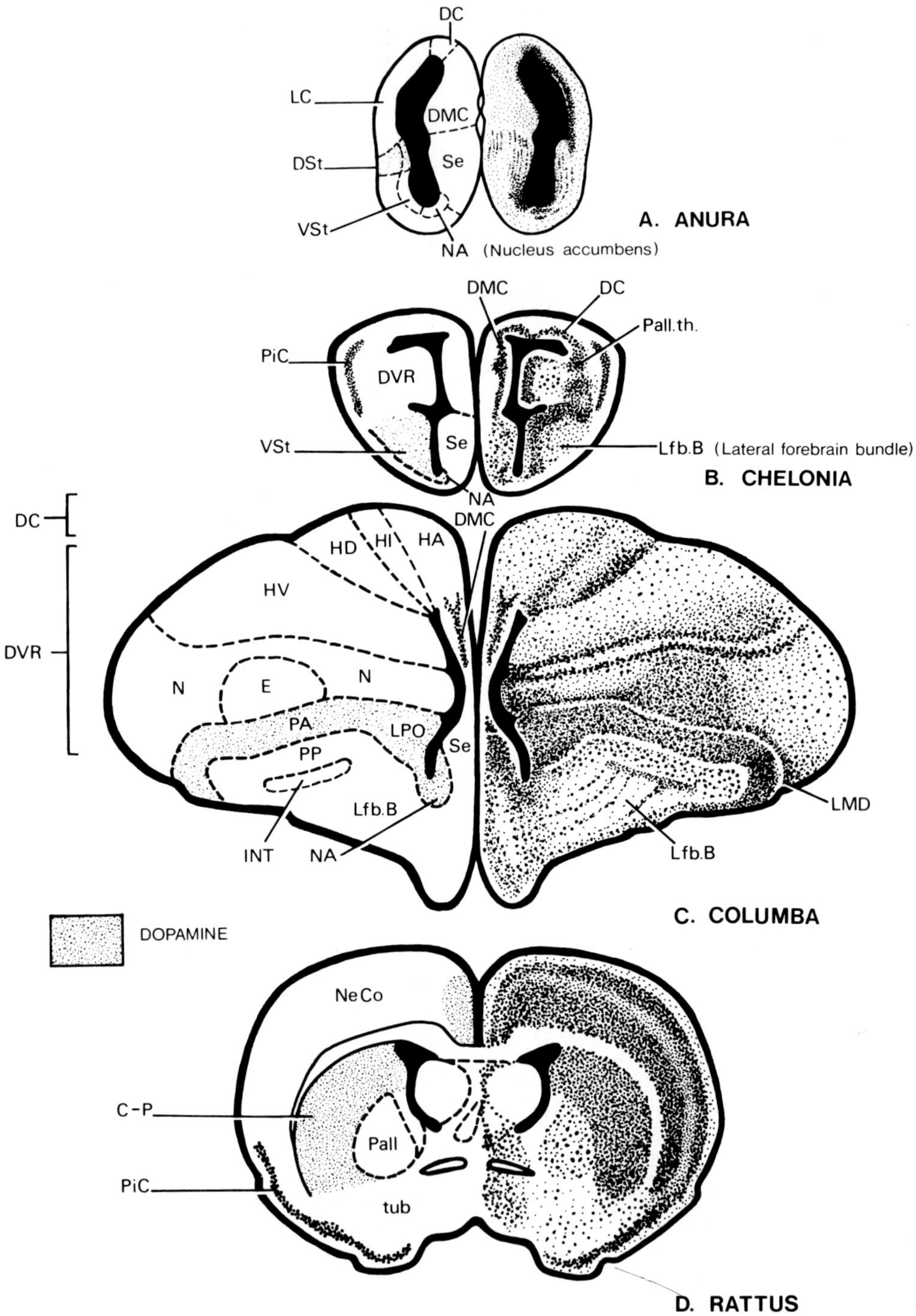

Fig.2. Semi-schematic representations of sections through the telencephalon from four groups of vertebrates. Modified from 117; reproduced by permission of the Zoological Society, London. It is not certain that the striatal monoamine in Anuran brains is dopamine.

At first sight, the condition in birds could not appear more removed from that in mammals, and also seems very different from that of reptiles. The telencephalon has a remarkably 'solid' appearance, apparently consisting of a mass of grey matter subdivided into many cytoarchitecturally distinct nuclei (5,38,45,61,95,96,104,109, 117 - see Figs. 1Dii and 2C). The plethora of descriptive analyses of the cyto-architecture may be conveniently dispensed with by parcelling the 'striatal' nuclei into two groups, dorsal and ventral. These groups are separated by a layer of fibres known as the dorsal medullary lamina (LMD, Fig. 2C). The ventral group is the more simple, containing no more than three nuclei [palaeostriatum augmentatum (PA), palaeostriatum primitivum (PP), and the intrapeduncular nucleus (INT)]. It will be shown later that it almost certainly corresponds to the reptilian ventral striatum. This group has also been referred to as the 'internal striatum' (81). Immediately anterior and medial to the palaeostriatum augmentatum is a region known as the paraolfactory lobe (LPO). The remaining (dorsal) division, which includes the neostriatum (N), ectostriatum (E), and ventral hyperstriatum (HV) is equivalent to at least part of the DVR. As in Crocodylia, the dorsal cortex is incorporated into the remaining DVR as the dorsal, accessory and intercalated hyperstriatum (HD, HA, HI) which together are sometimes known as the Wulst. Collectively the avian DVR and dorsal cortex comprise the 'external striatum' (81). The avian forebrain does contain some free dorsal cortex, particularly posteriorly, as well as a 'corticoid' rudiment in the superficial lateral parts of the DVR (20,95). Posteroventrally, the DVR contains an 'archistriatal' region, but only a small part of this can be compared with the mammalian amygdala (123).

The anamniote telencephalon is very variable in its morphology (5,27,96). In many animals it appears relatively simple (e.g. Figs. 1B and 2A), and the striatal thickening poorly developed - although dorsal and ventral divisions (DSt, VSt, Fig. 2A) have been described (66). In others (and especially in teleost fishes), the telencephalon is apparently composed almost entirely of corpus striatum (Fig. 1Fii), but this appearance is almost certainly illusory and due to developmental eversion of the hemispheres (83,84,85). Eversion is said to involve a thickening and lateral rotation of the hemisphere about its basal regions (1 and 2 in Fig. 1A) so that 'dorsomedial' (hippocampal) cortex (3 in Fig. 1A) comes to lie laterally, joined to the septum (2 in Fig. 1A) by an ependymal membrane that has all the appearance, from its position, of 'atrophic' cortex (Fig. 1Fi). However, as may be judged from the illustration, the real nature of the final arrangements is still disputed (c.f. 27 and 83,84,85; see also 96). This state of affairs underlies the dangers of treating comparative neuroanatomy as an exercise in topology rather than one of experimental investigation. Perhaps the most important general point to be made about the anamniote telencephalon is that it is not dominated by olfaction, as was once thought, but like that of other vertebrates, receives considerable ascending inputs from the diencephalon.

How, if at all, are these differing arrangements of topography and cytoarchitecture to be compared? The practice of naming subdivisions of the nervous system on the basis of cytoarchitecture alone is a useful, but hazardous, practice, since common names can have a peculiar potency in suggesting common functional significance (44). The nervous system is above all characterised as a mechanism in which neurons signal each other, generally through circuit-like arrangements, and any morphological analysis that ignores this is potentially sterile. Not that this 'connectivity'(or 'hodological') approach is itself without difficulties. The most notable pitfall (an example of which will be noted later) is the risk of circular argument: in a neuronal circuit the initial identification of one element (on what criteria?) prejudices the identities to be assigned to the remainder (86). In spite of this criticism (and Campbell's (12) refutation of it is not entirely convincing), the fact that the 'connectivity' approach reflects one of the fundamental features of the nervous system is a powerful argument for its continued use. Thus, in order to establish homologues of the basal nuclei, it is necessary to identify regions that have similar (if not identical) interconnections in various animals. (Needless to say, this starting point accepts that there is 'a mammalian corpus striatum' and not 'corpora striata' of fundamentally different configurations in different mammals - see 111).

In the present context, the key features of the connections of the mammalian corpus striatum (see 13,65,119 and Grofová in this volume; and Fig. 4A) appear to be (a) Afferents end in the striatum and are derived essentially from the intralaminar thalamus (99), the cerebral neocortex (37,64,69,115,116,121), and the substantia nigra pars compacta (14). The latter input is dopaminergic and can be identified by fluorescence histochemistry (2). (b) Efferents. The striatum has a projection to the substantia nigra, pars reticulata, but the major target of its outflow is the pallidum (19,110), which is the principal source of efferents from the basal nuclei as a whole. Pallidal efferents distribute largely to the part of the thalamus projecting upon the motor cortex, to the subthalamus, and to a lesser extent to the nucleus tegmenti pedunculopontinus of the brainstem tegmentum (15,70, 82). Let us consider these features in turn.

Afferent connections of the basal nuclei

This group of observations necessitates that we first determine homologues of (i) the substantia nigra, (ii) the intralaminar thalamus, and (iii) the neocortex, and then discover some common region of the telencephalon upon which they all project. The first exercise raises an immediate problem: unless some new and powerful criterion is introduced, the use of the 'connectivity' approach to determine homologies threatens a regression if not to infinity, then at least as far as peripheral receptors. As we shall see, this indeed occurs in the case of the search for a neocortical homologue. In the case of the substantia nigra, however, another factor emerges - the fact that the transmitter substance is the catecholamine compound dopamine.

(i) The substantia nigra. Both catecholamine and 5-HT containing perikarya, identified by histofluorescence techniques, occur in the brainstem tegmentum of amphibia (6,9,17), reptiles (90,93), and birds (35,12); but in no case is there a discrete nuclear aggregation in the midbrain readily identifiable (as by inspection) with the substantia nigra pars compacta (mammalian cell groups A8 and 9 - see 22). Many of these cells do, however, lie in the ventrolateral tegmentum and project upon the telencephalon, where, as shown in Fig. 2, catecholamines (probably dopamine) occur in particularly high concentration in the dorsal striatum (amphibia), ventral striatum and nucleus accumbens (reptiles), and palaeostriatum augmentatum, paraolfactory lobe and nucleus accumbens (birds). It is almost certain that this represents a dopaminergic innervation from the midbrain, corresponding to that of mammals (8,52,53,60,66,89,90,92). The shaded areas on the left of Fig. 2A,B,C are therefore prime candidates for homologues of the caudate-putamen. But since, in mammals, the midbrain dopaminergic cell group A10 projects to the prefrontal cortex (23,73), the telencephalic areas concerned cannot be unreservedly identified exclusively with the mammalian striatum. Catecholamine fibres have been shown to ascend from the brainstem to the dorsal cortex and DVR in both reptiles (90,94) and birds (8,111). Some of the aminergic cell groups concerned probably correspond to mammalian nuclei such as A10 and the locus caeruleus (22). One further reservation is that in many of these observations the catecholamine has not been rigorously identified as dopamine or noradrenaline.

(ii) The intralaminar thalamus. Attempts to identify intralaminar thalamic nuclei (IL) in non-mammals affords a good example of the regression problem taking the form of a circular argument: since most of the thalamus projects upon the telencephalon and the telencephalon is largely corpus striatum, then the dorsal thalamus must be largely intralaminar nuclei. The chief characteristics used to identify the intralaminar nuclei of the mammalian dorsal thalamus have been their physiological 'non-specificity' ('polymodality' involving polysynaptic pathways) and their reactions to telencephalic ablations. In the latter regard, unlike the neurons of the physiologically 'specific' principal nuclei, which degenerate dramatically after neocortical lesions, the perikarya of these nuclei are resistant to telencephalic damage. The observation that involvement of the striatum in the lesion is important, although not crucial*, has usually been interpreted as evidence for a major (if not exclusive) projection from intralaminar thalamus to striatum (99). One might predict that, if the telencephalon of non-mammals is largely striatum and the thalamus intralaminar, then almost any telencephalic lesion would lead to retrograde degeneration in the thalamus. This prediction is only partly true: with the exception of lesions in that part of the avian hyperstriatum known as the Wulst, retrograde degeneration is produced only by massive lesions of the DVR invading the ventral striatum or palaeostriatum (98,100). In other words, most of the dorsal thalamus does respond in a way strikingly similar to the 'non-specific' thalamus of mammals - as if it were indeed intralaminar. This is, as we shall see, a puzzling result, for other (more persuasive) evidence suggests much of the non-mammalian thalamus to be more akin to the 'principal' ('specific') nuclei of mammals. Another complication is that it is now certain that in mammals the intralaminar nuclei, in addition to their large projection upon the striatum, send some fibres to the cerebral cortex (49,68,78) - an observation which further limits the usefulness of this line of enquiry.

Anatomical criteria for discriminating between intralaminar and principal thalamic nuclei independently of their relationships with the telencephalon are by no means easy to apply - the characteristic (but small) ascending inputs from spinal cord and cerebellum are not universally distributed to all nuclei described as 'intralaminar', and their supposedly large reticular inputs are for the most part too ill-categorized and little investigated to be of use (see 29, 117, and Krauthamer, this volume, for discussion and references). But, as previously pointed out (26) and shown in Fig. 3, the most cursory examination reveals that the non-mammalian thalamus is apparently dominated not by somatic (S) and cerebellar (C) inputs (26,56), but by visual afferents that arrive through oligosynaptic pathways, whether from the retina (R) or optic tectum (T, Ts, Td). Indeed, thalamic afferents from brainstem auditory centres (A) complete the complement of sensory inputs in non-mammals, and add weight to the suggestion that whatever the status of each individual nucleus in comparison with possible mammalian counterparts (e.g. the avian nucleus ovoidalis might best be compared with the medial geniculate body), as a group they are apparently not 'intralaminar' but 'principal'. So complete has been this reappraisal of the non-mammalian thalamus that few, if any, authors will now hazard identifying the intralaminar nuclei in these animals (c.f. 29,91). There have followed, however, consequences more far-reaching than this.

(iii) Neocortex. The various nuclei which make up the major part of the avian and reptilian thalamus receive afferents that are physiologically and anatomically well-specified (including those from the retina and optic tectum), and project in turn to discrete regions of the DVR and/or dorsal cortex (39,46,58,59,62,79,80,81,91,102,117, 118, Lohman & Hoogland, personal communication). Such observations inevitably suggest comparison between the DVR/dorsal cortex and the mammalian neocortex. In the case of birds, they strengthen the assumption that the avian palaeostriatal complex alone is equivalent to the mammalian corpus striatum (c.f. Figs. 4A and B).

---

*One intralaminar group (nuclei centralis lateralis, medialis et paracentralis), rather like the posterior group of thalamic nuclei, shows some signs of (incomplete) degeneration after neocortical removal. But the changes become more marked when the lesion is widespread, and involves the striatum as well as cortex. The other group (nuclei centrum medianum et parafascicularis) degenerate only when massive ablations invade both cortex and corpus striatum.

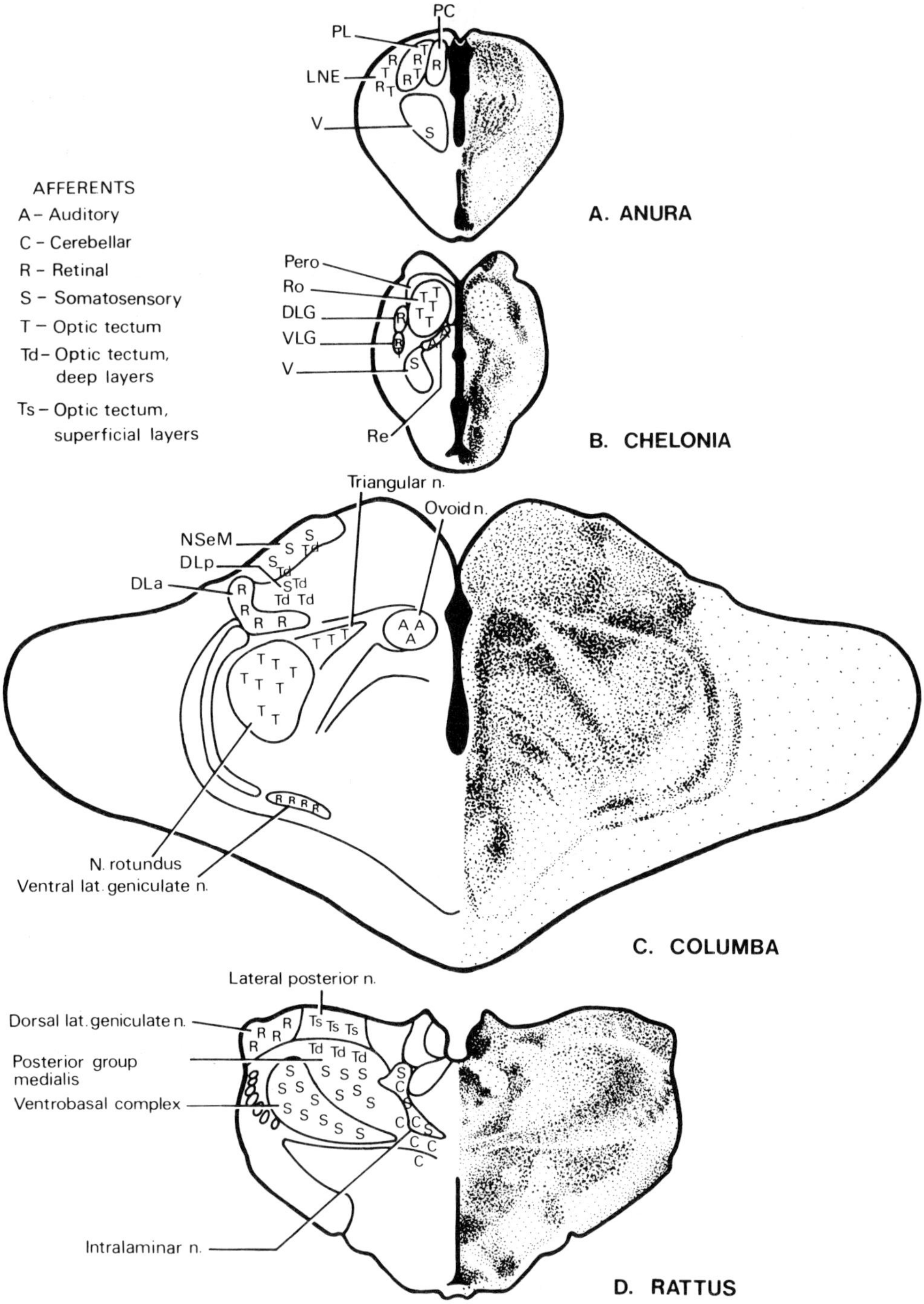

Fig. 3. Semi-schematic, simplified representations of the distribution of direct thalamic afferents in four groups of vertebrates. Data from: A. 25,107; B. 25, 33,39; C. 7,18,41,46,55,56,57,63, and Hunt & Webster, unpublished; D. 76,77, and Antonetti, Parnavelos, Price & Webster, unpublished. For further details see 26,116.

This assumption is further reinforced by the observation that the avian palaeostriatum augmentatum, like the mammalian striatum, is rich not only in dopamine, but also in acetylcholinesterase (8,60,92). Any hope that such a simple, clearcut distinction between the DVR ('neocortex') and ventral striatum ('corpus striatum') is a universal feature of the vertebrate forebrain has proved difficult to sustain. For example, the arrangements in reptiles are less easy to interpret. Both the tecto-recipient ('principal') nucleus rotundus (Ro) (38,39) and the perirotundal nuclei (which are perhaps intralaminar) (91) project not only to the DVR but also to the dopamine/acetylcholinesterase-rich ventral striatum, i.e. to the presumed homologue of the caudate-putamen (Fig. 5A). In amphibia, the picture is no less confusing (Fig. 5B) for there appears to exist an oligosynaptic pathway through the optic tectum for the transmission of visual information to the monoamine-rich dorsal striatum, and a channel superficially similar to the mammalian retino-geniculo-striate pathway, but ending in the medial cortex [which is in many ways reminiscent of mammalian limbic cortex (66,106,107)]. The work of Kicliter & Northcutt (67) has called into question the nature of the somatosensory pathways to these same parts of the amphibian telencephalon (113).

The 'homologous cell population' hypothesis of Nauta & Karten (81) was formulated in an attempt to elucidate problems of this kind. The hypothesis postulates that neuron 'pools', not cytoarchitecturally defined nuclei, are homologous when they are interconnected in the same way in different species. For example - the reptilian DVR could be said to contain a population of cells comparable with those in mammalian neocortex that receive the afferents from the principal thalamic neurons; and neuron pools equivalent to mammalian striatal and (as we shall see later) pallidal assemblies would be found in the ventral striatum. The rotund and perirotundal thalamic nuclei of reptiles might then simply be supposed to contain both 'principal' and 'intralaminar' cell populations intermixed (Figs. 5A and 6i). In that case, the homologues of the mammalian neocortex could be identified as cell populations in the DVR and dorsal cortex of reptiles and birds, and the medial pallium and dorsal striatum of amphibia - see Fig. 1Bi, Ci, Di, Ei. (This interpretation is reinforced by the embryological observations of Källen (54), who maintains that cell populations which invade the presumptive neocortex in mammalian embryos, give rise to the DVR in birds.)

If this thesis is accepted, then just as in mammals the neocortex projects upon the striatum in a topographically organised fashion (64,115,116) so should the avian DVR and dorsal cortex project upon the palaeostriatum augmentatum. There is indeed evidence that this is the case (1,60 - c.f. Figs. 4A and B). In reptiles, the dorsal cortex projects upon the DVR, and this in turn probably projects upon the ventral striatum (Lohman & Hoogland, personal communication, but see 69,70). The pallial thickening (which receives no thalamic afferents (39), and for which no homologue is known) also projects upon the ventral striatum. Thus far the analysis permits reasonable, not to say illuminating, comparisons between mammals and reptiles (c.f. Figs. 5A and 5A) - the dorsal cortex/DVR with neocortex and the ventral striatum with the corpus striatum. It is, however, based upon the assumption that the reptilian rotund and perirotundal thalamic nuclei are mixed 'principal' and 'intralaminar' populations. But once the 'cell population hypothesis' is accepted, there is no reason why it cannot be supposed that the mixed populations lie within the DVR and ventral striatum - i.e. that the 'neocortical' and 'striatal' populations are mixed, the thalamic populations homogeneous (see Fig. 6ii). In this case, striatal cells in the DVR would be separate from dopaminergic innervation* (90,91, 92,93,94). It can also be argued that the amphibian dorsal striatum likewise

*Lohman & Hoogland have recently described a catecholaminergic projection from midbrain to DVR in the Tegu lizard. The exact nature of the transmitter is, however, unknown. If noradrenaline, it could be compared with the NA innervation of neocortex from the locus caeruleus.

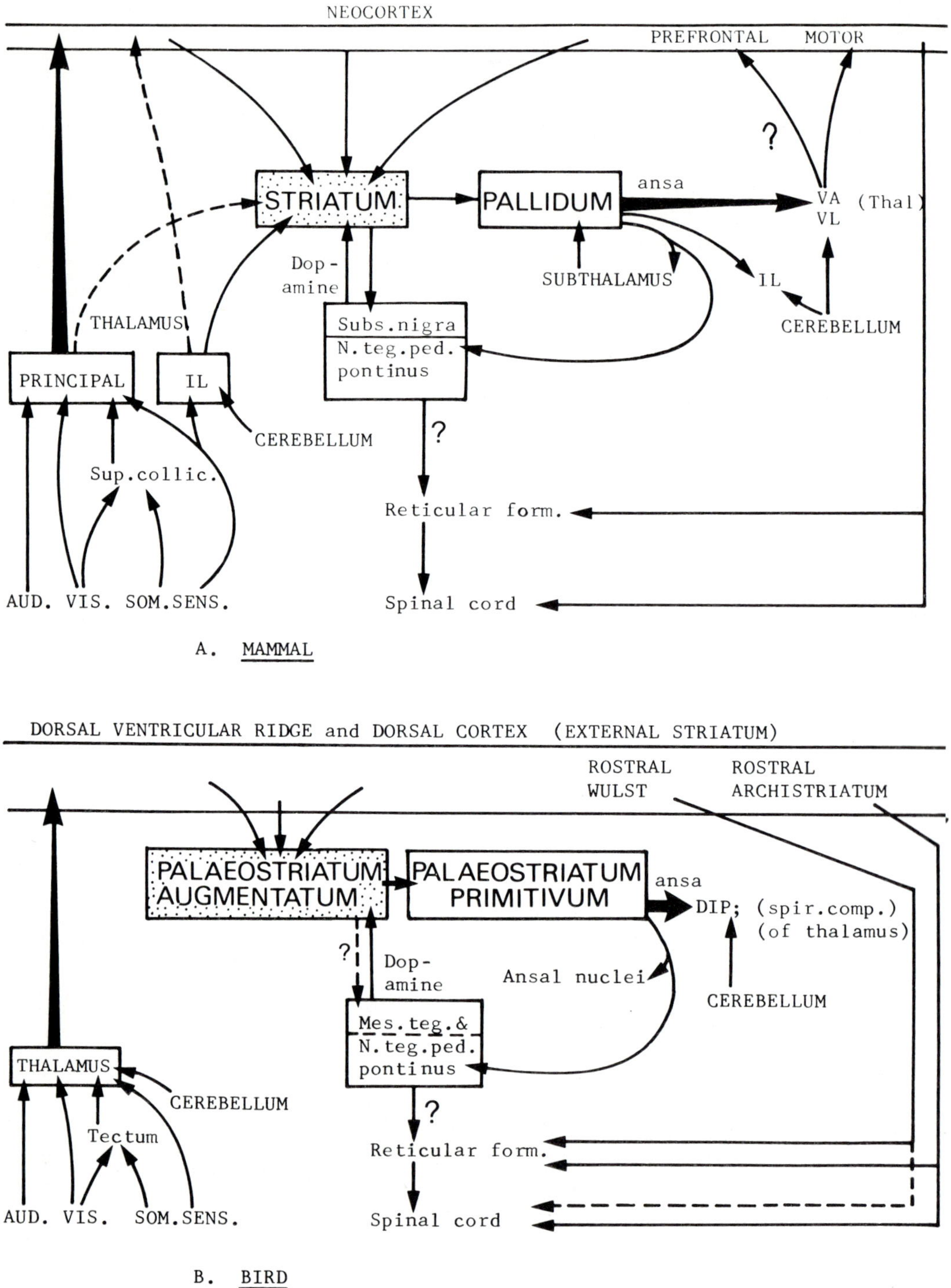

Fig. 4. A. Scheme of connections of mammalian basal nuclei (modified from 119; reproduced by permission of the Royal Society of Medicine, London). B. Some connections of the avian forebrain. AUD.,auditory; SOM.SENS.,somatosensory; VIS.,visual.

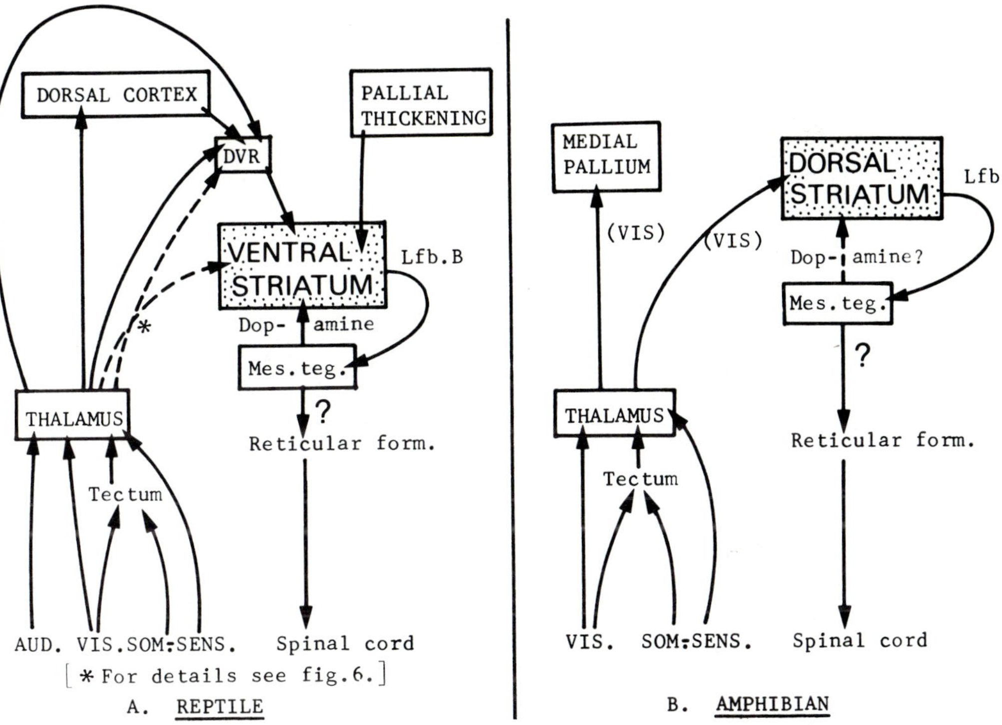

Fig. 5. Some connections of the forebrain in reptiles (A) and amphibia (B).

contains a 'neocortical' (visuo-receptive) cell population, and, as is the case in the reptilian ventral striatum and avian palaeostriatum augmentatum, because of the local concentration of dopamine, 'striatal' cell equivalents (c.f. Figs. 4 and 5).

It is interesting and pertinent that in at least some mammals some thalamic areas usually defined as 'principal'(parts of the posterior group nuclei that include tecto-receptive regions; the medial geniculate body; and the dorsomedial nucleus) as well as the intralaminar nuclei are said to project to both striatum and non-dopamine containing neocortex (28,38,42,49,72). These observations are clearly similar to those just described in relation to the reptilian forebrain. They pose similar questions of interpretation (Fig. 7). Are we to assume that these thalamic nuclei represent homogeneous populations (Fig.7i), or are the neocortex and striatum to be considered homogeneous (Fig. 7ii)? One may spin the web of complication almost indefinitely. For example, the deep (polysensory) layers of the mammalian superior colliculus project to both principal and intralaminar nuclei (Antonetti & Webster, unpublished observations; Krauthamer, this volume). Does this indicate the presence of a single cell population scattered through two or more thalamic nuclei? Or is it that here we are dealing with two thalamic cell populations, spatially separate, and distinguished also by the fact that save in this one respect their sources of afferents are quite different? At this point, it may be as well to summarize some little-explored problems raised by the cell population hypothesis and the connectivity approach in general.

1. The hypothesis cannot be directly tested: as may be judged from the present discussion, it has so far proved sufficiently procrustean to make possible an explanation for any and every set of observations.

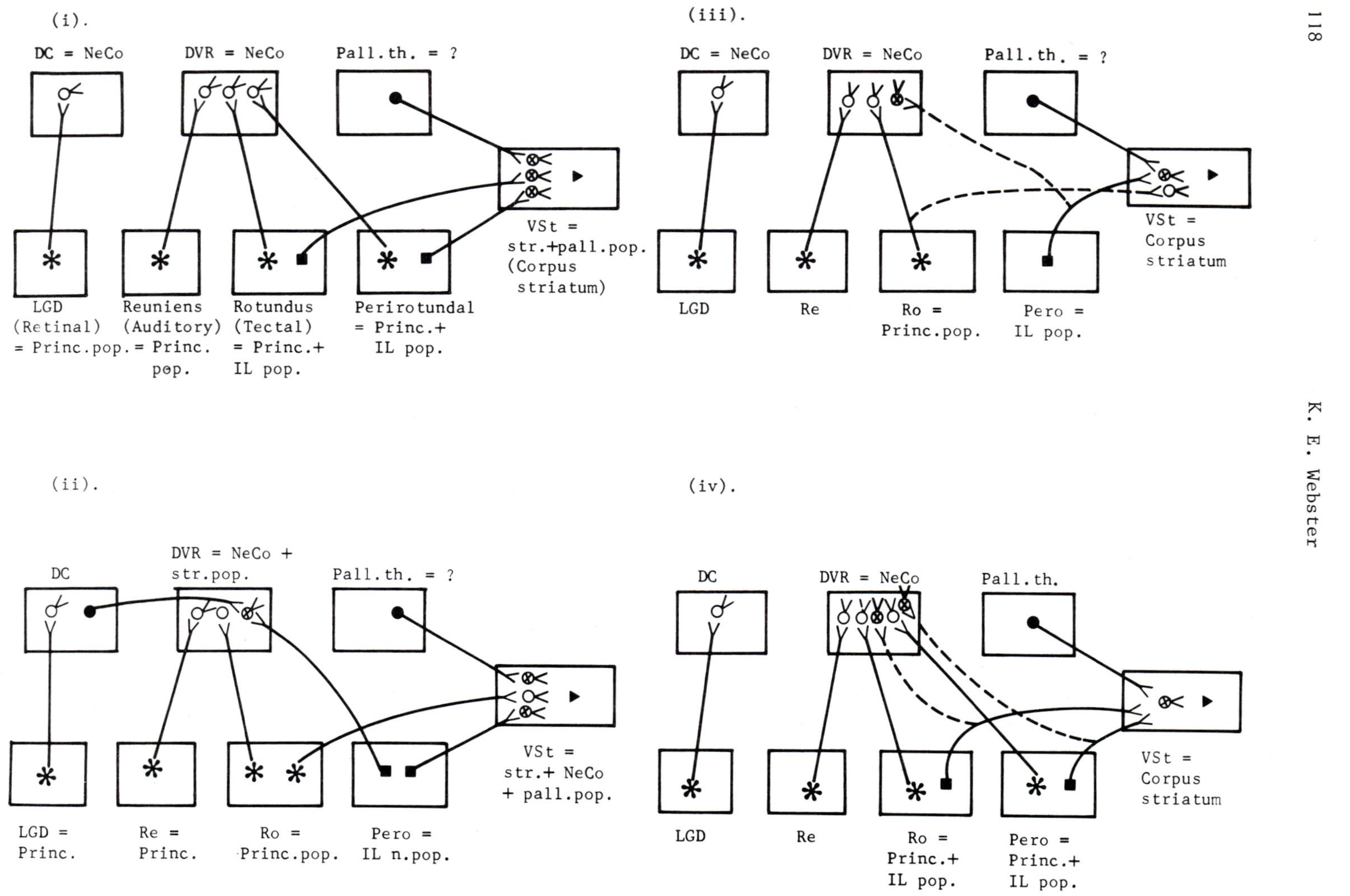

Fig. 6. Some of the ways of interpreting data on the connections of the reptilian ventral striatum and DVR using the homologous population thesis. See text and compare with Fig. 7.

2. At least a whole new system of nomenclature may be required to minimise conflict of concept and language - we still use nuclear nomenclature in the context of population concepts. Also, in mammals we have a strong tendency to think of a region (e.g. an area of neocortex) as a unit albeit one with multiple inputs (e.g. principal and intralaminar thalamus). In non-mammals multiple inputs are taken to imply the existence of multiple neuron populations comparable with aggregations (e.g. neocortex and striatum) that we commonly view as distinct parts of separate systems in mammals. If one adopts, for example, a 'reptilocentric' stance, and assumes 'by definition' that the rotund and perirotund nuclei are homogeneous populations of neurons, it is possible to conclude that the mammalian neocortex must contain two mechanisms that are spatially separate in reptiles. The significance (evolutionary, embryological and physiological) of the forms which neuronal aggregations take ('sheets', 'spheres') and the spatial coincidence or separation of neuron populations, is entirely unknown* (see also 3 below). The impact on our view of comparative structure-function relations is potentially enormous.

3. It is often simply not known if sets of afferents converge, or synapse with separate neuron populations in a nucleus. Similarly, precise details of the sources of efferent pathways are usually unknown. These problems are no less true of mammals (Fig. 7) and it is clear that analysis of observations on non-mammals will be influenced by this. The location of a particular population (or, indeed, whether it is believed to exist) and details of interconnections can be variously represented by different interpretations of experimental findings (Fig. 7). Analyses of the type upon which Figs. 6 and 7 are based involve, of course, great simplification. They ignore, for example, internal organisation (including interneurons) of the regions concerned. It is pertinent that although Pettigrew (97) maintains that in physiological terms the owl Wulst is remarkably similar to the supposedly homologous cat striate cortex, not only is the thalamic input bilateral (like that of most non-mammals - 27,46,62,79,80,102), but its internal structure is different. Similarly, there are probably fundamental differences between the internal organisation of the mammalian dorsal lateral geniculate body and its supposed homologue in the avian dorsolateral thalamus (Kemplay & Webster, unpublished observations).

4. Why the greater part of the avian dorsal thalamus reacts to telencephalic ablation in a way similar to the intralaminar or posterior group nuclei rather than principal nuclei remains puzzling, particularly since marked branching of axons and widespread distribution of collaterals is usually invoked to explain the characteristic reaction of these nuclei in mammals (Fig. 7i). There is at present little evidence indicating that organisation of this kind exists in the thalamo-telencephalic projection of birds. Collateral innervation of other regions, such as the bulbar brainstem, is a possible alternative, but would itself throw considerable doubt on the comparative interpretation of the thalamus outlined above.

---

*The columnar-organisation thesis of the cerebral cortex suggests one possibility: sheet-like aggregations are perhaps more readily formed from basic columnar building-blocks, each of which extends through the thickness of the sheet [D.H. Hubel & T.N.Wiesel, Proc. Roy. Soc. Lond. B 198, 1-59 (1977); R.A.Fisken, L.J. Garey and T.P.S. Powell, Phil. Trans. Roy. Soc. Lond. B 272, 487-536 (1975); J.J. Sloper, J. Neurocytol. 2, 351-359 (1973); A.J. Rockel, R.W.Hiorns and T.P.S. Powell, J. Anat. Lond. 118, 371 (1974) and personal communication]. Whatever its other significances, such an arrangement probably facilitates increase in overall size by the addition of more and more columns. From this 'modular' point of view, the expansion of 'spherical' accumulations would appear to present more problems. This supposes, among other things, that the 'spherical' aggregations are not in fact sheets so folded that the contiguities of opposing folds have become obscured.

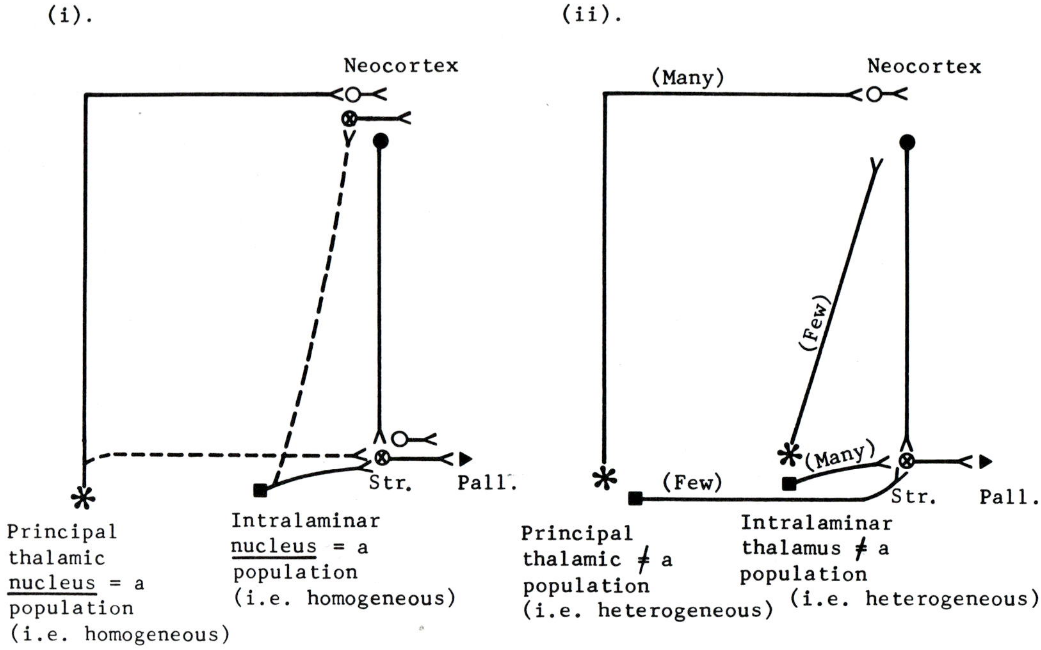

Fig. 7. Two alternative ways of applying the cell population thesis to connections of the mammalian corpus striatum.

In reptiles, however, this 'sustaining collateral' characteristic favours, but is not entirely explained by, an interpretation of thalamic projections such as shown in Fig. 6iii or iv. These figures represent 'collateralized' versions of the arrangements shown in Fig. 6i and ii - Fig. 6iii is a variant of 6ii (homogeneous thalamic nuclei), and Fig. 7iv derives from 6i (heterogeneous thalamic nuclei). Thus, in the organisations represented by both Fig. 6iii and iv, in order to damage or destroy most or all of the terminal arborisation of axons from cells in the rotund (Ro) or perirotund (Pero) nuclei - and thereby induce retrograde degeneration in such cells - the lesion would need to invade both the DVR and the ventral striatum (VSt). Such is the case (100). However, that the neurons of the reptilian lateral geniculate body respond in precisely the same way even in those species in which their projection axons end only in the dorsal cortex (39,95), suggests that factors other than the existence of collaterals (perhaps metabolic differences) may well be involved.

5. Many cell accumulations remain impossible to identify (e.g. the greater part of the avian DVR - see 60,81), or are not wholly comparable with apparent homologues (e.g. Ebner (27) argues cogently that the reptilian dorsal cortex corresponds at most to the subgranular layers of mammalian neocortex). This problem is at its most serious when connections just do not exist or are very poorly developed (as the spinothalamic systems of non-mammals). In the latter case, identities may be confusing, e.g. the intralaminar cell groups; neurons corresponding to the posterior group nuclei, particularly in relation to the differential distribution of axons from the laminae of the superior colliculus and optic tectum (16,41, Antonetti, Hunt & Webster, unpublished). Organisations of a kind quite alien to the mammalian brain [e.g. the direct inputs to telencephalon from the bulbar brainstem (114), which appear to include some from the optic tectum (10)] are impossible to deal with.

If, however, these difficulties are set to one side, it seems that there are parts of the non-mammalian telencephalon (the avian palaeostriatum augmentatum and para-olfactory lobe; the reptilian ventral striatum; and the amphibian dorsal striatum) which contain high levels of monoamine (in some instances identified as dopamine and most probably derived from cells in the midbrain) and which are therefore possible homologues of the mammalian striatum (caudate-putamen). Further, in birds and reptiles especially, there can be identified cell populations similar in some way to cells in the mammalian neocortex which project upon putative striatal neurons (the palaeostriatum augmentatum and ventral striatum). There is no evidence of such an arrangement in the case of amphibia. In reptiles, the value of the evidence lies between these extremes. Of the sources of striatal afferents known in mammals, only the intralaminar nuclei cannot be identified with any certainty in non-mammals.

We must now ask if such similarities as have been disentangled by examination of afferents are in any way strengthened by consideration of the pattern of efferent connections of the telencephalon.

Efferent connections of the basal nuclei

(i) The mammalian pallidum consists of large, multipolar neurons arranged in two groups - the inner and outer pallidal segments found in primates, or the globus pallidus and entopeduncular nucleus found in animals such as rat or cat. On these grounds, the avian palaeostriatum primitivum and intrapeduncular nucleus (PP, INT, Fig. 2C) are clear contenders for consideration as homologues. Support for this comparison is to be had from observation of the marked similarities of the ultra-structure of the globus pallidus and palaeostriatum primitivum (34). Further [like the mammalian caudate putamen, the avian palaeostriatum augmentatum, the reptilian ventral striatum, and the dorsal and ventral striata of amphibia (89,92)], the mammalian entopeduncular and avian intrapeduncular nuclei are rich in acetyl-cholinesterase. Unfortunately for this line of argument, this is not a feature of the primate internal pallidal segment (92). It should be noted that, just as the dopamine-rich paraolfactory lobe is distinguished from palaeostriatum augmentatum in receiving no input from the DVR, so does it not project upon the palaeostriatum primitivum, but rather to the hypothalamus (60). It will therefore be omitted from further discussion. The avian palaeostriatum augmentatum (putative caudate-putamen), however, projects massively upon the palaeostriatum primitivum and intrapeduncular nucleus (60). This suggests direct comparison with the striato-pallidal projection of mammals (70,82, c.f. Fig. 4). In the case of reptiles, the acetylcholinesterase concentration is again high in the dopamine-rich ventral striatum (52,53,92). A population of large cells which might be supposed to be 'pallidal' neurons is also to be found here, but any synaptic connection between them and the 'striatal' cells postulated to exist in the ventral striatum (v.s.) remains purely conjectural. Again, in amphibia, any connection between 'striatal' and 'pallidal' cells (which are, in any case, almost entirely postulates) is no more than speculation. In amphibia, the distribution of acetylcholinesterase does not correspond with that of dopamine (dorsal striatum) but extends into both dorsal and ventral striata (89).

(ii) In mammals, efferents from the pallidum project massively upon the ventral thalamus (VA,VL) and so upon the frontal cortex and, of especial importance in the present context, the motor cortices (Fig. 4A). This pathway represents a converg-ence of cerebellar and striatal influences also to be found in the intralaminar nuclei (IL). In birds, there appear to be two descending pathways (Fig. 4B), one from the rostral archistriatum, the other from the rostral Wulst, which are said to be related to the medial and lateral reticular formation and spinal cord (60,122, 123) in a way generally reminiscent of the medial and lateral components of the cortico-spinal and cortico-bulbar pathways of mammals (71). The rostral parts of the archistriatum and Wulst may therefore be homologues of the mammalian motor cortex. The palaeostriatum primitivum/intrapeduncular complex projects to the spiriform complex (spir. comp.) of the dorsal thalamus, including, most interestingly, the nucleus dorsointermedius (DIP) which also receives cerebellar afferents (56). The nucleus dorsointermedius, however, is not yet known to project specifically upon the presumptive somatic motor regions, which would need to be the case if it were to be compared with the more anterior parts of the mammalian ventral thalamic group, or indeed, if the avian somatic motor areas themselves are to be compared in a strict sense with mammalian motor cortex (Fig. 4). The nucleus dorsointermedius could, of course, correspond to an intralaminar neuron population. Whichever of these alternatives eventually proves the more satisfactory parallel, at least one other potential difficulty is raised by Karten & Dubbledam's (60) account. The somatic-motor archistriatum ('motor cortex') receives afferents from widespread regions of DVR ('neocortex'). In mammals such an arrangement exists only as between somatosensory and motor cortices (50). Also, it is proposed that in birds the descending ('pyramidal-like') pathways arise from a different population of DVR-dorsal cortex neurons from those projecting upon the presumptive caudate-putamen (i.e. palaeostriatum augmentatum). It was previously believed that this is not the case in mammals (101, 115), but more recent evidence suggests that even in this, mammals and birds may be very similar (51).

As in mammals, there appears to be an efferent connection with the brainstem nucleus tegmenti pedunculopontinus (N. teg. ped. pontinus - Fig. 4), but we are as ignorant of further connections of this nucleus in birds as in mammals. The projection of the palaeostriatum primitivum to the 'ansal nuclei' may correspond with that to the subthalamus, and the apparent projection to the midbrain tegmentum (Mes. teg.) could represent that to the substantia nigra pars reticulata from the mammalian striatum, since the lesions used by Karten & Dubbledam inevitably damaged the palaeostriatum augmentatum ('striatum') or its efferents.

In neither reptiles nor amphibia does there appear to be evidence of descending pathways from the forebrain comparable to the mammalian pyramidal system. The reptilian ventral striatum (postulated above to contain 'pallidal' cells) projects to the midbrain tegmentum (Mes. teg.) (91), as does the amphibian striatal region (40). Such a projection could represent the striatonigral pathway and/or that from pallidal neurons to the nucleus tegmenti pedunculopontinus (Fig. 5). No homologues of projections to the ventral and intralaminar nuclei of the dorsal thalamus, or the subthalamus, are known.

CONCLUSIONS

Forebrain organisation in which particular regions can with reasonable certainty be compared with the mammalian corpus striatum is to be found especially in birds, to a lesser extent in reptiles, and least in anamniotes (Figs. 4, 5 and 7). In general terms, however, it is clear that the interpretation of the non-mammalian hemispheres represented by Fig. 1B,Ci,Di and Fi better reflects the way in which the forebrain is organised than do the earlier accounts. Although the details are by no means certain, it seems likely that the forebrains of Holosteans and Teleosts evolved by a process that involved lateral bending (eversion), while those of amphibians remained relatively unspecialised. Källen's (54) observations of forebrain ontogeny in amniotes lend credence to the idea that, during amniote evolution, the hemisphere enlarged by proliferation, in the dorsolateral part of its wall, of neurons receiving 'specific' or 'principal' thalamic afferents, to produce either the DVR (X - Fig.1Ai) or expansion of the dorsal/neocortex (Y - Fig. 1Aii), or a combination of the two. In this limited sense, at least, the essential nature of the striatum proper remained intact. In all non-mammals, however (the evidence is particularly abundant for reptiles and amphibia), the possibility that homologous neuronal populations are intermixed in different ways must be borne in mind: neither the functional significance of this, nor the evolutionary mechanisms by which these different intermixings came into being are understood. It is also clear that the more removed, in a phylogenetic sense, is an organism from mammals, the more the striatal cell homologues become 'brainstem orientated', perhaps because the forebrain becomes more and more dissociated from direct somatosensory and cerebellar inputs, and from direct control of lower motor neurons*. Even in birds, this may prove a major difference from mammals. In birds, there is much evidence that implicates the forebrain in complicated learning and behaviour patterns, but largely in relation to the dorsal cortex and dorsal ventricular ridge (see 95,96,117,118 for reviews and references). Comparative functional studies of the presumptive homologues of the striatum proper are urgently needed: although the forebrain has been implicated in the control of movement, only in birds has a suggestion of the possible significance of the palaeostriatal complex been made - that it corrdinates overall motor functions (105). One obvious difficulty of comparison here is that there is little general agreement about the function of the mammalian corpus striatum (119, and Sections III and IV of this volume). Functional comparisons with mammals may, of course, prove futile, in as far as the different organisations of the remaining nervous system is ignored: the structure and function of the forebrain should not, and in the last analysis cannot, be made in isolation.

Acknowledgments
I wish to thank Lydia Malim for preparing the illustrations. Some of the original work referred to in the text was supported by a grant from the MRC.

Abbreviations
AUD, auditory; C-P, caudate putamen; DLG, dorsal lateral geniculate nucleus; DLp, posterior dorsolateral complex; DMC, dorsomedial cortex; DSt, dorsal striatum; DVR, dorsal ventricular ridge; E, ectostriatum; HA, hyperstriatum accessorium; HD, hyperstriatum dorsale; HI, hyperstriatum intercalatus; HV, hyperstriatum ventrale; IL, intralaminar nuclei; INT, intrapeduncular nucleus; LC, lateral cortex; LfbB, lateral forebrain bundle; LMD, dorsal medullary lamina; LNE, lateral neuropil; LPO, paraolfactory lobe; Mes.teg., mesencephalic tegmentum; N, neostriatum; NA, nucleus accumbens; NeCo, neocortex; N.teg.ped.pontinus, nucleus tegmenti pedunculopontinus; PA, palaeostriatum augmentatum; Pall, pallidum; Pall.th., pallial thickening; PC, posterocentral nucleus; Pero, perirotundal nuclei; PiC, piriform cortex; PL, posterolateral nucleus; pop., population; PP, palaeostriatum primitivum; Princ., principal; Re, nucleus reuniens; Ro, nucleus rotundus; Se, septum; SOM.SENS., somatosensory; spir.comp., spiriform complex; Str, striatum; Subs.nigra, substantia nigra; Sup.collic., superior colliculus; Thal., thalamus; V, ventral nucleus; VA, ventroanterior nucleus; VL, ventrolateral nucleus; VLG, ventral lateral geniculate nucleus; VIS., visual; VSt, ventral striatum; tub., olfactory tubercle.

References
1. N.J. Adamo, J. Comp. Neurol. 131, 337-355 (1967).
2. N.E. Andén, A. Dahlström, K. Fuxe and K. Larsson, Am. J. Anat. 116, 329-333 (1965).
3. J. Anthony, Traité de Zoologie, 14(2), 202-332, Masson, Paris (1970).
4. C.U. Ariens Kappers, Vergleichende Anatomie des Nervensystems, E.F.Bohn, Haarlem (1921).

---

*In this context the findings of Stephan & Andy (108) are of interest. They discovered that in mammals the most striking phylogenetic differences in relative volumes of various components of the brain are manifested by the neocortex. Although exhibiting considerably smaller differences, the next most important region in this respect is the basal nuclei - in other words the more neocortex a mammal possesses, the larger is its corpus striatum. Other authors have remarked the close interrelationship between the mammalian neocortex and corpus striatum (see 119 and Divac, this volume, for discussion and references).

5. C.U. Ariens Kappers, G.C. Huber and E.C . Crosby, The Comparative Anatomy of the Nervous System, Hafner, New York (1936).
6. W. Bartels, Z.Zellforsch.116, 94-118 (1971).
7. L.I.Benowitz and H.J. Karten, J. Comp. Neurol. 167, 503-520 (1976).
8. Å. Bertler, B. Falck, C.G. Gottfries, L.Ljunggren and E. Rosengren, Acta Pharmacol.(Kbh) 21, 283-289 (1964).
9. H. Braak, Z. Zellforsch. 106, 269-308 (1970).
10. P. Bradley and G. Horn, J. Physiol. Lond. 278, 46P (1978).
11. J. Cairney, J. Comp. Neurol. 42, 255-348 (926).
12. C.B.G. Campbell, in Evolution of Brain and Behaviour: Persistent Problems (R.B.Masterton, W.Hodos & H.Jerison, Eds.) pp.143-151, Halsted Press (1976).
13. M.B. Carpenter, in The Basal Ganglia (M.D.Yahr, Ed.) pp.1-35, Raven Press New York (1976).
14. M.B. Carpenter and P. Peter, J. Comp. Neurol. 144, 93-116 (1972).
15. M.B. Carpenter and N.L. Strominger, Am.J. Anat. 121, 41-72 (1967).
16. V.A. Casagrande, J.K. Harting, W.G. Hall, I.T. Diamond and F.G. Martin, Science (New York) 177, 444-447 (1972).
17. T. Chacko, M. Terlou and J. Peute, Cell Tiss. Res. 149, 481-495 (1974).
18. W.M. Cowan, L. Adamson and T.P.S. Powell, J. Anat. Lond. 95, 545-563 (1961).
19. W.M. Cowan and T.P.S. Powell, J. Neurol. Neurosurg. Psychiat. 29, 426-439 (1966).
20. E.H. Craigie, J. Comp. Neurol. 72, 299-328, 73, 179-234 (1940).
21. E.C. Crosby, J. Comp. Neurol. 27, 325-402 (1917).
22. A. Dahlström and K. Fuxe, Acta Physiol. Scand. 62 Suppl. 232, 1-55 (1964).
23. I. Divac, O. Lindvall, A.Björklund and R.E.Passingham, J. Comp. Neurol. (in press).
24. A. Durward, J. Anat. Lond. 65, 8-44 (1930).
25. S.O.E. Ebbesson, in Frog Neurobiology (R. Llinas and W. Precht, Eds.) pp.679-706, Springer Verlag (1976).
26. S.O.E. Ebbesson, J.A. Jane and D.M. Schroeder, Brain Behav. Evol. 6, 92-130 (1972).
27. S.O.E. Ebbesson and R.G.Northcutt, in Evolution of Brain and Behaviour in Vertebrates (R.B.Masterton, C.B.G.Campbell, M.E.Bitterman & N.Hotton, Eds.) pp.115-146, Halstead Press (1976).
28. F.F. Ebner, Ann. N.Y. Acad. Sci. 167(1), 241-257 (1969).
29. F.F. Ebner, in Evolution of Brain and Behaviour in Vertebrates (see ref. 27), pp.147-167 (1976).
30. L. Edinger, A. Wallenberg and G.M.Holmes, Abh. senkenb.naturforsch. Ges. 20 343-424 (1903).
31. G. Elliot Smith, Lancet 88, 1-16; 147-153; 221-227 (1910).
32. G. Elliot Smith, J. Anat. Lond. 53, 271-291 (1919).
33. R.E. Foster and T.L. Peele, Anat. Rec. 181, 530 (1975).
34. C.A. Fox, D.E. Hillman, K.A.Siegesmund and L.A. Sether, in Evolution of the Forebrain (R. Hassler & H. Stephan, Eds.) pp.237-248, G.Thieme Verlag, Stuttgart (1966).
35. K. Fuxe and L.Ljunggren, J. Comp. Neurol. 125, 355-382 (1963).
36. F. Goldby and H.J. Gamble, Biol. Rev. 32, 383-420 (1957).
37. P.S. Goldman and W.J.H. Nauta, J. Comp. Neurol. 171, 369-384 (1977).
38. A.M. Graybiel, Brain Res. 49, 229-244 (1973).
39. W.C. Hall and F.F. Ebner, J. Comp. Neurol. 140, 101-122 (1970).
40. M. Halpern, Brain Behav. Evol. 6, 42-68 (1972).
41. J.R. Hart, Ph.D. Thesis, University of Wisconsin (1969).
42. C.J. Heath, Brain Res. 21, 435-538 (1970).
43. L. Heimer, Ann. N.Y. Acad. Sci. 167(1), 129-146 (1969).
44. C.J. Herrick, The Brain of the Tiger Salamander, Univ. Chicago Press (1948).
45. G.C. Huber and E.C. Crosby, J. Comp. Neurol. 48, 1-225 (1929).
46. S.P. Hunt and K.E. Webster, Brain Res. 44, 647-651 (1972).
47. J.B. Johnston, J. Comp. Neurol. 26, 481-505 (1916).
48. J.B. Johnston, J. Comp. Neurol. 35, 337-481 (1923).

49. E.G. Jones and R.Y. Leavitt, J. Comp. Neurol. 154, 349-377 (1974).
50. E.G. Jones and T.P.S. Powell, Brain 93, 793-820 (1970).
51. E.G. Jones, J.D. Coulter, H. Burton and R. Porter, J. Comp. Neurol.173, 53-80, (1977).
52. A.V. Jourio, J. Physiol. Lond. 204, 503-509 (1969).
53. A.V. Jourio and M. Vogt, J. Physiol. Lond. 189, 489-518 (1967).
54. B. Källén, Ergebn. Anat. Entw. Gesch. 36, 62-82 (1962).
55. H.J. Karten , Int. Congr. Zool. 16(2), 23 (1963).
56. H.J. Karten, Anat. Rec. 148, 197-198 (1964).
57. H.J. Karten, Brain Res. 6, 409-527 (1967).
58. H.J. Karten, Brain Res. 11, 134-153 (1968).
59. H.J. Karten, Ann. N.Y. Acad. Sci. 167(1), 164-179 (1969).
60. H.J. Karten and J.L. Dubbledam, J. Comp. Neurol. 148, 61-90 (1973).
61. H.J. Karten and W. Hodos, A Stereotaxic Atlas of the Brain of the Pigeon, Johns Hopkins, Baltimore (1967).
62. H.J. Karten, W. Hodos, W.J.H. Nauta and A.M. Revzin, J. Comp. Neurol. 150 253-278 (1973).
63. H.J. Karten and W.J.H. Nauta, Anat. Rec. 160, 373 (1968).
64. J.M. Kemp and T.P.S. Powell, Brain 93, 525-546 (1970).
65. J.M. Kemp and T.P.S. Powell, Phil. Trans. Roy. Soc. Lond. (B) 262, 441-457 (1971).
66. E. Kicliter and S.O.E. Ebbesson, in Frog Neurobiology (R.Llinas & W. Precht, Eds.) pp. 946-972 (1976).
67. E. Kicliter and R.G. Northcutt, J. Comp. Neurol. 161, 239-254 (1975).
68. H.P. Killackey and F.F. Ebner, Science 179, 283-285 (1973).
69. H. Künzle, Brain Res. 88, 195-209 (1975).
70. Jon-Son Kuo and M.B. Carpenter, J. Comp. Neurol . 151, 201-236 (1973).
71. H.G.J.M. Kuypers, in New Developments in Electromyography and Clinical Neurophysiology 3 (J.E.Desmedt, Ed.) pp.38-63, Karger, Basel (1973).
72. C.M. Leonard, Brain Res. 12, 321-343 (1969).
73. O. Lindvall, A.Björklund, R.Y. Moore and U. Stenevi, Brain Res. 81, 325-331 (1974).
74. A.H.M. Lohman and G.M. Mentink, Brain Res. 45, 325-344 (1972).
75. A.H.M. Lohman and I. Van Woerden-Verkley, Brain Res. 103, 9-28 (1976).
76. R.D. Lund and K.E. Webster, J. Comp. Neurol. 130, 301-311 (1967).
77. R.D. Lund and K.E. Webster, J. Comp. Neurol. 130, 313-327 (1967),
78. G. Macchi, M. Bentiveglio, C. D'atena, P. Rossini and E. Tempesta, Neurosci. Lett. 4, 121-126 (1977).
79. R.E. Meier, J. Mihailovič and M. Cuénod, Exp. Brain Res. 19, 351-364 (1974).
80. D. Miceli, J. Peyrichoux and J. Repérant, Brain Res. 100, 125-131 (1975).
81. W.J.H. Nauta and H.J. Karten, in Neurosciences Second Study Programme (F.O. Schmitt, Ed.) pp.7-26, Rockefeller Univ. Press, New York (1970).
82. W.J.H. Nauta and W.R. Mehler, Brain Res. 1, 3-42 (1966).
83. R. Nieuwenhuys, Acta Morph.Neerl.Scand. 5, 65-78 (1962).
84. R. Nieuwenhuys, J. Morphol. 111, 69-88 (1962).
85. R. Nieuwenhuys, in Evolution of the Forebrain (R. Hassler & H. Stephan, Eds.) pp. 32-39, G. Thieme Verlag, Stuttgart (1966).
86. R. Nieuwenhuys and T.S. Bodenheimer, J. Morphol. 118, 415-450 (1966).
87. R.G. Northcutt, J. Comp. Neurol. 130, 109-147 (1967).
88. R.G. Northcutt, The Telencephalon of the Western Painted Turtle, Illinois Biol. Monogr. 43, (1970).
89. R.G. Northcutt, J. Comp. Neurol. 157, 379-390 (1974).
90. A. Parent, J. Anat. Lond. 114, 379-387 (1973).
91. A. Parent, Brain Res. 108, 25-36 (1976).
92. A. Parent and A. Olivier, J. Hirnforsch. 12, 73-81 (1970).
93. A. Parent and L.J. Poirier, J. Anat. Lond. 110, 81-89 (1971).
94. A. Parent and D. Poitras, Brain Res. 78, 345-358 (1974).
95. R. Pearson, The Avian Brain, Academic Press, London and New York (1972).

96. R. Pearson and L. Pearson, The Vertebrate Brain, Academic Press, London and New York (1976).
97. J.D. Pettigrew, Phil. Trans. Roy. Soc. Lond. (B) (In press - 1978).
98. T.P.S. Powell and W.M. Cowan, J. Anat. Lond. 95, 78-109 (1961).
99. T.P.S. Powell and W.M. Cowan, J. Neurol. Neurosurg. Psychiat. 30, 140-153 (1967).
100. T.P.S. Powell and L. Kruger, J. Anat. Lond. 94, 528-542 (1960).
101. S. Ramón y Cajal, Cellule 7, 125-148 (1891).
102. J. Repérant, J-P. Raffin and D.Miceli, C.R. Acad. Sci. Paris 279, 279-280 (1974).
103. W. Riss, M. Halpern and F. Scalia, Brain Behav. Evol. 2, 1-50 (1969).
104. M. Rose, J. Psychol. Neurol. Lpz. 21, 278-352 (1914)
105. E.A. Salzen and D.M. Parker in Neural and Endocrine Aspects of Behaviour in Birds (P. Wright, P.G. Caryl and D.M. Vowles, Eds.) pp.205-242, Elsevier, Amsterdam (1975).
106. F.Scalia, in Frog Neurobiology (R.Llinas & W.Precht, Eds.) pp.213-233, Springer-Verlag (1976).
107. F. Scalia, in Frog Neurobiology (R. Llinas & W.Precht, Eds.) pp.386-406, Springer-Verlag (1976).
108. H. Stephan and O.J. Andy, Ann. N.Y. Acad. Sci. 167(1), 370-387 (1969).
109. W. Stingelin, Vergleichend Morphologische Untersuchungen am Vorderhirn der Vögel, Helbin & Lichtenbahn, Basle (1958).
110. J. Szabo, Exp. Neurol. 27, 1-15 (1970).
111. C.F. Tyner, Brain Behav. Evol. 12, 75-96 (1975).
112. M. Tohyama, T.Maeda, J. Hashimoto, G.R. Shreshta, O. Tamura and N. Shimizu, J. Hirnforsch. 15, 319-330 (1974).
113. N.P. Vesselkin, A.L. Agayan and L.M. Nomokonova, Brain Behav. Evol. 4, 295-306. (1971).
114. A. Wallenberg, Anat. Anz. 24, 357-369 (1904).
115. K.E. Webster, J. Anat. Lond. 95, 532-544 (1961).
116. K.E. Webster, J. Anat. Lond. 99, 329-337 (1965).
117. K.E. Webster, Symp. Zool. Soc. Lond. 33, 169-203 (1973).
118. K.E. Webster, in Essays on the Nervous System, (R. Bellairs & E.G.Gray Eds.) pp. 258-298, Clarendon Press, Oxford (1974).
119. K.E. Webster, Proc. Roy. Soc. Med. 68, 203-210 (1975).
120. S.A.K. Wilson, Arch. Neurol. Psychiat. 11, 385-404 (1924).
121. E.H. Yeterian and G.W. van Hoesen, Brain Res. 139, 43-63 (1978).
122. A. Zecha, Acta Morph. Neerl. Scand. 5, 194-195 (1962).
123. H. Zeier and H.J. Karten, Brain Res. 31, 313-326 (1971).

Ontogeny of the Neostriatum

I. H. M. Smart and R. R. Sturrock

Department of Anatomy, The University, Dundee, Scotland

This chapter summarises the results of a series of studies on the development of the mouse neostriatum made in our laboratory. Although the mouse is not an ideal experimental animal for behavioural studies it so happens that it is the species whose neostriatal ontogenesis has been particularly well documented and is thus the best available starting point for a general introduction to the pattern of histogenesis of this area.

The Ventricular Elevations

The cell populations constituting the caudate-putamen or neostriatum originate from the germinal epithelium of the developing neuraxis in the region of the telo-diencephalic junction and the adjacent parts of the telencephalon. The first evidence of major differentiation in this area is a swelling appearing in the mouse at embryonic day 11 (E11) which straddles the telo-diencephalic boundary zone and, thus, protrudes into both the third and lateral ventricles. A day later at E12 a second more gently sloping swelling develops in the adjacent part of the telencephalic wall. These are respectively the medial and lateral ventricular elevations* (synonyms: ganglionic eminences, striatal elevations, Ganglionhugel, Streichhugel). As development proceeds the medial elevation comes to lie entirely within the lateral ventricle and in so doing becomes separated from the diencephalon; the lateral elevation becomes the larger and circumscribes the medial elevation rostrally, dorsally and caudally (Fig. 1). About E14 the sulcus between the two elevations begins to disappear and by E15 is barely distinguishable. By the end of gestation (E19) both elevations have receded into the lateral wall of the lateral ventricle. The graph in Fig. 2 describes the change in surface area of the two elevations during the prenatal period; maximum area is reached at E15 to E16 after which there is a steady decline consistent with the recession of the eminences into the ventricular wall. A similar sequence is described by Lammers (15) in the golden hamster (Cricetulus griseus).

---

*We have come to prefer Lammer's (15) term 'ventricular elevation' as it does not prejudge the issue by implying any necessary connection with the neostriatum or any other of the basal ganglia and leaves the term 'ventricular ridge' as a term in reptilian neuroanatomy for the similarly located permanent protrusions in reptiles which may not be fully comparable to the transient elevations of mammals.

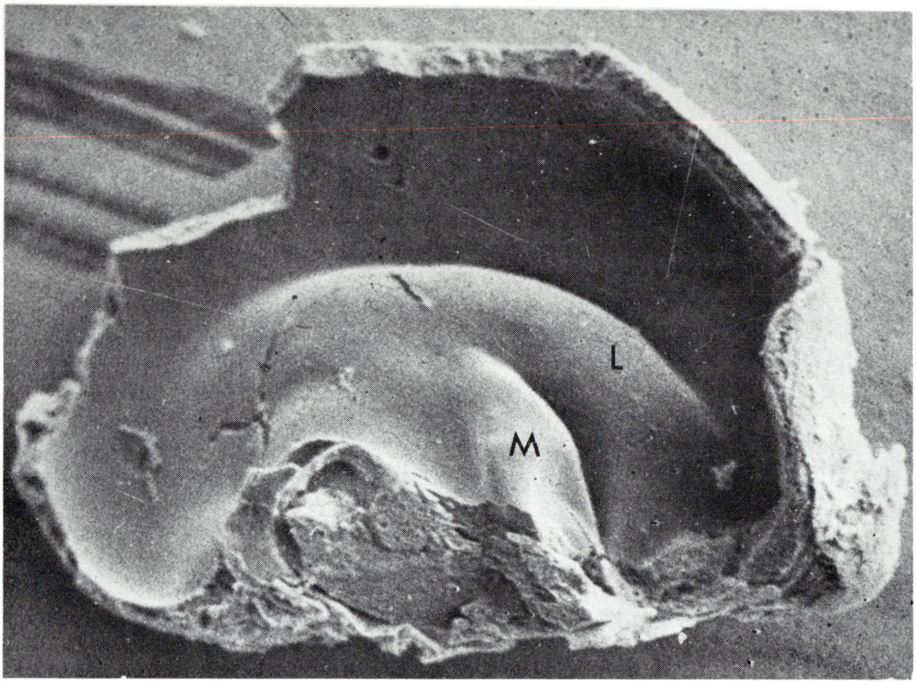

Fig. 1 Scanning electron micrograph of lateral ventricular wall of E14 mouse brain. Note large lateral elevation L, circumscribing smaller medial elevation M. Occipital pole of hemisphere to left; olfactory bulb to right.

The most recent account of the development of these features in the human is given by Hewitt (11) who has also confirmed the existence in the human of a third elevation which appears later than the other two and lies between their rostral poles. A third elevation is not present in either the mouse or golden hamster (Refs. 35 & 15).

Proliferative Histology

The ventricular elevations are the sites of production of the neurons and neuroglia of the basal parts of the telencephalon. The histological structure of each elevation is similar and reflects an enhanced proliferative capacity of the germinal epithelium. At E10 the presumptive sites of the elevations are composed of pseudostratified cells of the ependymal layer (synonyms: ventricular layer, matrix layer). An important feature of the ependymal layer, in this as in other locations, is that on entering the S phase of DNA synthesis the cell nuclei migrate to the ventricular surface where mitosis occurs. Thereafter the postmitotic daughter nuclei return to the non-ventricular aspect of the layer (Refs. 28, 29). In areas of the ependymal layer with a high mitotic index, competition for space at the ventricular surface results in a state in which nuclei about to enter the visible stages of mitosis are obliged to wait for a vacancy at the ventricular surface (Refs. 33, 34). In such locations 'subsurface prophases' are found belonging to cells which have entered visible mitosis before a vacancy at the ventricular surface has become available. The appearance of subsurface prophases can be taken to represent a choking of the proliferative mechanism. Ependymal choke occurs in most parts of the developing mouse neuraxis (Refs. 33,34, 35) and is relieved to a greater or lesser extent by the appearance of non-surface mitotic figures occurring in nuclei which have been freed from the interkinetic migratory cycle (Ref. 35). The ventricular elevations from their first appearance at E11 and E12 are associated with such a population of nuclei which do not migrate during mitosis (Ref. 35). These cells in all probability have lost cytoplasmic contact with the ventricular surface. This population lies within the concavity of the elevations adjacent to the nonventricular surface of the ependymal layer from which it is derived, and is termed the subependymal layer (synonyms: subventricular or submatrix layer). The subependymal layer becomes a major feature of the elevations and by E14 to E15 takes over from the ependymal layer as the major site of mitotic activity (Ref. 35). The subependymal layer increases in volume up to E16 and then declines progressively (Fig. 3). A remnant, however, persists into adult life (Refs. 32, 16), and even senescence* as a site of undifferentiated mitotically active cells.

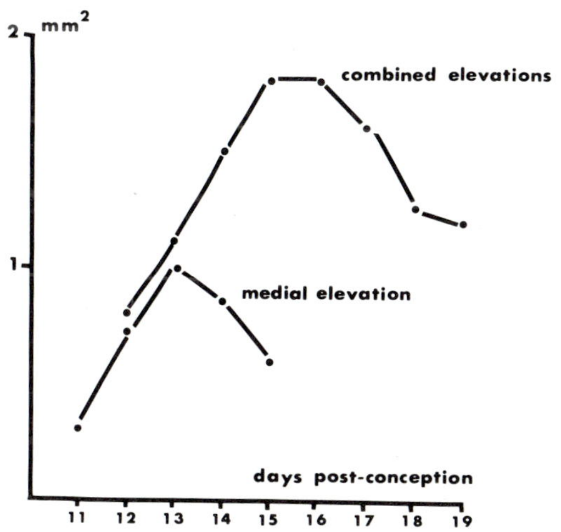

Fig. 2 Estimated change in surface area of medial and lateral elevations during prenatal development in the mouse.

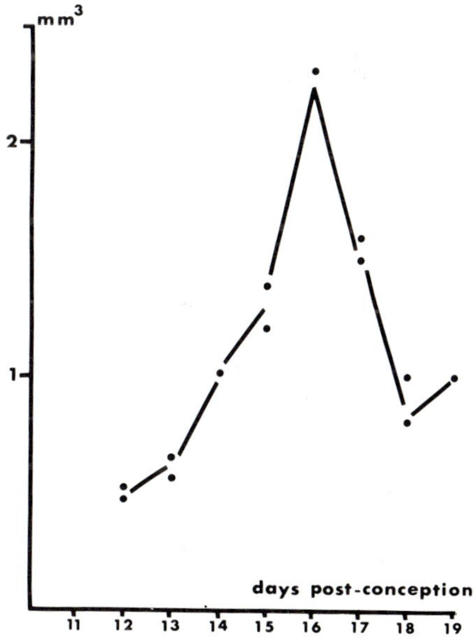

Fig. 3 Estimated change in volume of combined EL and SEL of ganglionic eminences during prenatal development in the mouse.

The subependymal layer although making its first appearance within the ventricular elevations is not confined to them. It accompanies the outgrowing olfactory ventricle and also extends into the pallial parts of the hemisphere. The

*Manuscript in preparation.

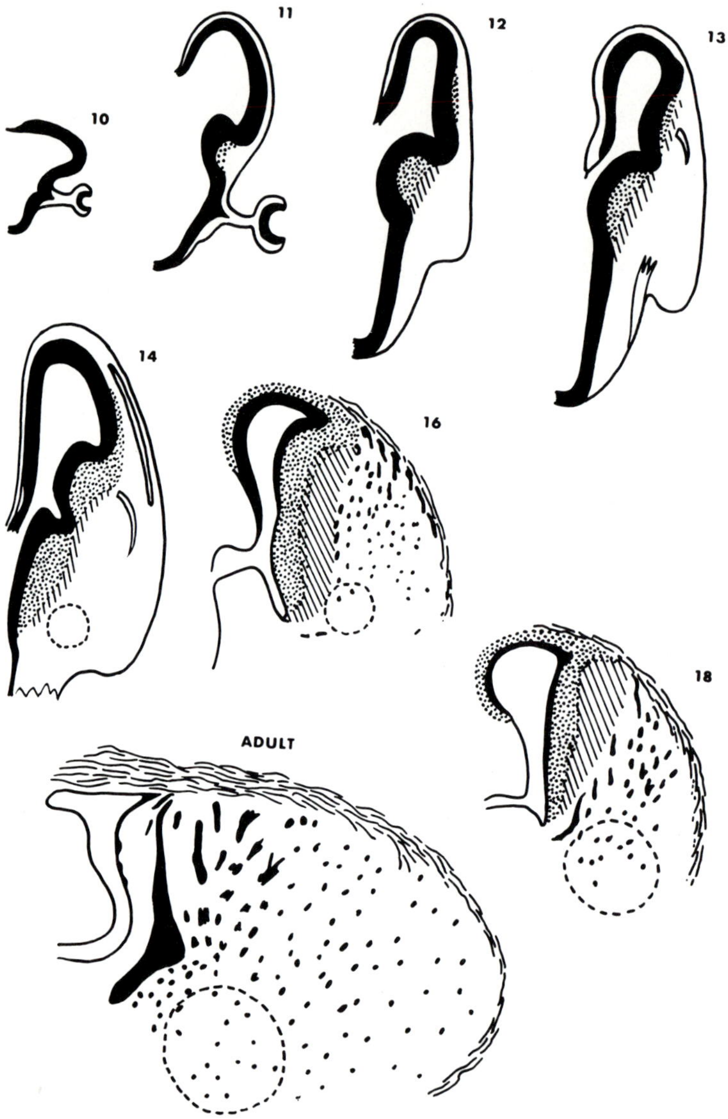

Fig. 4. Diagrams of coronal sections through mouse brains at different embryonic ages. The numbers indicate the post-conceptional age. Solid black - ependymal layer; stipple - subependymal layer; cross hatching - medial denser part of striatal anlage. Comma-shaped area at E13 and 14 represents plane of cleavage between the lateral boundary of neostriatal anlage and the circumferentially-running ependymoglial fibres emanating from the caudato-pallial angle (Cf. Fig. 5). From E15-16 onwards this plane is increasingly occupied by the corticofugal axons of maturing neurons and will eventually be the site of the corpus callosum in the adult. Broken circle indicates position of globus pallidus anlage.

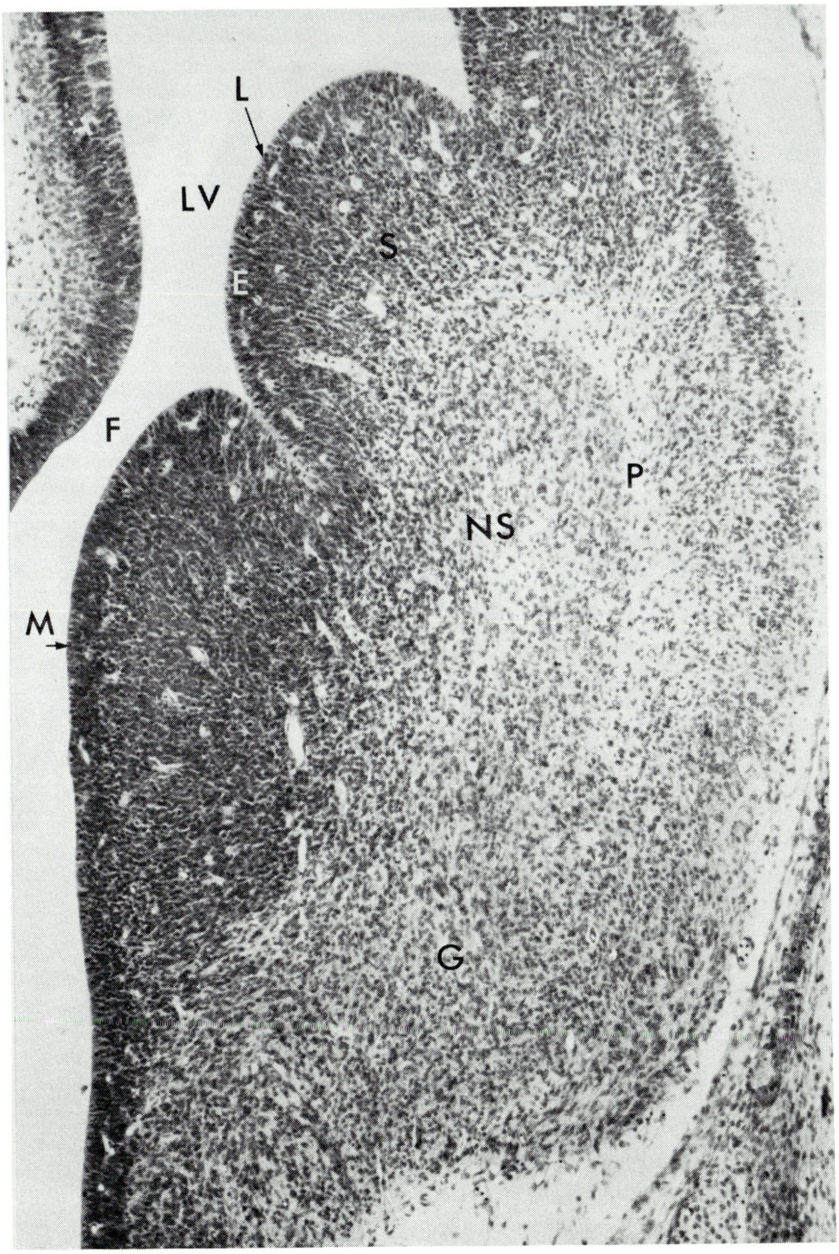

Fig. 5. Coronal section through E13 mouse brain at level of interventricular foramen: LV - lateral ventricle; F - interventricular foramen; M - medial ventricular elevation; L - lateral elevation; E - ependymal layer; S - subependymal layer; NS - neostriatal anlage; G - anlage of globus pallidus: P - plane of cleavage between neostriatal anlage and cortex (Cf. Figs 4 and 7).

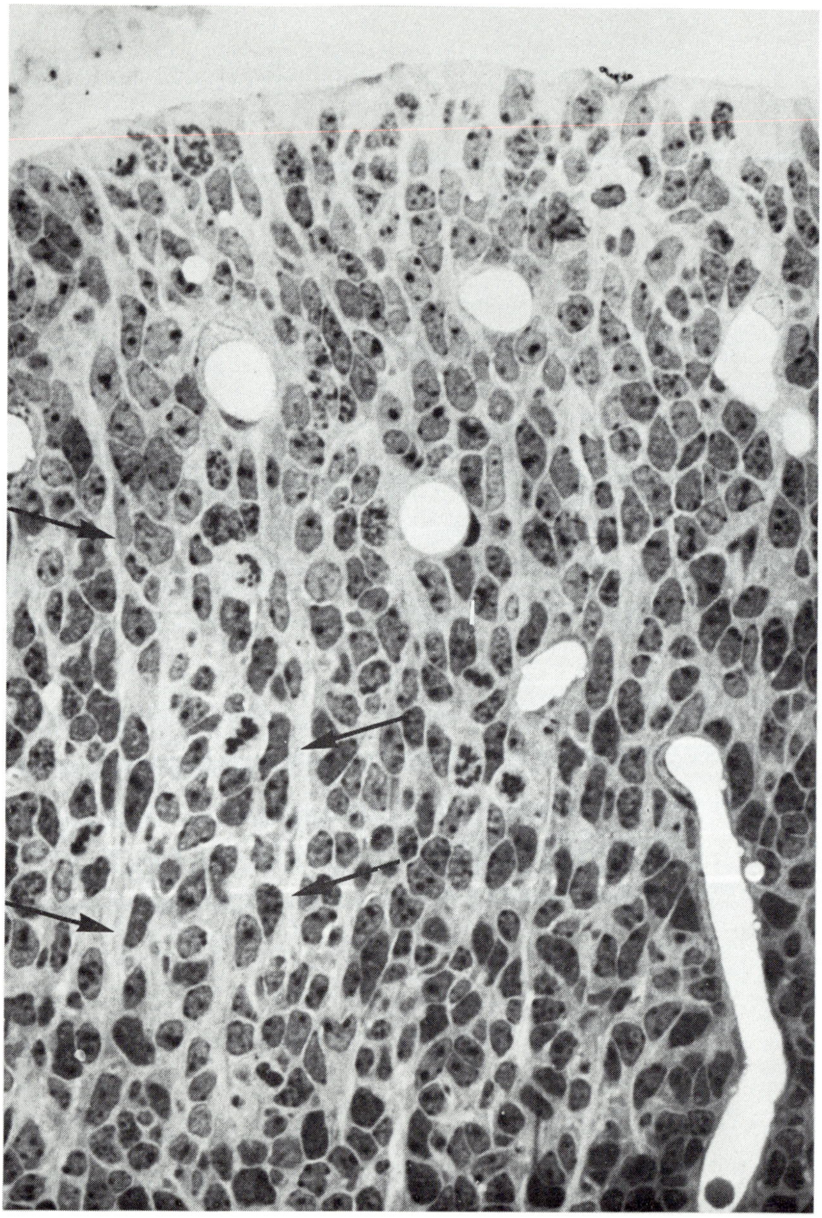

Fig. 6. Toluidine blue stained semithin section of ependymal and subependymal layers of lateral elevation at E15. The arrows mark the pale staining ependymoglial fibres extending deep into the subependymal layer (x1200). Note also numerous mitotic figures in subependymal layer and relatively few figures at ventricular surface.

subependymal layer can be taken to represent the evolution of a more 'efficient' mechanism of cell production fundamental to the generation of the increased cell populations of the mammalian brain.

The Neostriatal Primordium

The aggregation of neuroblasts and glioblasts forming the neostriatal primordium is first clearly distinguishable at E12 lying lateral to the subependymal layer of the ventricular elevations and medial to the site of the future corpus callosum (Figs. 4, 5). Two parts can be distinguished: a medial part where nuclei are densely packed and a lateral part with more widely separated nuclei interspersed with fibre bundles (Figs. 4, 5). The volume of the primordium increases progressively from its appearance at E12 reaching half its adult volume about 6 days after birth (Ref. 14). The final adult volume is greater on the left (Ref. 14).

Identification of Cell Types

The following descriptions are based on an extensive series of 'semithin' (Refs. 18, 37, 38) and ultrathin sections of the developing mouse telencephalon. A combination of light microscopic studies of semithin sections and electron microscopy enables a large area to be studied in the semithin section while details of differentiation and myelination can be observed with the electron microscope.

Ependymal layer. In early embryonic stages the ependymal layer of the ventricular elevation consists of a pseudostratified columnar epithelium, made up of cells in which nuclei and cytoplasm are pale staining. Bundles of pale staining fibres can be traced from this layer for varying distances into the subependymal layer (Fig. 6). These fibres probably correspond to the ependymal cell processes seen in Golgi preparations (Figs. 7, 8, 9).

Subependymal layer. The cells of the subependymal layer have small, randomly orientated pleomorphic nuclei and scanty cytoplasm. Their characteristics are indeterminate and no classification into neuroblasts or glioblasts can be made (Fig. 10).

Neostriatal primordium. The aggregation of cells forming the neostriatal primordium appears lateral to the adjacent half of each elevation at E12 to E13. It is initially composed of immature cells of the neuron and neuroglia series. Neuroblasts are first present at E13 and differentiated neurons by E14. A mixture of all stages of neuron differentiation can be found throughout the developing neostriatum from E14 to postnatal day 5 (P5). Neuroblasts can no longer be identified after P5. The substantial increase in volume of the neostriatum occurring thereafter (Ref. 14) can be attributed to the increase in neuropil and the formation of myelin.

Microglia are the first differentiated representatives of the neuroglia series to be identified appearing at E13 and reaching adult numbers of P15, but these probably arise from intraventricular macrophages rather than from the ependymal or subependymal layers (Refs. 17, 39).

Glioblasts appear in the neostriatal primordium at E13. The first differentiated astrocytes are present at birth and reach adult numbers at P5. Both degenerating and dividing astrocytes can be observed at this time suggesting that the astrocyte population is undergoing continuous renewal. Oligodendrocytes, first observed at

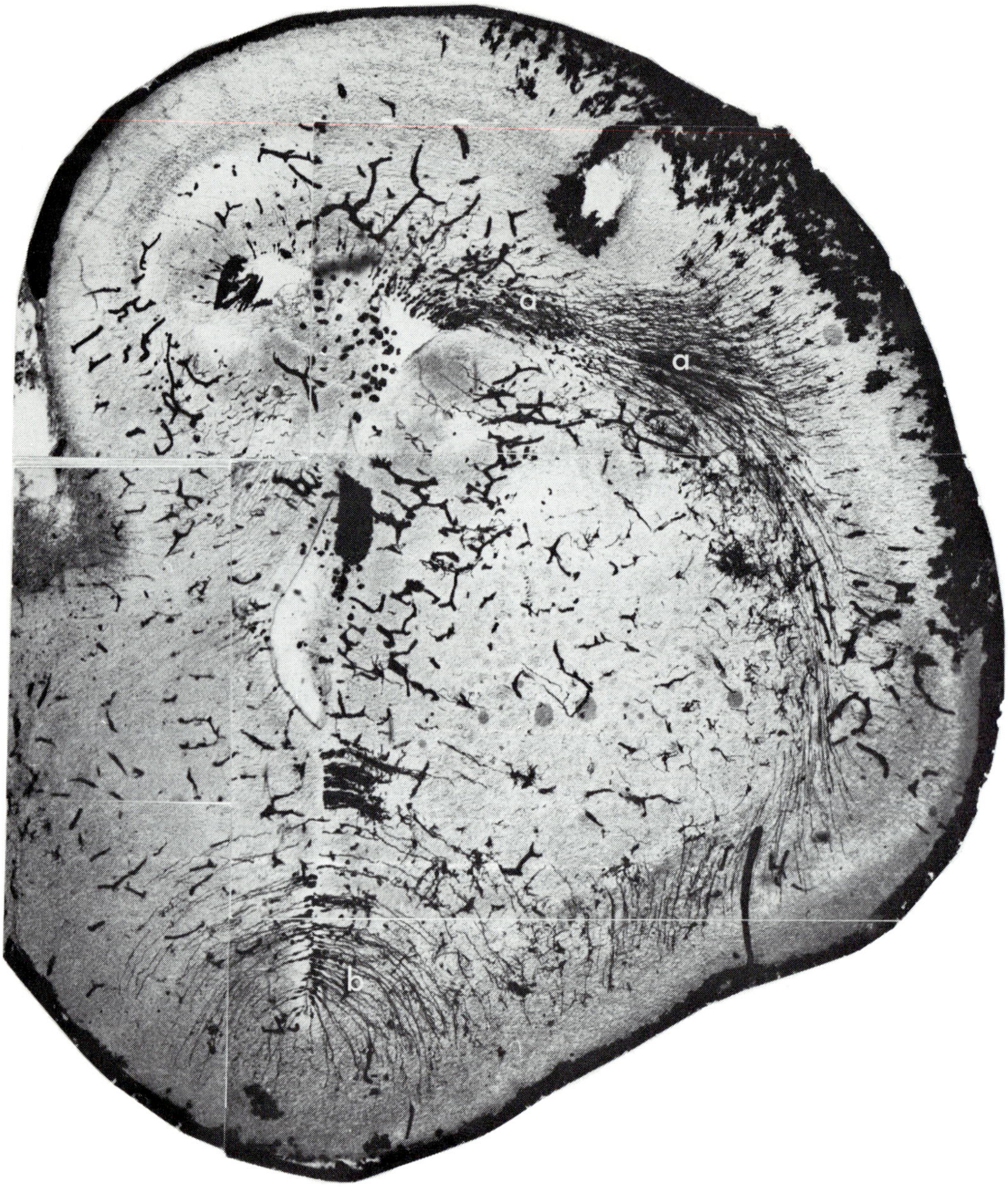

Fig. 7. Photocollage of coronal section through rostral part of mouse telencephalon at E16 impregnated by rapid Golgi method. Note (a) radiation of cell processes from caudato pallial angle circumscribing neostriatum and (b) curving path of ependymoglial cell processes from ventral part of lateral ventricle.

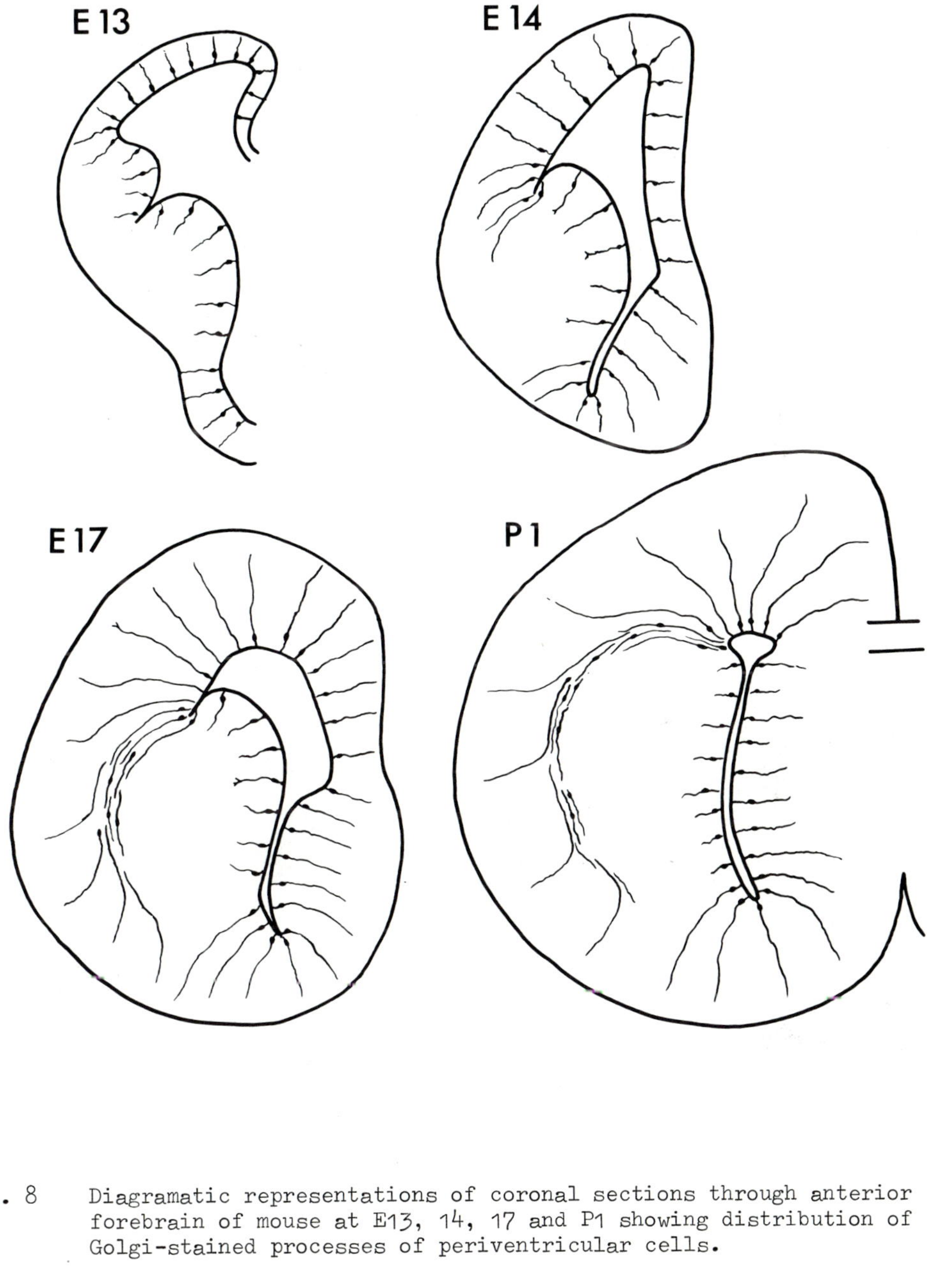

Fig. 8 Diagramatic representations of coronal sections through anterior forebrain of mouse at E13, 14, 17 and P1 showing distribution of Golgi-stained processes of periventricular cells.

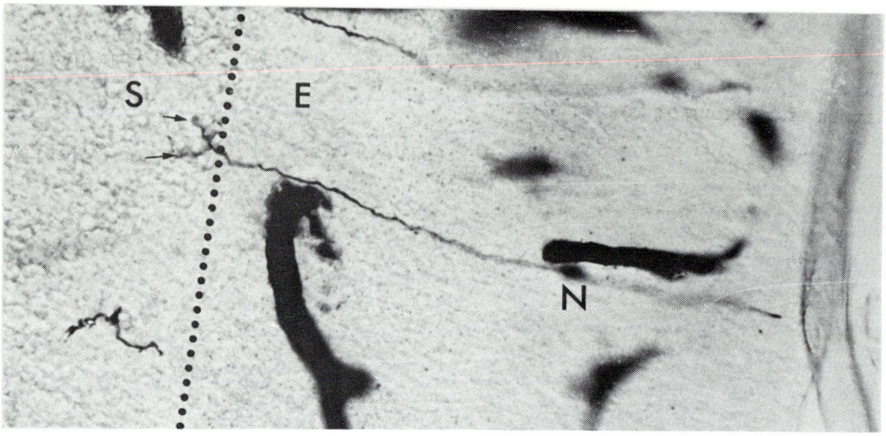

Fig. 9 Golgi-stained section through lateral elevation at E14. Note ependymal cell with terminal arborisation (arrows) just lateral to the boundary between ependymal and subependymal layer which is marked by dotted line. N, ependymal cell nucleus; E, ependymal layer; S, subependymal layer.

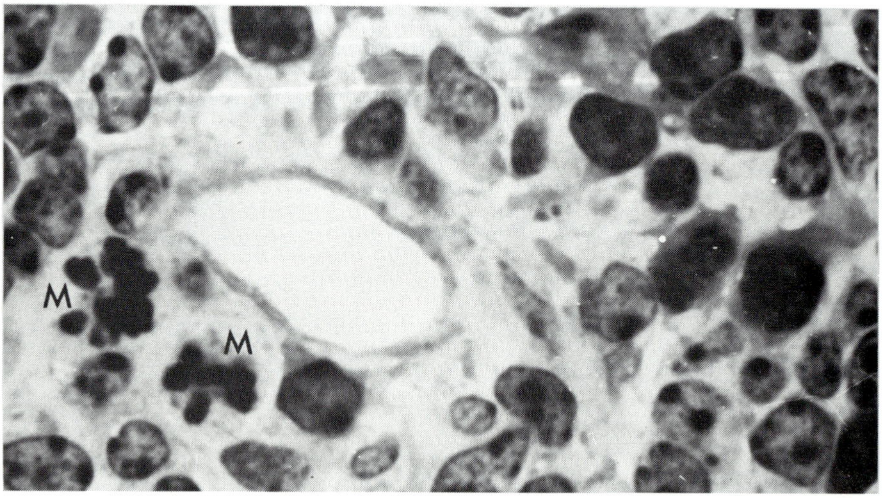

Fig. 10. Toluidine blue stained 1 µm section of the subependymal layer at E15 showing the cells with small nuclei and scanty cytoplasm. Both dark and light cells are present. Note the two mitotic figures (M).

P5, increase in number rapidly up to P20 and then more slowly up to P90. Myelination in the fibre bundles dispersed throughout the neostriatum begins about P10 at which time 1.3 % are myelinated. The percentage of myelinated axons rises rapidly at first to 6.4 % at P15 and 12.0 % at P20, then more slowly to 28.9 % at P90 and reaching at P180 a final level of 36.1 %.

Two different types of striatal neuron, a large and a small type, have been described in the mouse but the large neurons form only 3 % of the neuron population (Ref. 21). In the rat Lu and Brown (20) have identified no less than seven morphological types of neuron in the caudate nucleus using the size of the neuronal soma and the morphology of axons and dendrites (demonstrated by metallic stains) as criteria. In other animals, e.g. man and monkey (Refs. 42, 4) large and small neurons can also be identified with routine stains such as haematoxylin and eosin.

Autoradiographic Studies

Our mouse autoradiographic material consists of autoradiographs of serially sectioned brains of adult mice which had received a single injection of 3H-thymidine during prenatal life. Thymidine, a specific DNA precursor, is taken up by cells during the DNA synthetic phase which precedes the onset of visible mitosis and is subsequently retained in the nucleus. Nuclei which have taken up labelled thymidine can thus be detected autoradiographically. The label becomes progressively weaker after each subsequent division. Thus, heavily-labelled nuclei belong to cells which have been 'born' about the time of injection of 3H-thymidine while more weakly labelled nuclei belong in general to cells which have undergone subsequent divisions. In our material, nuclei with 25 or more overlying silver grains were considered to be 'heavily labelled'. The minimum label was taken as 10 overlying silver grains. Since silver grains can only be counted under high power magnification, and preferably oil immersion, some form of mapping is necessary to gain an accurate idea of the distribution of labelled cells over a wide area. Our maps were prepared by marking each labelled nucleus on a greatly enlarged photograph of the coronal section under study.

In Fig. 11 the distribution of cell nuclei with more than 10 overlying silver grains is shown in mice injected with 3H-thymidine at E11, E14 and E18. In this case all intensities of label are recorded on the same map to accentuate the numbers and give a better idea of the trend of distribution. It is probable that the majority of precursor cells underwent one or two divisions after exposure to 3H-thymidine and then differentiated. Thus the cells marked in Fig. 11 were probably born within three days of labelling. The first neurons to be born (E11 - E13) are more numerous in the lateral half of the caudate/putamen; those born in the middle period (E14 - E16) are fairly evenly scattered and the last born (E18 - P1) lie predominantly in the medial half. The maps in Fig. 12 are drawn from one horizontal and four coronal levels and show the distribution of heavily labelled nuclei in mice injected at E16. In the coronal sections the density of labelled cells seems to be greatest rostrally and least caudally; such a gradient, however, is not apparent in the horizontal section. Also, in the most rostral of the coronal sections, labelled cell nuclei appear to be more frequent ventrally while in coronal sections caudal to the interventricular foramen they appear to be more densely distributed dorsally.

The first workers to use 3H-thymidine labelling in striatal histogenesis were Sidman and Angevine (31) who found that in the mouse neostriatal neurons were produced from E10 to E15. In a later report, also on the mouse, Angevine and McConnell (1) described lateromedial and caudorostral gradients in neostriatal neuron production which began at E12, reached a peak at E14 and was largely complete by E17, although proliferation continued until one day postnatum. In a recent personal communication Angevine and McConnell (Angevine J.B. & McConnell J.A.-

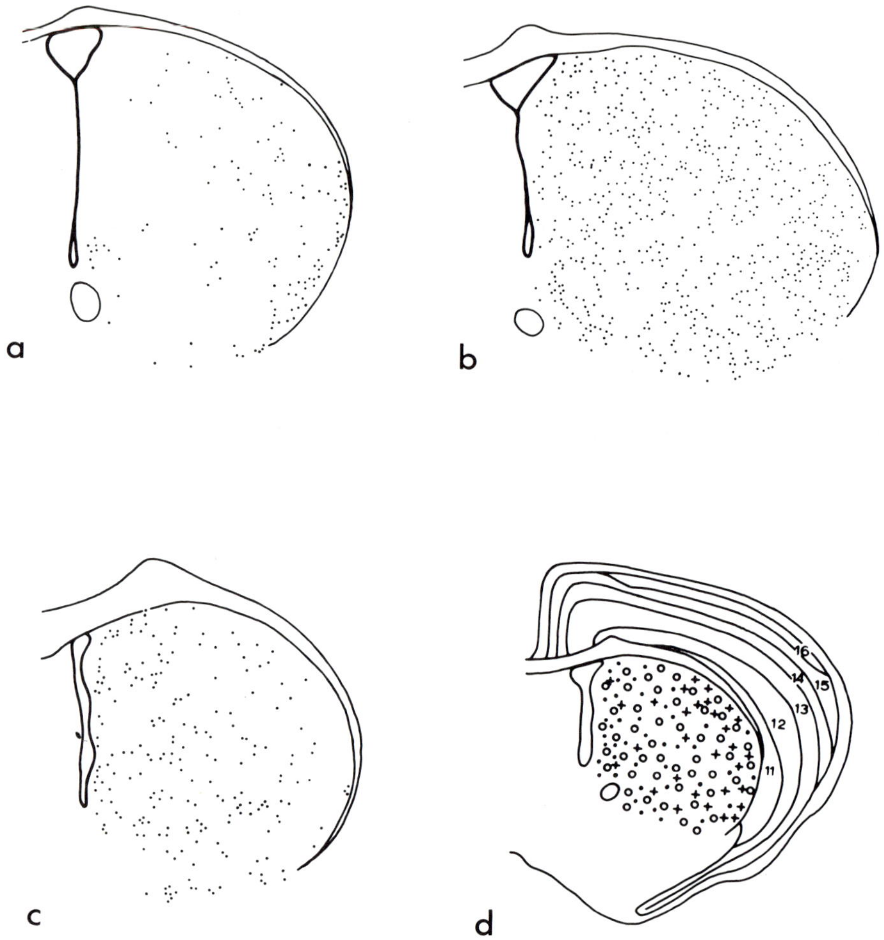

Fig. 11 a - c Distribution maps of nuclei in adult mouse neostriatum with $>$ 10 overlying silver grains. Mice received a single pulse of [^{3}H] thymidine at (a) E11, (b) E14 and (c) E18. In (d) the numbers in the cortical layers indicate the embryonic day on which the majority of the cells in each layer are produced; these do not coincide with the histological layers. The symbols within the neostriatum indicate the time of production of striatal cells: E11-12, crosses; E14-15, circles; E18-19, dots.

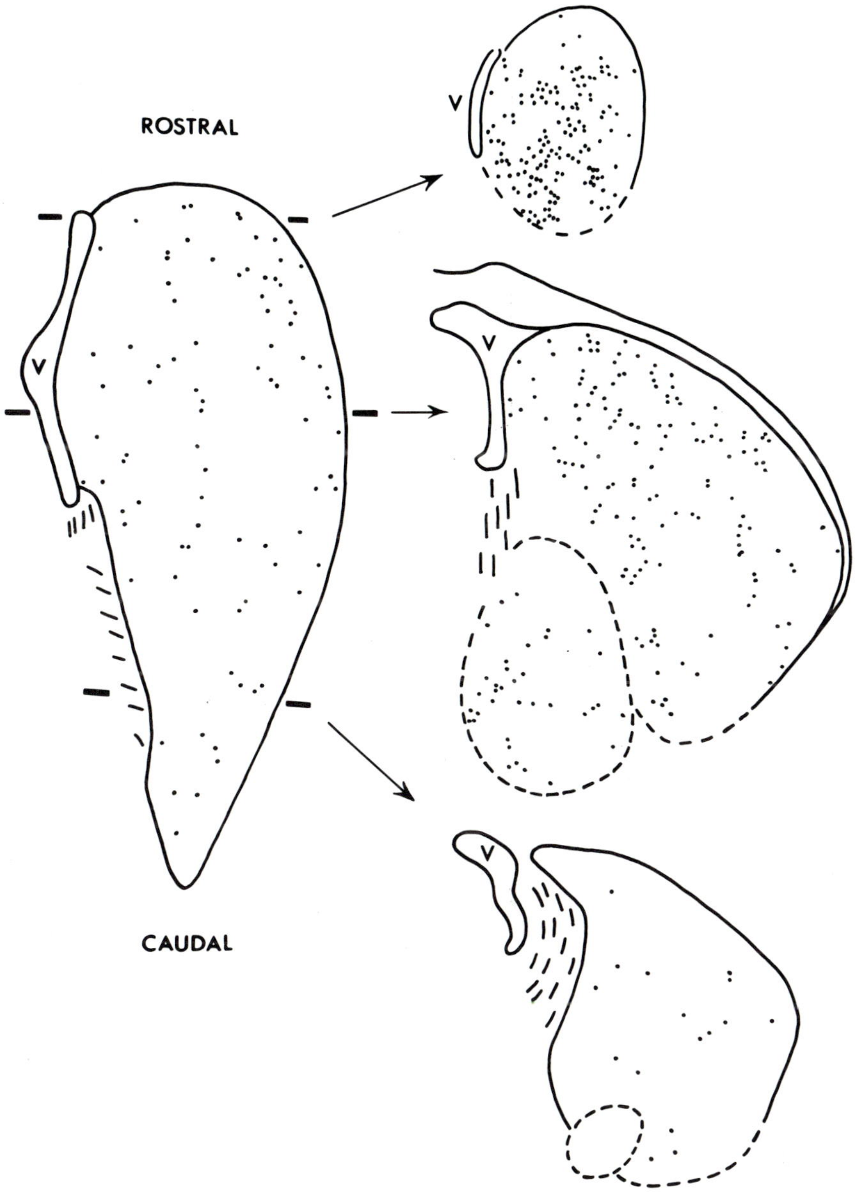

Fig. 12 Distribution of heavily-labelled nuclei (i.e. those with 25 or more overlying silver grains) at different levels in the corpus striatum of adult mice injected with [^{3}H]thymidine at E16 - on the left is a horizontal section and on the right coronal section taken at three different levels; the latter are taken from the same brain.

personal communication) confirmed that neuron origin throughout the neostriatum began at E10 at which time labelled cells were more numerous posteriorly. The number of labelled neurons peaked at E14 in all parts of the neostriatum with the building to this peak being most gradual posteriorly and most rapid anteriorly. Neuron production ceased at E17 posteriorly, at E18 in the middle, and at P1 anteriorly.

It thus seems possible to conclude in the case of the mouse that (1) the majority of the early produced cells are located laterally and caudally, (2) the last cells to be produced tend to lie rostrally and medially, (3) these two extreme populations overlap, (4) cells formed between these two extreme times of cell production are more uniformly distributed throughout the caudate-putamen. The autoradiographic picture is consistent with the histological findings presented earlier in the chapter.

An additional feature noted by Angevine and McConnell (1) in the mouse was the presence of small clusters of cells with labels of similar intensity suggesting that cells of the group were born contemporaneously. Angevine and McConnell (Angevine, J. B. & McConnell, J. A. - personal communication) found that these cell clusters, too, had a noticeable gradient; at E10 they were found to lie posteriorly and laterally at E10 to E13 at the middle level and laterally and at E17 to P1 in the middle region and medially. Brand and Rakic (4) also describe similar cell clusters occurring in the monkey which were formed by 15 - 20 neurons bearing a similar intensity of label.

One autoradiographic study has demonstrated different times of origin of cell types. Brand and Rakic (4) reported that in the Rhesus monkey (gestation period 165 days) large neurons were produced at the beginning of striatal neurogenesis (E36 - E43) while the second population of smaller cells was produced during the whole period (E36 - E80). These authors also state that no gradient was observed in their material.

In the rat Bisconte and Marty came to slightly different conclusions (Ref. 3). Using a semi-automated counting method and computer-aided statistical treatment they found that in the rat (gestation period 21 days) neostriatal neurons were produced mainly between E13.5 and E15.5 and that they were distributed uniformly with no rostro-caudal gradient. The last neurons to be labelled (E16.5 to E17.5), however, lay dorso-laterally. In both the rat and monkey, therefore, neostriatal cells are produced in a relatively short time span within the gestation period whereas in the mouse neostriatal neuron production continues up to the first day after birth. That there may be species differences is suggested by Bisconte and Marty's (3) finding that the last neurons to be formed in the rat neostriatum lie laterally and dorsally whereas in the mouse the last formed cells lie medially and rostrally.

We have in this laboratory recently completed an extensive autoradiographic study of cortical histogenesis in the mouse, the results of which are summarised in Fig. 11 alongside the neostriatal findings. The earliest formed cortical cells are found principally ventrally and medially in the deep cortical layers and the later formed cells lie superficially and dorsally.

The Neostriatum and the Ependymoglia System of the Forebrain

Ependymoglia cells are non-neuronal cells with cell bodies located in the ependymal layer and long processes which extend through the mantle layer towards the pial surface. In sub-mammals ependymoglia cells persist throughout life as a major component of the neuroglial population. In mammals cells of this type are present transiently in the early stages of development and except in a few locations have disappeared by the time of birth. During the embryonic period in mammals the ependymoglial cell processes seem to act as 'guide fibres' conducting neuroblasts from their site of formation adjacent to the ventricles to their eventual location in the grey matter (Refs. 22, 24, 25, 10).

Nieuwenhuys has generalised this idea by regarding the ependymoglial processes as a 'natural' system of coordinates whose course indicates the route followed by migrating neurons (Ref. 23). Nieuwenhuys used this principle in mapping the submammalian brain stem; cranial nerve nuclei were referred back along the glial coordinates to their putative site of origin at the surface of the fourth ventricle. The results proved useful in homologising cranial nerve nuclei which had undergone different degrees of migration in different species.

Our Golgi studies of the developing mouse forebrain have demonstrated that a well-developed ependymoglial fibre system exists in relation to those areas of the ventricle giving rise to cortical cells but is conspicuously lacking in the ventricular elevations (Fig. 7, 8). In the ventricular elevations ependymoglial cell processes lose contact with the pial surface about E12 and in our material could not be traced much beyond the ependymal layer. At least some of these abbreviated ependymoglial fibres terminate as arborisations at the boundary between the ependymal and subependymal layers (Fig. 9). By contrast, the ependymoglia fibres originating from cells dorsal and ventral to the ventricular elevations are well developed and pursue curving courses which circumscribe the neostriatal primordium (Fig. 8). Distribution of cortical neurons, therefore, appears to take place by the 'conventional' guide fibre systems whereas the neostriatal cell population makes use of some other mechanism. As has been seen in the autoradiography section of this chapter, a given neuron generation becomes widely dispersed in the adult neostriatum. How this dispersal is brought about and controlled is a problem yet to be solved.

Origin of Neostriatal Cells

As there are two ventricular elevations the question arises as to the destination of the cells produced by each. The neostriatal primordium of the mouse lies opposite adjacent halves of each elevation and, if proximity is any guide, is receiving cells from the subependymal layer of each elevation. The ventral part of the medial elevation is in close proximity to the primordium of the globus pallidus with which it is coextensive and there is, thus, a case for considering that cells of the globus pallidus may arise from this source.

The dorsal part of the lateral elevation seems to have cortical affinities as the extended ependymoglial system persists here; processes of ependymoglial cells from this area arch laterally round the neostriatal primordium to reach the ventrolateral cortex (Fig. 8). Radiating fibres with a similar origin and course have also been described in the fetal sheep (Ref. 2).

The possible origin of cells from the dorsal part of the lateral elevation may have some comparative anatomical interest as an area similarly located in the reptile has been mentioned by some investigators (see review by Goldby & Gamble (8)) as a possible site of neurons forming the reptilian equivalent of 'general cortex'.

The origin of the basal nuclei cells has received considerable attention in the developing human brain. Major interest in the area arose following the demonstration by Vogt and Vogt (42) of the different functional properties and pathological susceptabilities of the caudate-putamen and the globus pallidus. An embryological explanation for this difference was sought and a series of studies over the years described the caudate-putamen as developing from the ventricular elevations and the globus pallidus from the basal plate of the diencephalon. The pallidal primordium was described as being brought into the telencephalon by differential growth at the telo-diencephalic junction which displaced the pallidal cells into the basal telencephalon by a movement analogous to inussusception. One of the earliest proponents of this concept was Spatz (36) and of the later writers Schneider (30) and Richter (26) have written detailed works on the same theme. Schneider (30) also described an additional migration of cells from the thalamic diencephalon into the ventricular elevations, implying that some neostriatal cells were also of diencephalic origin. In the mouse the only definite evidence for basal ganglion cell populations originating from the diencephalon is the fact that the medial ventricular elevation incorporates part of the wall of the third ventricle. This diencephalic part of the medial elevation lies medial to and coextensive with the primordium of the globus pallidus, and by its proximity is the most likely source of pallidal cells. There is no convincing evidence in the mouse for extensive cell immigrations from the more ventral parts of the diencephalon.

The Reptile-Mammal Transition

In the forebrain of adult reptiles there are well-developed ridges which protrude from the lateral wall of the lateral ventricles and thus occupy a position similar to the ventricular elevations found in embryonic mammal brains. The cell populations of the reptilian ventricular ridges appear to originate directly from the embryonic ependymal layer. In the embryonic reptilian material at our disposal we have not seen mitotic figures at any location other than at the ventricular surface, nor are any mentioned in Kirsche's (13) detailed account of the development of the turtle brain. (The Chelonians are thought to be the same evolutionary stem that gave rise to mammals.) The subependymal layer thus seems to be a feature which evolved during the reptile-mammal transition and its appearance represents a breakdown of the mitotic monopoly of the ependymal layer and its restrictive practice of interkinetic nuclear migration. This change in the mechanism of cell production reflects the generation of larger cell populations of both neurons and neuroglia which underlie the increase in size of the mammalian telencephalon.

Many attempts have been made to homologise nuclear groups in the reptilian ventricular ridges with the mammalian basal nuclei (see review by Ris, Halpern & Scalia (27)). This exercise may be even more difficult if the changes in histogenesis are also taken into account. In mammals the subependymal layer preempts the position within the concavity of the ventricular elevations occupied by basal ganglion cells in the reptile and there is consequently a shift laterally of the greatly increased numbers of maturing neurons and their fields of synaptogenesis, both of which are themselves responding to the changing input from the new cortical areas. Comparing the striatum of reptiles and mammals may be analogous to comparing a medieval town with the modern city which has replaced it; the change is not one of simple multiplication of medieval (or reptilian) institutions but the installation of a different order of complexity. A possible way of formulating an approach to the reciprocal growth of neocortex and neostriatum is suggested in the recent paper by Katz and Lasek (12).

Biochemical Development

The onset of neuronal differentiation in the mouse neostriatum at E13 coincides with the appearance in the neostriatum of fluorescent axons which have been shown by microspectrofluorimetry to contain dopamine (Ref. 9).

While caution must be exercised in comparing results obtained from different species due to the differences in the length of gestation and differences in the stage of brain development at birth, the brains of rats, kittens, rabbits and mice are probably at a similar stage of development at birth. The proportion of dopamine in the neostriatum of neonatal rats, rabbits and kittens is remarkably similar with 13 % of the adult value present in neonatal rats (Ref. 7), 15 % in rabbits (Ref. 41) and 15 % in kittens (Ref. 5). The postnatal increase in neostriatal dopamine appears to continue for a considerable time. Loizou (19) has suggested that dopamine containing axon terminal proliferation is highest during the first four postnatal weeks. At first the areas of fluorescence are arranged in islands (Ref. 19) and gradually become diffuse. Tennyson *et al.* (41) observed a similar pattern in the rabbit. The one month period over which the number of axon terminals appears to increase in the rat coincides with the period of neuronal branching and development of dendritic spines (Ref. 20).

A detailed correlation of structural and biochemical development in the neostriatum has been carried out in the rabbit (gestation period 30 days) by Tennyson *et al.* (40, 41). Dopamine containing terminals were found in the ventromedial putamen at E19 and in the late fetal period a second ingrowth of dopamine containing terminals occurred in both the putamen and caudate. The appearance of fluorescence first in the lateral striatum coincides with the latero-medial gradient of cell production and cell differentiation described earlier. Tennyson *et al.* (40, 41) also found a caudorostral gradient in time of appearance of fluorescent axons. Islands of fluorescence were observed in the late fetal and early postnatal period similar to those observed in the rat by Loizou (19). It would be interesting to know if these islands correspond with the grouping of contemporaneously produced cells as found in autoradiographic studies (Ref. 4 and personal communication from Angevine & McConnell).

The appearance of the uptake mechanism for radioactive dopamine in the rabbit coincided with the first appearance of fluorescent axons at E19. Endogenous dopamine levels increased significantly from E19 to E28 and gradually rose from 15 % of the adult value at birth to 62 % at 45 days postnatum. The uptake levels of radioactive dopamine, however, reached adult levels much earlier at P25 suggesting that the mechanisms for dopamine uptake matures before dopamine is present at adult levels (Ref. 40).

Tyrosine hydroxylase, the enzyme which converts tyrosine to DOPA is present at 8 % of adult levels in neonatal rats. During fetal development there is a shift of this enzyme from the region containing dopaminergic cell bodies to the region containing the axons terminals i.e. from the substantia nigra to the neostriatum (Ref. 6).

Summary

The ventricular elevations which dominate the basal parts of the mouse telencephalon during development represent a modification of the basic proliferative mechanism of the neural epithelium. Associated with this is an early abbreviation of the embryonic ependymal cells which lose contact with the pial surface and which in this region do not appear to extend beyond the subependymal layer. There is strong circumstantial evidence that elsewhere in the developing brain, including the cortical areas, these processes are involved in some way as 'guide fibres' leading neuroblasts to their peripheral destination. Their absence suggests that the distribution of neuroblasts from the elevations to the striatal area takes place by a different method. While contemporaneously produced neurons may settle in all parts of the adult neostriatum, there is a tendency for neurons born early in the period of neostriatal histogenesis to lie predmoninantly in the lateral and caudal parts of the neostriatum and for those born at the end of the period to lie medially and rostrally. The neocortex in the mouse matures along a dorsocaudal gradient but there is as yet insufficient data to demonstrate whether or not there is any developmental correlation between the neocortex and the neostriatum.

The cortical areas are generated in a much more conventional way than the cells originating from the ventricular elevations proper and it may be worth considering that the primary evolutionary innovation occurring between reptile and mammal in this area is the production of much more populous basal nuclei and that the cortical areas, including the neocortex, evolve by riding on the back of the neostriatum both literally and figuratively.

References

(1) J. B. Angevine Jr, & J.A. McConnell, Time of origin of striatal neurons in the mouse, Anat. Rec. 178, 300 (1974).

(2) K.-E. Astrom, On the early development of the isocortex in fetal sheep. In Progress in Brain Research, 26 (Eds C. G. Bernhard & J. P. Schade) 1 - 59. Elsevier, Amsterdam, London, New York (1967).

(3) J.-C. Bisconte, & R. Marty, Analyse chronoarchitectonique du cerveau de rat par radioautographie. I. Histogenese du telencephale. J. Hirnforsch. 16, 55-74 (1975).

(4) S. Brand & P. Rakic, Time of origin of neurons in primate neostriatum: H3-thymidine autoradiographic analysis in the Rhesus monkey, Anat. Rec. 190, 345-346 (1978).

(5) J. D. Connor & N. H. Neff, Dopamine concentrations in the caudate nucleus of the developing cat, Life Sci. 9, 1165-1168 (1970).

(6) J. T. Coyle & J. Axelrod, Tyrosine hydroxylase in rat brain: developmental characteristics, J. Neurochem. 19, 1117-1123 (1972).

(7) J. T. Coyle & D. Henry, Catecholamines in fetal and newborn rat brain, J. Neurochem. 21, 61-67 (1973).

(8) F. Goldby & H. J. Gamble, The reptilian cerebral hemispheres, Biological Reviews, 32, 383-420 (1957).

(9) G. S. Golden, Embryological demonstration of a nigro-striatal projection in the mouse. Brain Res. 44, 278-282 (1972).

(10) C. K. Henrikson & J. E. Vaughn, Fine structural relationships between neurites and radial glial processes in developing mouse spinal cord. J. Neurocyt. 3, 659-675 (1974).

(11) W. Hewitt, The development of the human caudate and amygdaloid nuclei, J. Anat. (Lond.) 92, 377 (1958).

(12) M. J. Katz & R. J. Lasek, Evolution of the nervous system: role of ontogenetic mechanisms in the evolution of matching populations. Proc. Natl. Acad. Sci. USA. 75, 1349-1352 (1978).

(13) W. Kirsche, Die entwicklung des Telencephalons der Reptilien und deren Beziehung zur Hurn-Bauplanlehre, Nova Acta Leopoldina 37/2, 9-78 (1972).

(14) H. -J. Kretschmann & F. Wingert, Biometrische Analyse der Volumino des Striatum einer ontogenetische Reihe von Albinomausen, Z. Anat. Enwickl.-Gesch. 128, 85 (1969).

(15) G.J. Lammers, On the development of the strio-amygdaloid complex in the Chinese hamster, Cricetulus griseus, Ph.D. thesis, Katholieke Universiteit Nijmegen, (1976).

(16) P. D. Lewis, The fate of the subependymal cell in the adult rat brain, with a note on the origin of microglia. Brain, 91, 721-735 (1968).

(17) P. D. Lewis, Glial reaction to cerebral injury. Proc. Roy. Soc. Med. 67, 130-131 (1974).

(18) E. A. Ling, J. A. Paterson, A. Privat, S. Mori & C. P. Leblond, Investigation of glial cells in semithin sections. I. Identification of glial cells in the brain of young rats. J. Comp. Neurol. 149, 43-72 (1973).

(19) L. A. Loizou, The postnatal ontogeny of mono-amine containing neurons in the central nervous system of the albino rat. Brain Res. 40, 395-418 (1972).

(20) E. J. Lu & W. J. Brown, The developing caudate nucleus in the euthyroid and hypothyroid rat. J. Comp. Neurol. 171, 261-284 (1977).

(21) P. L. Mensah, Neuronal cell death in the neostriatum of the perinatal mouse. Anat. Rec. 190, 479-480 (1978).

(22) D. Kent Morest, A study of neurogenesis in the forebrain of Opossum pouch young, Z. Anat. Entwickl.-Gesch. 130, 265-305 (1970).

(23) R. Nieuwenhuys, Topological analysis of the brain stem of the lamprey Lampetra fluviatalis, J. Comp. Neurol. 145, 165-178 (1972).

(24) P. Rakic, Mode of cell migration to the superficial layers of fetal monkey neocortex, J. Comp. Neurol. 145, 61-84 (1972).

(25) P. Rakic, Genesis of the dorsal lateral geniculate nucleus in the Rhesus monkey: site and time of origin, kinetics of proliferation, routes of migration and pattern distribution of neurons. J. Comp. Neurol. 176, 23-52 (1977).

(26) Richter, E. Uber die Entwicklung des Globus Pallidus und des Corpus Subthalamicum bein Menschen. In: R. Hassler & H. Stephan (Eds) Evolution of the Forebrain. G. Thieme, Stuttgart (1966).

(27) W. Ris, M. Halper, & F. Scalia, The quest for clues to forebrain evolution - the study of reptiles. Brain Behav. Evol. 2, 1-15 (1969).

(28) F. E. Sauer, The interkinetic migration of embryonic epithelial nuclei, J. Morphol. 601, 1-11 (1936).

(29) M. E. Sauer & B. E. Walker, Radioautographic study of interkinetic nuclear migration in the neural tube. Proc. Soc. Exp. Biol. and Med. 101, 557-560 (1959).

(30) R. Schneider, Ein Beitrag zur Ontogenese der Basalganglien des Menschen, Anat. Nachr. 1, 115-137 (1949).

(31) R. L. Sidman & J. B. Angevine, Autoradiographic analysis of time of origin of nuclear versus cortical components of mouse telencephalon, Anat. Rec. 142, 326-327 (1962).

(32) I. Smart, The subependymal layer of the mouse brain and its cellular production as shown by radioautography after thymidine-H3 injection, J. Comp. Neurol. 116, 325-349 (1961).

(33) I. H. M. Smart, Proliferative characteristics of the ependymal layer during early development of the mouse diencephalon, as revealed by recording the number, location and plane of cleavage of mitotic figures, J. Anat. 113, 109-129 (1972).

(34) I. H. M. Smart, Proliferative characteristics of the ependymal layer during the early development of the spinal cord in the mouse. J. Anat. 111, 365-380 (1972).

(35) I. H. M. Smart, A pilot study of cell production by the ganglionic eminences of the developing mouse brain, J. Anat. 121, 71-84 (1976).

(36) H. Spatz, Zur Ontogenese des Striatums und des Pallidum, Dtsch. Z. Nervenheilk 81, 185-188 (1924).

(37) R. R. Sturrock, Light microscopic identification of immature glial cells in semithin sections of the developing mouse corpus callosum. J. Anat. 112, 521-537 (1976).

(38) R. R. Sturrock, Development of the indusium griseum. II. A semithin light microscopic and electron microscopic study, J. Anat. 125, 433-445 (1978).

(39) R. R. Sturrock, A developmental study of epiplexus cells and supraependymal cells and their possible relationships to microglia, Neuropath. & Appl. Neurobiol. 4, 307-322 (1978).

(40) V. M. Tennyson, R. E., Barrett, G. Cohen, L. Cote, R. Heikkila & C. Mytilineou, The developing neostriatum of the rabbit: correlation of fluorescence histochemistry, electron microscopy, endogenous dopamine levels, and 3H-dopamine uptake. Brain Res. 46, 251-285 (1972).

(41) Tennyson, V. M., Barrett, R. E., Cohen, G., Cote, L., Heikkila, R. & Mytilineou, C. Correlation of anatomical and biochemical development of the rabbit neostriatum. In: D. H. Ford (Ed.) Progress in Brain Research. 203-217 Elsevier, Amsterdam, London & New York (1973).

(42) C. Vogt & O. Vogt, Zur Lehre der Erkrankungen, des Striaren Systems, J. psychol. Neurol. 25, 628-846 (1920).

Ventral Striatum

Lennart Heimer* and Gary van Hoesen**

**Department of Anatomy, University of Virginia, Charlottesville, Virginia*
***Department of Neurology, Beth Israel Hospital, Boston, Massachusetts*

The dorsal portion of the striatum is a prominent and easily localized component of the basal telencephalic gray matter in all mammals. While the lateral ventricle and subcallosal fasciculus form clear boundaries dorsally, few anatomical landmarks demarcate striatal boundaries ventrally. In fact, at many points in the basal forebrain it is nearly impossible to distinguish striatal elements from modified regions of the basal telencephalon like the nucleus accumbens septi and the olfactory tubercle. In this chapter evidence will be reviewed which indicates that the latter structures may more appropriately be viewed as specialized portions of a ventral striatum which share many features with the main dorsal part of striatum.

Nucleus accumbens and olfactory tubercle

Some authors (29,8,33,9) have indicated that accumbens may be part of septum, since it receives precommissural fornix fibers. Ariens Kappers (25), on the other hand, who introduced the term nucleus accumbens septi, meaning literally a nucleus leaning against the septum, considered it to be part of striatum, and many other authors (15,43,51,45) have had similar opinions. In 1923, Johnston (24) published a major article on the evolution of the forebrain, where he championed the notion that the accumbens, although striatal in nature, nevertheless should be distinguished from the rest of the striatum on the basis of its seemingly strong relations with the olfactory system. Subsequent normal anatomical studies appeared to concur with this opinion and the concept of "olfacto-striatum" (22,35) gradually became part of the neurological terminology. However, when modern experimental anatomical tract-tracing methods came into use in the fifties, it became obvious that no primary or secondary olfactory connexions reached the nucleus accumbens and the concept of "olfacto-striatum" gradually lost its appeal.

When it became apparent that the nucleus accumbens, like the main body of the neostriatum, receives a dopaminergic input from the midbrain (14), interest was again focused on this structure. Anatomists studied its connexions and its synaptology (27,49,34,38,6,16,5,48) and physiologists and pharmacologists became interested in its functional and histochemical characteristics (30,40,44,1,11,36,37,39,4,13). The results of these studies however are often contradictory, especially in regard to the extrinsic connexions of the accumbens.

If the classification of nucleus accumbens has been controversial, the situation has been no less puzzling in regard to the olfactory tubercle, especially in primates. Whereas several authors (2,3,47,32) have indicated that the striatum reaches all the way to the ventral surface in the region of the olfactory tubercle or the anterior perforated space (fig. 1A and B), other scientists, notably Crosby and Humphrey (7) have maintained that a laminated cortical region separates the striatum from the basal surface, not only in nonprimates but also in primates including the human. The situation in macrosmatic laboratory animals, such as rats, rabbits and guinea pigs seems, at least at first sight, to be considerably less controversial. In such animals, a more or less laminated olfactory tubercle is in most places separated from the main part of striatum by the prominent tracts of the medial forebrain bundle, (figs. 2 and 3A and B), and the olfactory tubercle has almost without exception been considered to be a somewhat modified, but nevertheless integral part of the olfactory cortex rather than of striatum. In several recent papers, however, we have emphasized the kinship between the striatum and the olfactory tubercle, not only in the microsmatic primates, but also in macrosmatic laboratory animals such as for instance the rat (19,20,21).

In spite of the fact that the more or less laminated olfactory tubercle of macrosmatic mammals is reminiscent of and continuous with the laterally located olfactory cortex, there are in reality few similarities between the two structures, and the boundary between them is remarkably sharp, both in Nissl-preparations (see arrow in figure 2) and in AChE-stained preparations, where the heavily positive olfactory tubercle contrasts sharply with the very lightly stained olfactory cortex (fig. 3B). The dorsal boundary of the tubercle, however, is considerably more elusive, especially at more rostral levels where many cell bridges provide a continuity between the deep part of the olfactory tubercle and the accumbens (fig. 2). Indeed, instead of trying to define a boundary between the tubercle and the accumbens, it seems appropriate to propose that the accumbens reaches the surface between the fiber bundles of the deep olfactory radiation. Further caudally the bundles of the deep olfactory radiation, which connect rostral olfactory areas with the diencephalon, become more prominent, thereby gradually reducing both the number and the extent of "striatal" cell bridges. However, occasional cell bridges can be seen also at more caudal levels where they emanate from the main body of the striatum rather than from the accumbens (fig. 3A). These striatal cell bridges are well displayed also in AChE-preparations (fig. 3B).

With the aid of modern experimental techniques, many of the extrinsic connexions of both the primate and the nonprimate olfactory tubercle have now been traced, and the results obtained indicate that the basic pattern of connectivity is principally the same regardless of whether the olfactory tubercle is laminated, as in many of the small macrosmatic laboratory animals, or mostly non-laminated, as in the microsmatic monkey. Furthermore, there are striking similarities in connexional pattern between the olfactory tubercle and the accumbens on one hand, and between these two structures and the main body of the neostriatum on the other hand. On the basis of these hodological considerations, the concept of the ventral striatum* was recently introduced (21,18).

---

*NOTE:

The ventral striatum includes the nucleus accumbens, the medium-celled parts of the olfactory tubercle and the ventral subcommissural part of the main body of the striatum. Although these structures, which are continuous with each other through s.c. "striatal" cell bridges, are characterized by a "striatal" pattern of connexions, they are also characterized by some hodological features which distinguish them from the rest of the striatum. Note that Webster (the present volume) uses the term ventral striatum in a different meaning.

Ventral Striatum

The ventral striatum, i.e. the accumbens, the main part of the olfactory tubercle and the subcommissural part of the main body of the striatum, are characterized by the same pattern of extrinsic connexions as the main, dorsal part of the striatum (fig. 4). In other words, major afferent pathways to the accumbens and the olfactory tubercle originate in cerebral cortex, in nonspecific thalamic nuclei and in dopaminergic cells of the midbrain. The apparent absence of efferent accumbens or tubercle projections to cerebral cortex would be another feature of ventral striatum anatomy comparable to the absence of dorsal striatal projections to cerebral cortex. Furthermore, projection fibers from both accumbens and olfactory tubercle terminate in a region, which apparently constitutes a ventral extension of globus pallidus, i.e. ventral pallidum (49,18). Whereas the presence of a tubercle projection to the ventral pallidum (10,18) does not seem to have been mentioned earlier, the accumbens projection to the ventral pallidum was mentioned several years ago in a normal anatomical study (43), and has been confirmed in several more recent experimental studies in the rat (45,49,6,48).

Although this short review of the connexions of the ventral striatum has emphasized the "striatal" pattern of connexions, many of the accumbens and olfactory tubercle pathways described in the literature are not easily characterized as "striatal" in nature. However, it should be emphasized that there is little consensus regarding these connexions. This is especially the case in regard to the accumbens projections which have recently been traced both in the rat and the monkey. For instance Conrad and Pfaff (6), who used the autoradiographic technique in the rat, traced accumbens projections to a variety of regions including not only globus pallidus and substantia nigra, but also septum, anterior hippocampal rudiment, frontal cortex, preoptic area, hypothalamus, dorsomedial thalamic nucleus and zona incerta. Unfortunately, the authors do not provide an adequate description of the extent of the injections. Swanson and Cowan (45), who also studied the rat, were apparently somewhat more restrictive with their injections, and they did not confirm any of the accumbens projections described by Conrad and Pfaff besides the ones to globus pallidus and substantia nigra. Williams et al (48), who used degeneration methods in the rat, described accumbens projections to cingulate gyrus, lateral hypothalamus and intralaminar thalamic nuclei in addition to globus pallidus and substantia nigra. A recent autoradiographic study in the monkey (38), finally, revealed yet another picture of the accumbens projections. According to the latter results, the majority of the accumbens projection fibers terminate in lateral hypothalamus, whereas other fibers terminate in cingulate gyrus, septum, diagonal band, midline thalamic nuclei, habenula, caudate nucleus and substantia nigra. Clearly, except for the projections to globus pallidus and substantia nigra, there is hardly any consensus regarding the accumbens projections.

As already indicated, we have good reason to propose that the olfactory tubercle projects to the ventral pallidum. Two other olfactory tubercle projections, one to the nuclei gemini (31,42) and the other to the mediodorsal thalamic nucleus (17), seem to indicate that the olfactory tubercle contains several different functional anatomical units. Whereas the main, medium celled part of the tubercle is undoubtedly striatal in character, many of the larger cells in the deeper part of the tubercle may well belong to a ventro-rostral extension of the globus pallidus. As a matter of fact, initial HRP studies in our laboratory seem to indicate that some of these larger cells give rise to the tubercle projection to the medio-dorsal thalamus (see also 41) and the suggestion has been offered that the projection is part of a pallido-thalamic system (21). Other large "extra-striatal" cells among the fascicles of the medial forebrain bundle apparently represent the origin for the tubercular projection to nuclei gemini. The habenular complex and the septum, finally, have traditionally been thought of as being closely related to the olfactory tubercle. There is, however, very little evidence for either tuberculo-habenular or tuberculo-septal fibers in the most recent experimental-anatomical literature.

Telencephalic input to the ventral striatum

As already indicated, there are striking similarities between the dorsal and the ventral striatum. For instance, both parts of the striatum receive one of their most important inputs from the cerebral cortex. Whereas the dorsal striatum receives afferent fibers from extensive regions of neocortex, ventral striatum receives fibers from allocortex as well as from temporal association cortex (20,46). However, there are also significant differences. For example, whereas there are no apparent amygdaloid projections to dorsal striatum one of the most distinctive features of the ventral striatum, is its close relation to the amygdaloid body (fig. 4). The amygdaloid projection to ventral striatum apparently originates in the basolateral and basomedial amygdaloid nuclei (28). Part of ventral striatum, furthermore, is characterized by its close relation to the olfactory system. This relation, however, seems to vary with the importance of the sense of smell. Macrosmatic mammals are characterized by a comparatively large direct olfactory bulb input to the most superficial part of the ventral striatum, i.e. the olfactory tubercle (figs. 5A and B). In most microsmatic primates, on the other hand, only a restricted ventrolateral part of the ventral striatum receives olfactory-bulb projection fibers (fig. 6).

There seems to be an interesting correlation between the laminar arrangement of cells in the ventral striatum and the degree of olfactory bulb input. The olfactory tubercle of the anosmic dolphin and certain whales is characteristically non-laminated (12,23). In the rhesus monkey there is a tendency towards lamination only in the rostro-lateral part of the tubercle where olfactory bulb projection fibers invade the surface. In the rat, on the other hand, where the whole extent of the tubercle receives olfactory bulb input (fig. 5B) there is a tendency towards lamination of superficial cell elements throughout the olfactory tubercle (fig. 5A). It seems reasonable to suggest that the laminated regions of the mammalian olfactory tubercle represent a modified portion of striatum, organized to receive primary olfactory input.

In a recent report, Yeterian and van Hoesen (50) have noted that cortico-striate projections in the rhesus monkey are closely linked to the pattern of cortico-cortical connections. Although Yeterian and van Hoesen confirmed Kemp and Powell's (32) cortico-striate topography, as indicated with fiber degeneration methods, autoradiographic observations revealed that reciprocally interconnected areas of the cerebral cortex project, not only to their own unique locus within the striatum, but to all other striatal loci which are the targets of cortical areas with which they have intrahemispheric connections. Thus, a given location in the striatum receives a primary projection from a proximally located cortical area, as well as secondary projections from numerous distally located areas. While Yeterian and van Hoesen did not extend this principle to subcortical portions of the telencephalon which also project to the striatum, it is conceivable that this principle of organization may apply to these connections as well. For example, with regard to the ventral striatum, and especially those areas designated olfactory tubercle, it may not be a coincidence that the olfactory bulb and primary olfactory cortex, two areas heavily interconnected, also project powerfully to the olfactory tubercle component of the ventral striatum. A similar pattern may exist in regard to the relation between the basal amygdaloid complex and that portion of the inferior temporal cortex immediately lateral to the rhinal sulcus (areas 35 and 36 in the monkey). These areas are reciprocally interrelated, and both regions project to the ventral striatum in the monkey. In total, these observations may imply that all telencephalic areas which project to striatum are characterized by the principle described by Yeterian and van Hoesen (50).

REFERENCES

1. ANDEN, N.E. and JACKSON, D.M., J. Pharm. Pharmacol. 27, 666-670 (1975).
2. BECCARI, N., Archives of Italian Anatomy and Embryol. 9, 173-220 (1910).
3. BECCARI, N., Archives of Italian Anatomy and Embryol. 10, 261-328 (1911).
4. BROWNSTEIN, M. et al., Brain Research 79, 431-436 (1974).
5. CHUNG, J.W. et al., Cell Tiss. Res., 172, 1-14 (1976).
6. CONRAD, L.C.A. and PFAFF, D.W., Brain Research, 113, 589-596 (1976).
7. CROSBY, E.C. and HUMPHREY, T., J. Comp. Neurol., 74, 309-347 (1941).
8. CROSBY, E.C. et al., Correlative Anatomy of the Nervous System, Macmillan, New York, pp. 360 and 419 (1962).
9. DeFRANCE, J.F. and YOSHIHARA, H., Brain Research 90, 159-163 (1975).
10. DeOLMOS, J.S. et al., Neuroscience Abstract 1, 680 (1975)
11. ELKHAWAD, A.O. and WOODRUFF, G.N., Brit. J. Pharmacol. 54, 107-114 (1975).
12. FILIMONOFF, I.N., J. fur Hirnforschung 8, 1-23 (1965).
13. FONNUM, F. et al., J. of Neurochem. 29, 221-230 (1977).
14. FUXE, K., Acta Physiol., (Scand. Suppl.), 64, 243-252 (1965).
15. GURDJIAN, E.C., J. Comp. Neurol. 45, 249-281 (1928).
16. HASSLER, R. and CHUNG, J.W., Cell Tiss. Res., 168, 489-505 (1976).
17. HEIMER, L., Brain Behav., Evol. 6, 484-523 (1972).
18. HEIMER, L. and WILSON, R.D., In: Golgi Centennial Symposium, ed. M. Santini, Raven Press, pp. 177-193 (1975).
19. HEIMER, L., ROBARDS, M. and SWITZER, R., Exp. Brain Res., In: Afferent and Intrinsic Organization of Laminated Structures in the Brain, ed. O. Creutzfeldt, pp. 141-148 (1976).
20. HEIMER, L., Van HOESEN, G. and ROSENE, D., Int. J. of Neurol., in press (1978).
21. HEIMER, L., In: The Limbic System, eds. K. E. Livingston and O. Hornykiewiez, Plenum Press (1978).
22. HERRICK, C.J., Brains of Rats and Men, University of Chicago Press, Chicago, Illinois (1926).
23. JACOBS, M.S. et al., J. Comp. Neurol., 141, 205-271 (1971).
24. JOHNSTON, J.B., J. Comp. Neurol., 35, 337-481 (1923).
25. KAPPERS, A., In die physogenese des Rhinencephalons, des Corpus striatum und der Vorderhirnkommissuren, (p. 173-288), Folia Neuro-biologica, 1 Band, Leipzig (1908).
26. KEMP, J.M. and POWELL, T.P.S., Brain 93, 525-546 (1970).
27. KOIKEGAMI, H., Folia Psychiat. Neurol., 21, 151-180 (1967).
28. KRETTEK, J.E. and PRICE, J.L., J. Comp. Neurol., 178, 225-253 (1978).
29. LAUER, E.W., J. Comp. Neurol., 82, 215-254 (1945).
30. LORENS, S.A. et al., J. comp. physiol. Psychol., 73, 284-290 (1970).
31. LUNDBERG, P.O., Acta physiol. scand., 49, suppl. 171 (1960).
32. MACCHI, G., J. Comp. Neurol., 95, 245-305, (1951).
33. NAUTA, J.H. and HAYMAKER, W., Hypothalamic Nuclei: and fiber connections in: The Hypothalamus, (Haymaker, Anderson and Nauta, editors), Charles C. Thomas, Springfield, Illinois (1969).
34. NIMI, K. et al., Brain Research 21, 327-343 (1970).
35. PAPEZ, J.W., Comparative Neurology, Crowell-Collier Publishing Company, New York (1929).
36. PIJNENBURG, A.J.J. et al., Psychopharmacologia (Berl.) 41, 87-95 (1975a).

37. PIJNENBURG, A.J.J. et al., Psychopharmacologia, (Berl.) 41, 175-180 (1975b).
38. POWELL, E.W. and LEMAN, R.B., Brain Research 105, 389-403 (1976).
39. PYCOCK, C. and HORTON, R., Brain Research 110, 629-634 (1976).
40. ROLLS, E.T., Brain Research 31, 275-285 (1971).
41. SAPAWI, R.R. and DIVAC, I., Neuroscience Letters 7, 183-189 (1978)
42. SCOTT, J. W. and LEONARD, C.M., J. Comp. Neurol., 141, 331-344 (1971).
43. SMITH, O.C., J. Comp. Neurol., 51, 65-127 (1930).
44. STEVENS, J.R., Psychomotor epilepsy and schizophrenia: a common anatomy, In: M.A.B. Brazier (ed.) Epilepsy: Its Phenomena In Man, Academic Press, New York (pp. 190-214) (1973).
45. SWANSON, L.W. and COWAN, W.M., Brain Research 92, 324-330 (1975).
46. VAN HOESEN, G., MESULAM, M.M. and HAAXMA, S.R., Brain Research 109, 375-381 (1975).
47. VonECOMOMO, C.F. and KOSKINAS, N., die cytoarchitektonik der hirnrinde des erwachsenen Menschen, Wien und Berlin: Springer (1925).
48. WILLIAMS, D.J. et al., Brain Research, 130, 217-227 (1977).
49. WILSON, R.D., The neural associations of nucleus accumbens septi, Masters Thesis, M.I.T., Cambridge, Massachusetts (1972).
50. YETERIAN, E.H. and Van HOESEN, G., Brain Research 139, 43-63 (1978).
51. YOUNG, M.W., J. Comp. Neurol., 65, 295-401 (1936).

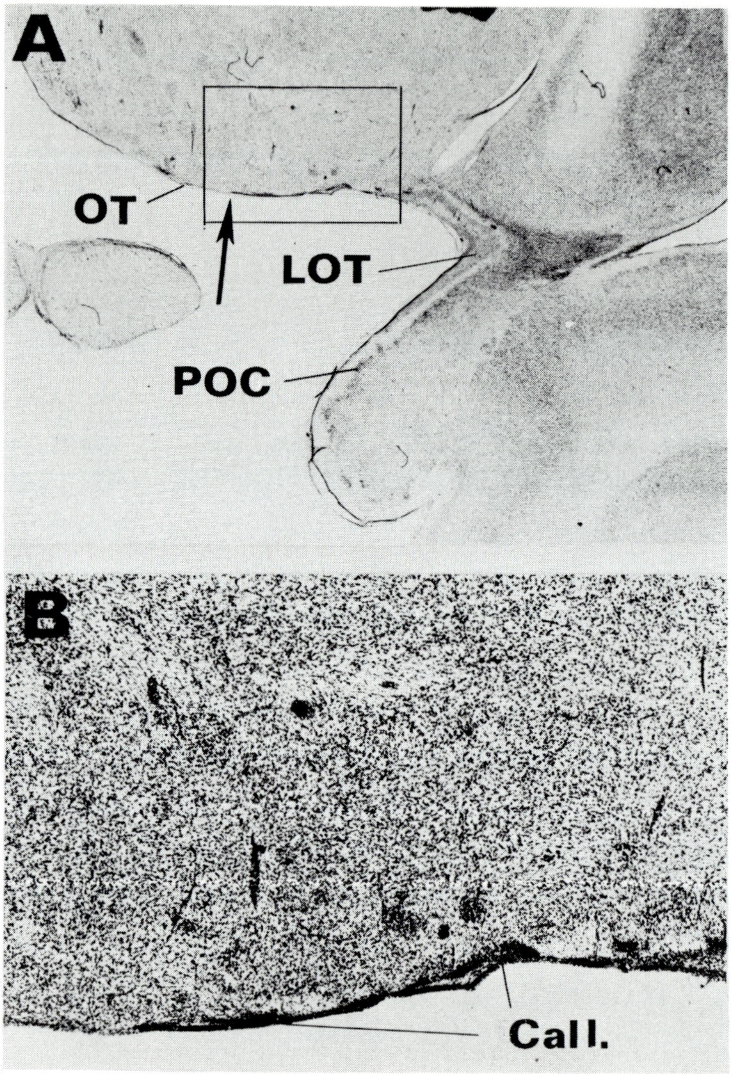

Fig. 1 Cross-section through the basal forebrain and rostral uncal cortex of the rhesus monkey. Note that the olfactory tubercle [OT] contains no conspicuous evidence of lamination. Lateral olfactory tract [LOT] fibers are distributed to the primary olfactory cortex [POC] and to the superficially located cells in the lateral part of the OT. Call = Islands of Calleja. (From Heimer 1978 with permission from Plenum Press).

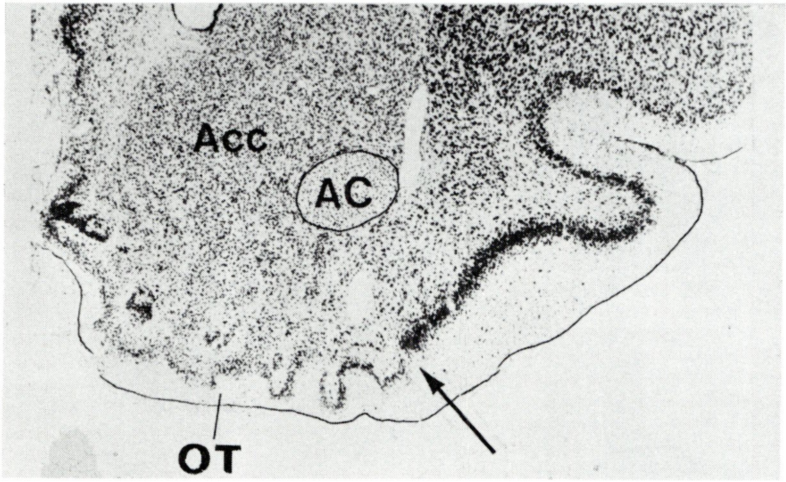

Fig. 2 Nissl stained frontal section through the rostral part of the basal forebrain in the rat. The arrow denotes the obvious junction between the primary olfactory cortex and the olfactory tubercle (OT). Note the absence of any clear boundary between the deep portions of the olfactory tubercle and the nucleus accumbens [Acc]. AC = anterior commissure. (From Heimer 1978 with permission from Plenum Press).

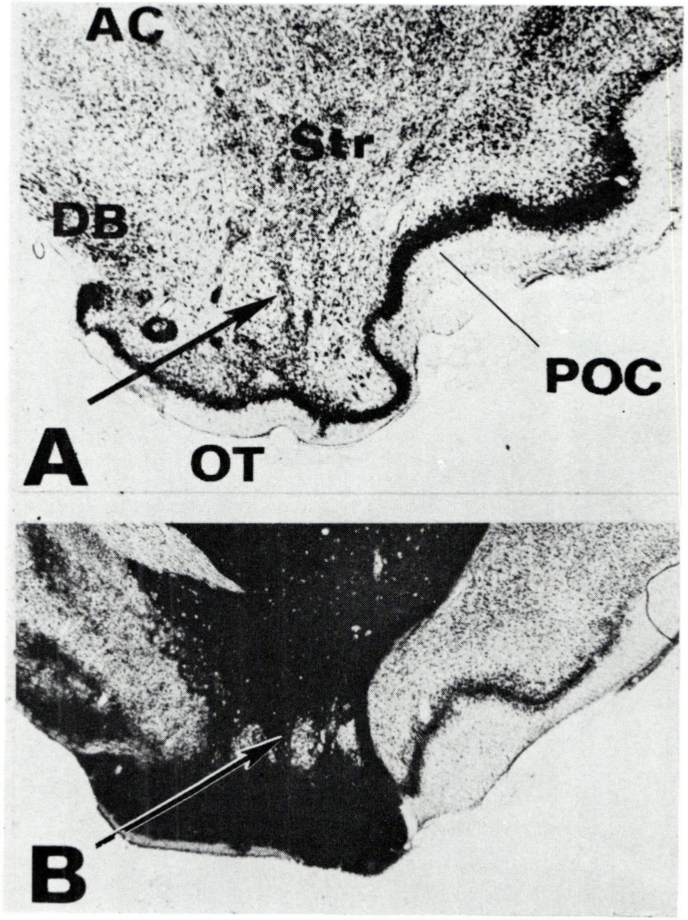

Fig. 3

Fig. 3 Nissl-[A] and AChE-[B] stained sections through the caudal part of the olfactory tubercle (OT) of the rat. Note the presence of "striatal" cell bridges (arrow) in the Nissl-section (A) and in the complementary AChE-section (B). The AChE section was counterstained with cresyl violet. (From Heimer 1978 with Permission from Plenum Press).

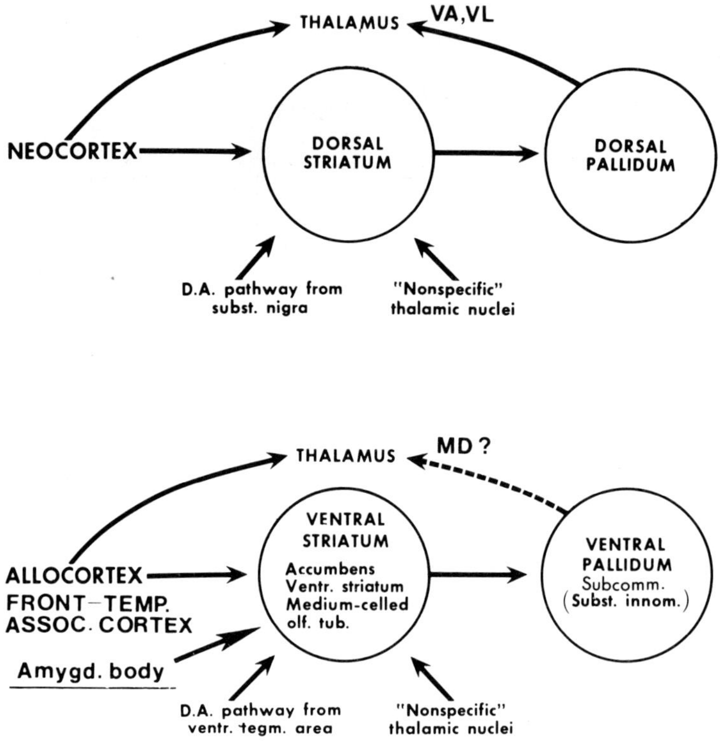

Fig. 4 Schematic figures depicting the major connexional relationships of the dorsal and ventral striatum in mammals. (From Heimer 1978 with permission from Plenum Press).

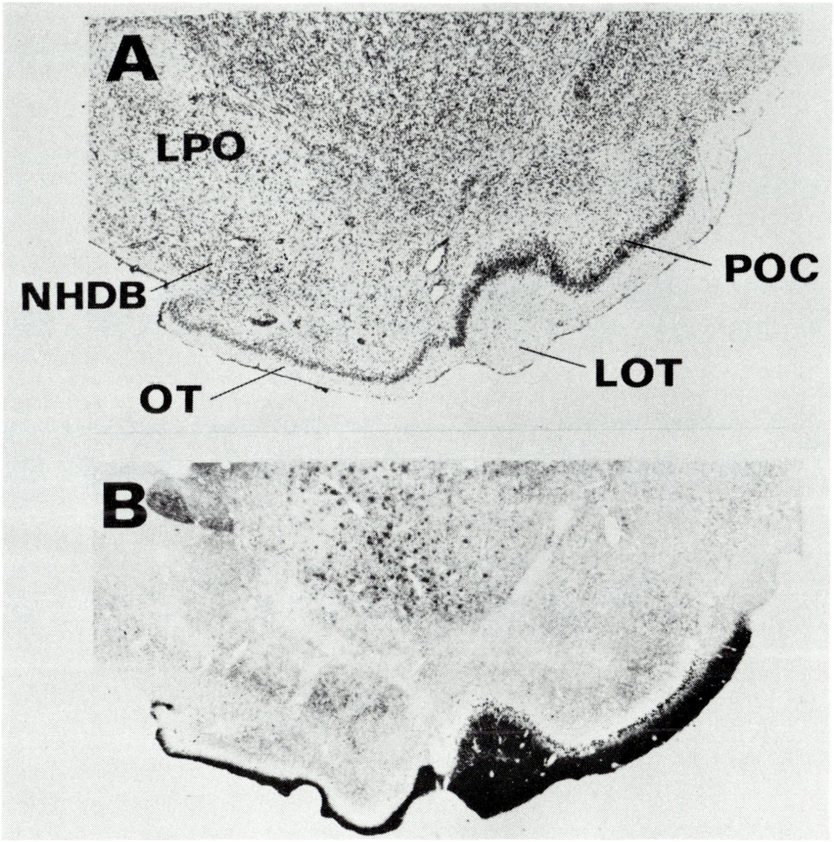

Fig. 5 Nissl-[A] and silver-[B] stained frontal sections through the olfactory tubercle [OT] and the primary olfactory cortex [POC] in the rat. Note the dense argyrophilia in the degenerating lateral olfactory tract [LOT], as well as in the olfactory projection areas of the OT medially, and the POC laterally. LPO = lateral preoptic area. NHDB = nucleus of the horizontal limb of the diagonal band. (From Heimer 1978 with permission from Plenum Press).

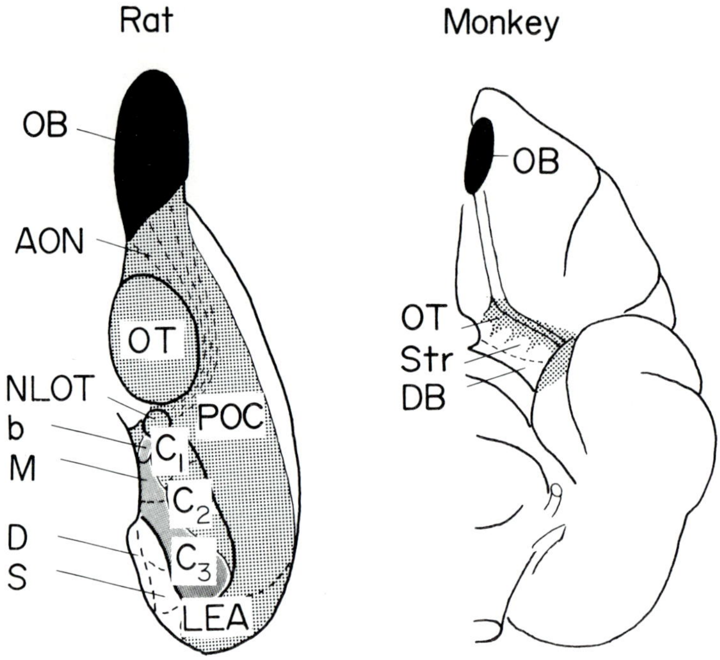

Fig. 6 The projection fields [stippled area] of the olfactory bulb projection fibers on the basal surface of the rat and the monkey brain. The dark stippling in the rat brain denotes the location of accessory olfactory bulb terminal fields. AHA = Amygdalo-hippocampal transition area; AON = Anterior olfactory nucleus; b = bed nucleus of the accessory olfactory tract; C_1, C_2, C_3 = Cortical amygdaloid nuclei; D = Dentate gyrus; DB = Diagonal band; LEA = Lateral entorhinal area; M = Medial amygdaloid nucleus; NLOT = Nucleus of the lateral olfactory tract; OB = Olfactory bulb; OT = Olfactory tubercle; POC = Primary olfactory cortex; S = Subiculum; Str = Striatum. (From Heimer 1978 with permission from Plenum Press).

Neurobehavioral Approaches

Introduction

I. Divac and R. G. Öberg

The behavioural roles of the NS have been studied with different methods of brain manipulation (mainly lesions and reversible electrical or pharmacological treatments) and by recording of neural events in behaving animals. Each of these techniques has serious shortcomings, not least when applied to a subcorical structure. Thus, both destruction of NS tissue and chemical and electrical interference with NS mechanisms tends to involve fibers of passage or, occasionaly, even adjacent cell groups. Gross recordings are similarly nonselective, while single-unit recordings may be hampered by sampling errors; either recording technique encounters the difficulties of correlating neural processes with any of many simultaneously occurring environmental and behavioural events. The research on the functions of NS has had its share of poorly controlled experiments and prejudged conclusions, many of which were awarded a surprising longevity in the literature. Yet, in the hands of careful investigators, these techniques have also yielded some valuable information, some of which has recently been substantiated by the superior (but still imperfect) techniques of inactivation by cooling, lesioning with selective neurotoxins, 2-DG tracing of neural activation, and microstimulation.

The aim of the following section was to review the dependable behavioural data obtained with each of the main treatment and recording techniques. Regrettable, Dr. Ungerstedt, who reviewed the pharmacological approach to the study of NS functions, did not find time to submit a chapter to this book.

Activity of Neurones in the Neostriatum and Related Structures in the Alert Animal

E. T. Rolls, S. J. Thorpe, S. Maddison, A. Roper-Hall, A. Puerto and D. Perret

University of Oxford, Department of Experimental Psychology, South Parks Road, Oxford OX1 3UD, England

To investigate the function of the neostriatum, the activity of single neurones has been recorded in the caudate nucleus and putamen of the alert animal in a number of studies. In the review of these investigations which follows, it will become clear that analysis of neuronal activity in the neostriatum of the alert animal is particularly appropriate, because although the activity of some neostriatal neurones is not unconditionally linked to sensory or motor events, such neurones may nevertheless be involved in the performance of particular tasks by the animal. Tasks in which the activity of neurones in the neostriatum has been investigated include the initiation and performance of movements. One reason for this is that anatomically the neostriatum projects to motor structures in the globus pallidus and substantia nigra, which in turn influence the thalamus and thus the motor cortex. Additionally, damage to the basal ganglia has been associated clinically with disorders of movement, and in both man and animals damage to the nigrostriatal system which depletes the neostriatum of dopamine produces a failure to initiate movement normally. The activity of neurones in the globus pallidus and substantia nigra will also be considered in this review, because these structures receive the major efferent projections of the neostriatum and provide evidence on its function.

1. CAUDATE NUCLEUS AND PUTAMEN

a) Movement-related activity

De Long (1972) recorded unit activity in the putamen in monkeys trained to make arm movements, and found some units which discharged in relation to these movements. To investigate Kornhuber's (1971) suggestion that the basal ganglia function primarily in the generation of slow, "ramp" movements, whereas the cerebellum functions largely in the pre-programming and initiation of rapid, "ballistic" movements, De Long (1973) and De Long and Strick (1974) recorded from neurones in the putamen and other structures during ramp and ballistic arm movements. For the ramp movement the monkey slowly pushed or pulled a lever between two 1 cm zones separated by a distance of 5 cm. For the ballistic movement the monkey rapidly moved the lever across the distance separating the zones and was permitted to hit against a distant mechanical stop. Movement times between the zones had to be >600 msec for the ramp movement and <140 msec for the ballistic movement in order to obtain a juice reward. A unit in the putamen with activity which was particularly related to the ramp movement rather than to the ballistic movement is shown in Fig. 1. Of 187 units in

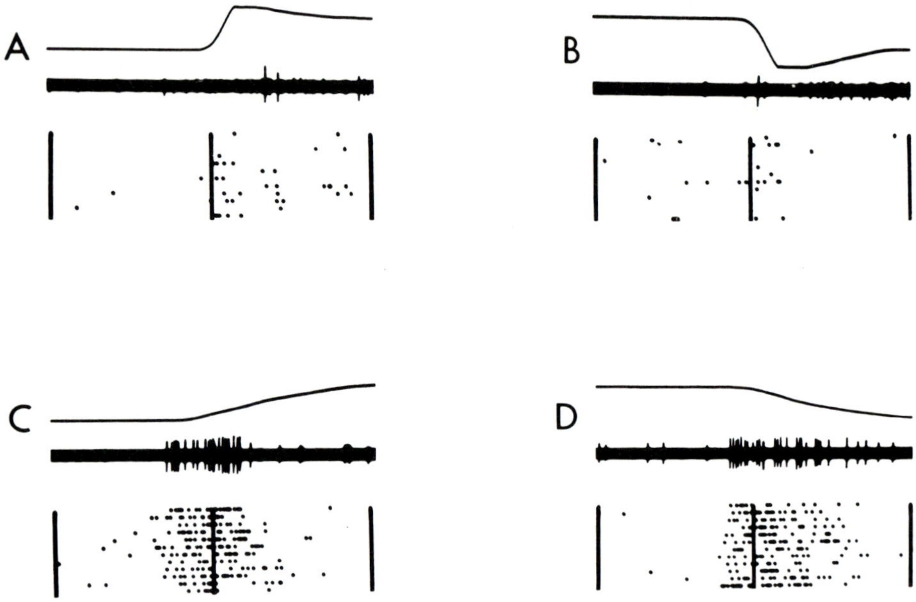

Fig. 1

Activity of a ramp-related unit from the putamen during pushing and pulling ballistic (A and B) and ramp (C and D) movements. For A-D the upper trace represents the position of the lever during a single trial; the middle trace, the unit discharge for the same trial; and the lower portion, the activity of the unit during 12 successive trials shown in raster form. Each trial is aligned on the time of leaving the zone (centre bar). The interval from the center to the margin of the raster is 1 sec. The unit discharged strongly in relation to the ramp movements (C and D) but only weakly in relation to ballistic movements (A and B). (Reprinted with permission from DeLong, M.R. and Strick, P.L., 1974, Brain Research, 71, 327-335)

the putamen related to these movements, approximately 45% were preferentially ramp-related, less than 10% were related preferentially to the ballistic movement, and the remainder were related to both movements. For comparison, in the globus pallidus approximately 17% of the units with activity related to these movements were preferentially ramp-related. The corresponding figure for the cerebellum was less than 3%, and in the motor cortex no units were found which discharged preferentially in relation to ramp movements as opposed to ballistic movements (De Long and Strick, 1974). These findings are consistent with the hypothesis that there are neurones in the putamen involved in the control of movement, and particularly in the generation of slow movements rather than rapid movements. Nevertheless, the fact that many of the movement-related neurones fired in relation to both the ramp and ballistic movements, was taken as evidence that the putamen is not solely involved in the control of ramp movements. It was noted in this work that the firing of at least some neurones in the putamen was associated with activity in the axial musculature. It is possible that the activity of at least some of the ramp-related neurones in the putamen resulted from postural adjustments made by the monkey. Activity related to posture has also been described by Anderson (1978), who reported that a small number of neurones in the putamen fired in relation to postural adjustments made to maintain the position of the head when the monkey's chair was tilted. In another study Dolbakyan, Hernandez-Mesa and Bureš (1977) found neurones in the caudate nucleus of the rat which altered their activity just before the rat made a forelimb movement, and surmised that the neurones were involved in the control of postural reactions stabilising the body of the rat in front of the feeder to which the forelimb movement was made.

Movement-related neuronal activity has been found in the caudate nucleus of the monkey by Buser, Pouderoux and Mereaux (1974, 1975). Three types of upper limb movement were studied, (a) reaching for a peanut and bringing it to the mouth, (b) finger and wrist movements involved in pressing a lever to obtain fruit juice, and (c) arm and shoulder movements involved in (isotonic) pulling on a string to lift a weight to obtain fruit juice. Of 460 cells recorded, approximately 40% altered their activity in relation to one (or more) of these movements. The alteration was usually an increase in firing rate of 30 to 100% from a relatively low baseline rate of less than 10 spikes/sec. In task (a), 21% of the caudate neurones recorded had activity related to grasping with the fingers and flexion of the arm. A further 6% of the neurones had activity which increased during the chewing. For the lever-pressing task (b), 43% (39/90) of the neurones recorded altered their activity in relation to the movements made by the distal parts of the limb. Approximately 15% of the neurones recorded altered their activity in relation to the movements made by the arm and shoulders in the lever-pulling task (c). Passive manipulation of the limb failed in the several cases in which it was tested to affect the activity of cells which fired during active movement. It has also been noted by Matsunami and Cohen (1975) that passive limb movements generally failed to affect neurones in the caudate nucleus.

Niki, Sakai and Kubota (1972) have also found movement-related neuronal activity in the caudate nucleus of the monkey, but only in a small proportion of the neurones recorded. In 14 caudate neurones of thousands tested, the neuronal activity occurred in relation to arm movements made in delayed spatial alternation and single alternation performance. The neuronal activity sometimes preceded the onset of EMG in the triceps brachii by 200 msec (but this was not necessarily the first EMG produced in the task). No units were found which had activity related to the delay part of the delayed spatial alternation, and this is different to neuronal activity in the prefrontal cortex (Kubota and Niki, 1971). This difference is of interest in that lesions of the prefrontal cortex or the part of the caudate to which it projects both impair performance on a spatial delayed alternation task (Divac, 1972). This suggests that either the caudate nucelus has a different function in the task to the prefrontal cortex, or that the deficit after caudate lesions was due to the interruption of other pathways passing through the caudate nucleus.

In recordings made in the caudate nucleus and putamen of the monkey, we have also found that the activity of some neurones is related to movements made by the monkey, although it is usually more difficult than with neurones in the globus pallidus to demonstrate a strong relation between neuronal activity and a particular movement. An example of the activity of a caudate neurone with movement-related activity is as follows. The neurone had a slow spontaneous firing rate of 2 spikes/sec, and increased its rate when the monkey extended his contralateral arm to depress an illuminated panel (Fig. 2). The unit responded during extension of the arm to reach the panel, or during pressing the panel with the fingers if the arm was

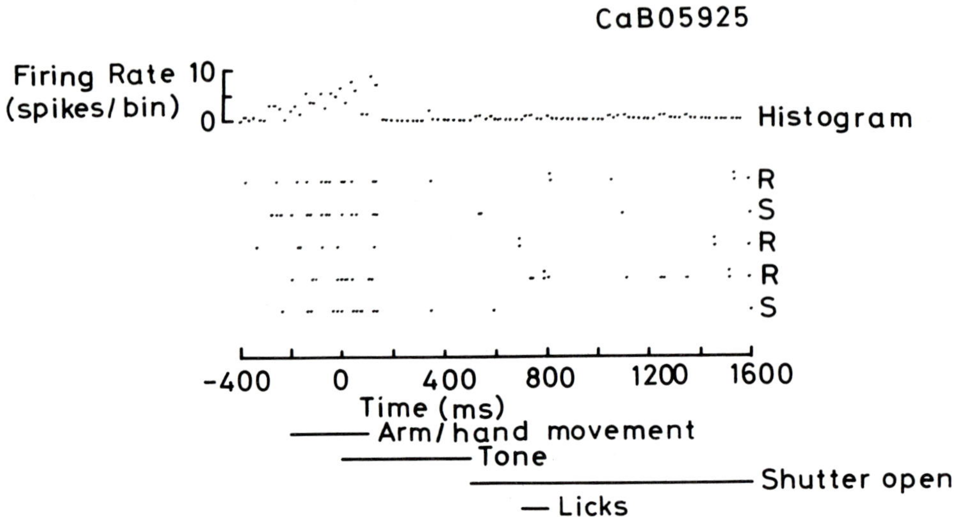

Fig. 2

Activity of an arm-movement related unit in the caudate nucleus. On each trial the monkey extended his arm and pressed an illuminated panel (at time 0 in the diagram) to start the trial. The activity of the unit was associated with this movement. The panel press started a 500 msec signal tone after which a shutter in front of the monkey opened to reveal one of two stimuli. If it was a food-related visual stimulus, the monkey could lick (see double dots in the dot displays) to obtain fruit juice (R trials). If it was a saline-related visual stimulus, the monkey obtained aversive hypertonic saline if he incorrectly made a lick (S trials). The activity of the unit was not related to the visual discrimination part of the task.

already extended. The response illustrated was related to movement of the contralateral arm, and similar but smaller responses were obtained with movements of the ipsilateral arm and hand. The activity of the neurone was not affected by other aspects of the task, which included the sounding of a signal tone for 500 msec at the press of the panel, followed by the opening of a shutter in front of the animal who then performed a visual discrimination as shown in Fig. 2. Of our present total sample of 307 quantitatively studied and anatomically verified neurones in the caudate nucleus and putamen, 21 (6.8%) had activity which was related to movements of the hand, arm, or mouth, or to general body movements. The activity was often not related only to one particular movement (e.g. the unit in the example above fired to both arm and hand movements), and in many cases the responses were relatively small changes in firing rate, compared to the large changes in firing rate of pallidal neurones (see below).

Thus neurones with activity related to movements have been found in the caudate nucleus and putamen of the alert monkey. Although the activity of some neurones is particularly related to slow, ramp movements, this could be because of postural adjustments, since these affect the activity of some neostriatal units. Further, the exact relation of the activity of neostriatal neurones to the movements studied has been difficult to analyse. For instance, De Long and Strick (1974) noted that units in the putamen often fired before movement of a lever could be detected, but were not able to demonstrate the timing relations between neuronal activity and muscle responses because of the difficulty of relating neuronal activity to activation of a particular muscle involved in the movement.

b) Activity related to sensory stimulation

Sensory stimulation can also produce responses in some neostriatal neurones. For example, Buser et al. (1974, 1975) noted that some caudate units which were not influenced by the movements being studied altered their activity when noises such as the voice of the experimenter occurred in the laboratory, or when a reward such as fruit was shown to the monkey. Buser et al. also thought that the signal used to start a trial, a light, might influence the activity of some caudate neurones. Vinogradova and Sokolov (1975) described units in the caudate nucleus of the unanaesthetised rabbit with responses to flashes and clicks. They noted that these units often showed habituation and dishabituation, or sometimes enhancement of the response with repeated presentation of the stimuli.

In our sample of 307 units recorded in the head of the caudate nucleus and the putamen of the alert monkey, a number of neurones with responses classified as sensory were found. For example, 8 units had visual responses which occurred both when an object such as food was shown directly to the monkey and also in a test for visual responses in which a shutter opened to reveal the same object. If a unit responded in only one of these tests, as was the case for many neostriatal units, then such a unit was not classed as having a purely sensory response, in that other factors such as the performance of the monkey must have been important. The latencies of the visual responses to these stimuli were 100 msec or longer, measured from the time the shutter opened. Similarly, 6 units responded unconditionally to auditory stimuli in that they responded both to the noise of the signal tone or the shutter opening in the shutter discrimination shown in Fig. 2, and also to noises such as clicks made in the laboratory. The latencies of these auditory responses were usually between 20 and 30 msec. A further two units were classified as polysensory, having responses for example associated both with the sight of a squeeze bulb from which air was blown gently onto the monkey's face and with the noise made by the squeeze bulb. Thus in our study relatively few (16/307 or 5.2%) neostriatal units were classified as having simple unconditional auditory or visual sensory responses which occurred independently of the performance of the animal and of the situation in which the stimuli were presented. Many more units which did respond in association with stimuli or events which were relevant to the performance of a particular task, but did not respond to the same stimuli outside the task, were found as described below.

c) Other types of response of neostriatal neurones

Although the activity of some units in the neostriatum is related to movements, and of others to sensory stimulation independently of the situation in which the stimulation is given, the activity of many others is related to other aspects of an animal's behaviour.

For example, Soltysik, Hull, Buchwald and Fekete (1975) have found that the activity of some neurones in the caudate and putamen was related to the performance of a simple delayed response task in the monkey. In the task, a light came on to start a trial, and if the monkey waited for a delay of more than 6.5 sec before pulling a response lever by hand, a fruit-juice reward was obtained. Neurones were found which fired in relation to different aspects of the task, for example to the light onset, during the delay period, in association with the behavioural response, and when the pulses of fruit juice were delivered. Seventy-three percent of the caudate neurones recorded, and 56% of the neurones recorded in the putamen, fired in relation to one or more aspects of the task, but it is not easy to determine to which part of the monkey's performance the activity was related. Relatively gradual changes in neostriatal unit firing which could not be related to movements of particular muscles were sometimes observed, and it was speculated that these changes could be related to the enabling of classes of responses (Soltysik et al., 1975).

In our investigations we have recorded the activity of single neurones in the caudate nucleus and putamen while monkeys initiate behavioural responses such as feeding. One reason for investigating the initiation of feeding is that damage to the nigrostriatal system which depletes the neostriatum of dopamine results in a failure to initiate behavioural responses, in that lesioned animals show catalepsy, a failure to orient to stimuli, and a failure to initiate feeding and drinking (Marshall, Richardson and Teitelbaum, 1974). A second reason is that Buser et al. (1974, 1975) have reported that some caudate neurones responded when food was shown to the monkey, and we wished to compare the activity of these neurones with that of neurones in the lateral hypothalamus and substantia innominata which have responses associated with the sight of food (Rolls, Burton and Mora, 1976; Rolls, Sanghera and Roper-Hall, 1978; Rolls, 1978). The methods were similar to those used for the analysis of the activity of hypothalamic neurones described in detail elsewhere (Rolls et al., 1976, 1978). Briefly, the activity of neostriatal neurones was measured in two situations. The first was a "clinical" testing situation in which the animal was shown food and fed by the experimenter, allowing the experimenter to obtain preliminary information about the factors which influenced the firing rate of the neurone. The second was a visual discrimination task of the type shown in Fig. 2 in which the time course of neuronal responses could be determined (with a PDP-11 computer providing peristimulus time histograms) relative to the different aspects of the task which included a tone which signalled the start of the trial, the sight of the stimulus which indicated that a feeding response would be rewarded, and the lick movement itself. It was found that many neurones in the neostriatum responded in one of the test situations but not in any simple comparable way in the other, so that the responses were situation-dependent. For this reason, relatively few neostriatal neurones were classified in the categories described above, as being movement-related or as having a simple sensory response independent of the situation.

Firstly, neurones with activity which increased after the monkey was shown food, or in some cases as the food was brought towards the monkey's mouth, were found (see example in Fig. 3). Comparable responses did not occur after the monkey was shown non-food objects (see Fig. 3). It was found that 13 of the 15 neurones in this category failed to respond either to the food-related visual stimulus in the visual discrimination task, or to food objects shown through the shutter. It is interesting to compare these units with the units in the lateral hypothalamic area which show maintained responses to the sight of food in the clinical feeding situation and also respond to the sight of food-related stimuli in the shutter-controlled visual discrimination. This difference in response properties in the two areas indicates that the neostriatal neurones have a different function to the hypothalamic neurones. The activity of the neostriatal neurones could be related to responses which the animal makes only in the clinical feeding situation, or could be complex sensory responses which depend on the testing situation.

Secondly, a considerable proportion (54 of 307, or 17.6 %) of neostriatal units were classified as task-related, in that they showed distinct time-locked changes in activity during the visual discrimination task, but could not be placed in the movement-related or unconditional sensory classes described above, or in the other classes described below. An example of the activity of one of these units is shown in Fig. 4. In Fig. 4A it is shown that the neurone increased its firing rate approximately 160 msec after the shutter opened on both reward (R)

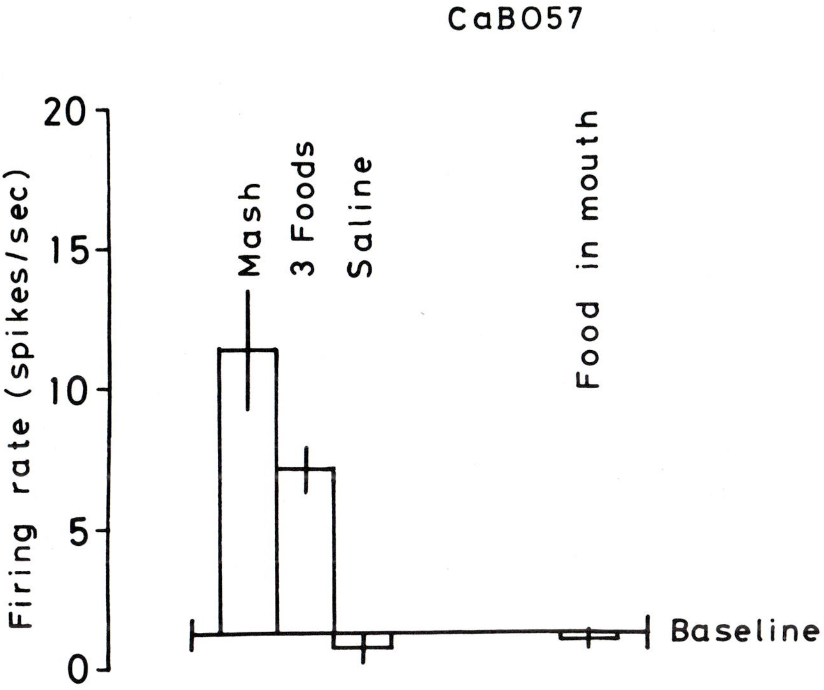

Fig. 3

Firing rate (mean and S.E.M.) of a neurone in the caudate nucleus when mash, 3 different foods, and aversive hypertonic saline were shown to and brought towards the monkey's mouth for feeding. The lack of response associated with food in the mouth is also shown.

and saline (S) trials, but whereas on reward trials there was a further increase of firing rate at approximately 250 msec, on saline trials the unit stopped firing at about the same time. Fig. 4B illustrates that when a 500 msec tone signalled the start of each trial, the unit had already increased its rate before the shutter opened (at about 100 msec after the tone started, although this is not shown in the diagram). It should be emphasized that this unit, and the others in this category, were not found to have responses in the clinical situation which were associated with particular movements of for instance the mouth, or sensory events.

Of the 54 units classified as task-related, 16 had activity (generally an increase in rate) which preceded the opening of the shutter. This was either when a tone was used to signal the start of a trial, or, as in some series of trials, when the animal was allowed to initiate a trial himself by depressing an illuminated panel. None of these units whose activity changed during the tone period responded with latencies of less than 100 msec, so it was unlikely that these responses were simple sensory responses to the sound of the tone. The possibility that these responses are simple sensory ones is made even less likely by the fact that if no tone cue was present, a number of these units showed increases in response to the shutter opening. Not all task-related units responded in this fashion. Some 38 of the 54 units classed as task-related showed no increase in activity during the tone period, but changed their activity (generally by

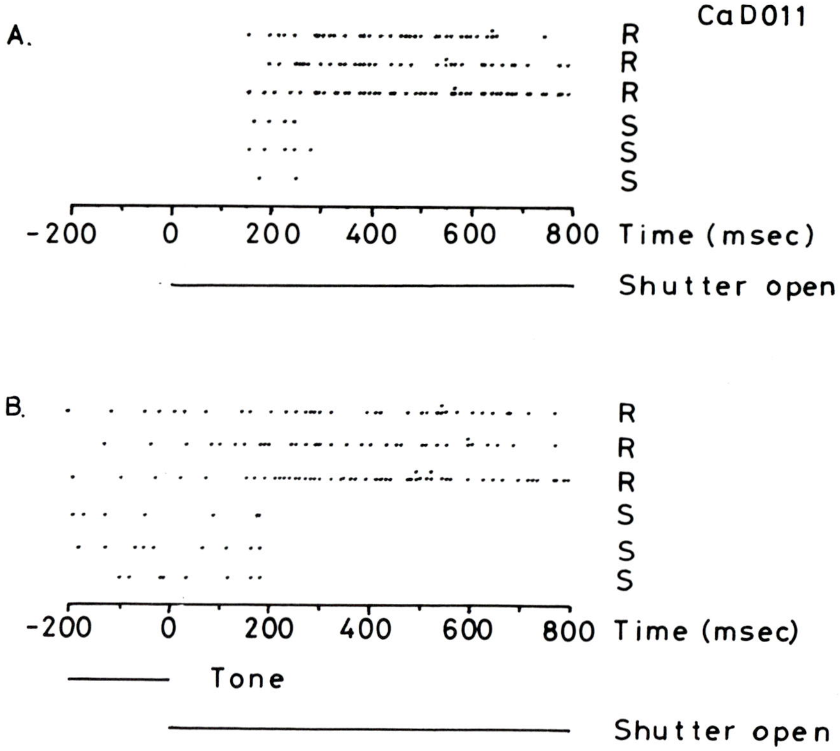

Fig. 4

Activity of a caudate unit during visual discrimination under two conditions. A. The opening of the shutter (at time zero) was unsignalled. B. The shutter opening was preceded by a 500 msec tone cue. Reward (R) and saline (S) trials are as described in Fig. 2.

increasing their firing rate) after the shutter had opened. Some of these units responded only if the monkey could see the shutter opening, and the associated response latencies were usually 100 msec or more. Other units responded to the noise of the shutter opening (in that a response occurred if the shutter was blocked from view), generally with latencies of 30-50 msec.

The responses of many units in this category were closely related to the animals' performance, as shown by the following examples. First, the neuronal and behavioural response latencies often became shorter relative to the shutter opening if the start of the trial was cued with a tone, or a self-start procedure was used. Second, some of the units fired only while the monkey was actively engaged in the performance of each trial, that is from the first cue (e.g. the tone if present, or the shutter opening) until lick responses were made, or until the monkey had prevented making a response on saline trials (see e.g. Figs 4 and 6). Conversely, other units fired steadily until the first cue for the start of a trial, and remained quiet until the end of the trial. Clearly, such units could not be responding simply to sensory events occurring during the task. Although the activity of these units could simply result from movements made by the animal during the task, the observation that their activity was usually not influenced when the animal made a wide range of movements in the test situation, suggests that their activity was not simply motor. Rather, these neurones may be involved in enabling the animal to make responses, and this possibility is considered below, in the Synthesis.

A small number of neurones responded differentially in the visual discrimination on reward as compared with saline trials. Examples of such differential responses are illustrated in Figs 4 and 6. The differential responses of such neurones occurred with latencies of 150-400 msec after the shutter opened, but it was characteristic of these units, and especially those with short latencies, that either their differential responses were not maintained for more than a few blocks of 15 trials, or were not well correlated with the animal's responses. This is in contrast with the consistent differential responses with latencies of 150-200 msec of hypothalamic food-related neurones in the same situation (Rolls et al., 1978).

Thirdly, another class of units (41 or 13.4% of the total sample) was unresponsive in the visual discrimination task, or when other objects were presented through the shutter, but responded during clinical testing. For example, 25 units responded when the animal was shown various objects, with most excited but several inhibited. As described above, some of these only responded when food was shown (e.g. Fig. 2) while the others responded to both food and non-food objects.

Other units changed their activity in relation to other events. For example, 5 units increased their firing rate when objects were removed from close to the animal, and others responded in situations similar to those described previously by Buser et al. (1974, 1975), for example when an experimenter moved in the laboratory, or the door was opened.

Fourthly, a relatively large population of neostriatal neurones (34 of the sample of 307 neurones, or 11.1%) responded in a previously undescribed way. All these neurones increased their firing rate when the experimenter reached to a tray out of the sight of the monkey in order to obtain a food or non-food object, which was then presented to the animal, and fed to him if it was food. All of these neurones responded either when the experimenter reached towards the tray, or when he brought his arm back, or to both movements. Many of these neurones fired as the monkey saw the experimenter's arm emerging from behind the screen which hid the tray, and stopped firing as soon as the monkey saw what was in the experimenter's hand (Fig. 5). As it appeared that these neurones responded to cues which predicted the presentation of food or non-food objects, we were very interested to determine whether these neurones would respond to cues which predicted other events such as the start of a trial in the visual discrimination. Twelve of the 34 units which responded to pre-presentation arm movements also responded (with an increase of rate for all except 1 unit) during the tone period which preceded the shutter opening in the visual discrimination. Figure 6 shows an example of a unit which increased its firing after the shutter opened if no tone was present (Fig. 6A), but which responded before the shutter opened if a tone signalled the start of a trial. The neurone shown in Fig. 6 also responded before the tone if the animal could reach for a panel to start the trial himself (Fig. 6C).

A further 8 of the 34 units did not respond to the tone, but increased their firing rate after the shutter opened, an event which presumably could also act as a cue that a discrimination was now required. This possibility receives support from the responses of the neurone illustrated in Fig. 7. The neurone responded with a 200 msec burst of firing starting approximately 120 msec after the opening of the shutter (top 4 trials, unsignalled start), which was not signalled by a tone in this series of trials. However, if the monkey was able to initiate a trial himself by pressing an illuminated panel, but otherwise the discrimination remained

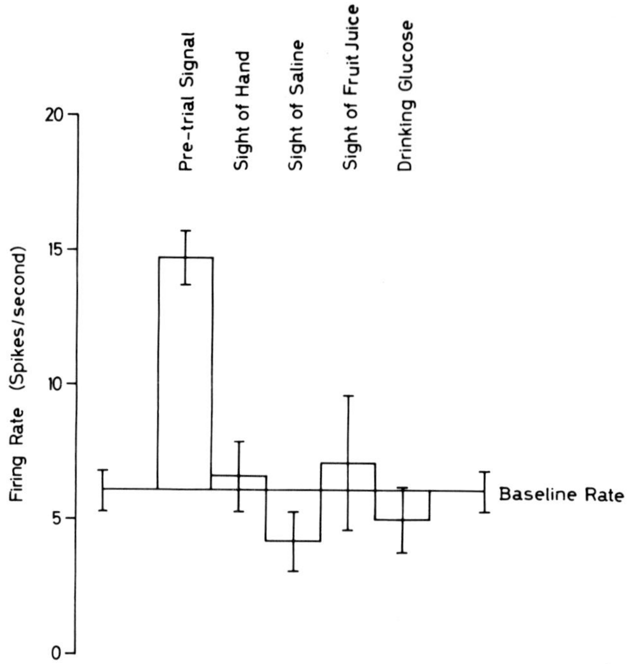

Fig. 5

The activity of this caudate neuron was associated with a signal which immediately preceded each trial, but there was no response during the trial to the sight of the syringe from which the monkey was fed hypertonic saline, of another, different, syringe from which the monkey was fed fruit juice, or of the empty hand, nor was there any neuronal response associated with drinking. The pre-trial signal in use was simply the movement of the experimenter's arm as it moved to bring the food or non-food syringe from behind a screen.

the same, there was no neuronal response associated with the shutter opening, now that in the self-started task the shutter did not provide the only cue about the start of the trial (lower 8 trials of Fig. 7, self-start). This suggests that the unit did not respond to the shutter opening per se, but only if it was important as a cue for the start of a trial.

Thus some of the neurones which responded to pre-presentation arm movement cues also responded in a different situation when other cues indicated the start of a visual discrimination trial. It is possible that these neurons respond to particular stimuli because they are important as cues to the animal. Consistent with this possibility, it was found that some neurones which initially did not respond to arm movements in directions other than towards the food tray, did start to respond to movements in another direction if these were regularly followed by the presentation of food. However, other neurones did not show such changes in responsiveness, and thus would not respond to all cues which predicted the presentation of food. This conclusion, that the neurones would not fire simply because the animal expected food, is supported by the observation that each neurone appeared to respond to different aspects of the stimulus situation. For example, some of the neurones responded only when the experimenter reached to the tray, whereas

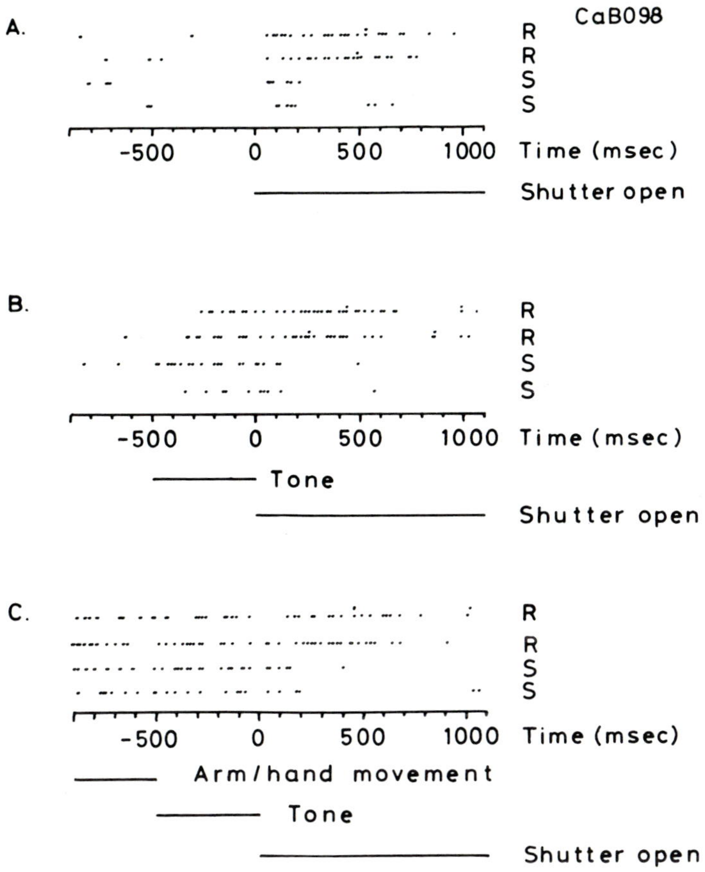

Fig. 6

Activity of a unit responding to pre-presentation arm movements during visual discrimination under three conditions. A. The shutter opening (at time zero) was unsignalled. B. The shutter opening was preceded by a 500 msec tone cue. C. The animal was allowed to initiate the trials himself by pressing an illuminated panel which started the 500 msec tone. Reward (R) and saline (S) trials are as described in Fig. 2.

others responded only when he brought his arm back, and others were strongly excited only during the last small hand-movement before the object became visible round the edge of the screen. Other neurones responded when the experimenter merely looked in the direction of the food tray as if he was about to select an object to show to the monkey. Further, whereas most of the neurones were direction-specific in that they responded only when arm movements were made in the direction of the food tray, some responded to arm movements made to other locations. Additional evidence for specificity has been described above, in that in the visual discrimination task, some of the neurones responded to the tone cue, some to the shutter opening, and some to neither of these cues. Because in a given situation the responses of individual neurones show great diversity, it is unlikely that the activity of these neurones simply reflects particular movements made by the animal. Further, it

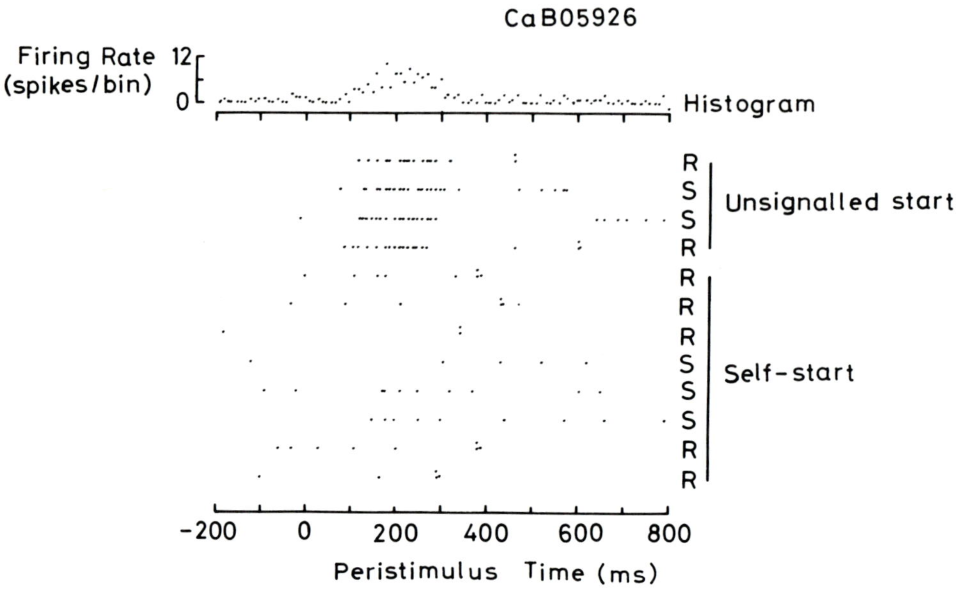

Fig. 7

Activity during visual discrimination of the caudate neurone illustrated in Fig. 5. Its firing rate increased approximately 120 msec after the shutter opened at time 0 on trials on which the shutter opening was unsignalled (upper 4 trials). If the monkey initiated each trial by pressing an illuminated panel to start the trial (self-start, lower 8 dot displays), then there was no neuronal activity correlated with the task. Reward (R) and Saline (S) trials are as described in Fig. 2. The histogram is the sum of 12 trials.

should be emphasized that, although these neurones responded to cues which predicted the presentation of the object, most stopped firing when the animal saw the object, regardless of whether it was food or not. Therefore these neurones did not fire simply because the animal expected food or reward. Rather it appears that individual neurones provide information about a great variety of events which are directly relevant or potentially relevant to the initiation of behaviour (see Synthesis below).

2. GLOBUS PALLIDUS

Because the globus pallidus is a major recipient of efferent projections of the caudate nucleus and putamen the activity of neurones in the globus pallidus of the alert animal provides information on the output of neostriatal neurones, and is therefore considered here.

Most investigators agree that the activity of at least some neurones in the globus pallidus can be related to particular movements made by the animal. For example, De Long (1971) found that 19% of a sample of 551 pallidal units in the monkey discharged in consistent temporal relation to rapidly alternating movements of the arm. Neuronal activity was usually related to movements of the contralateral arm. Other neurones with activity related to leg movements were found. Also, Anderson (1978) has described neurones in the globus pallidus with activity related to postural adjustments made by monkeys. Rolls, Burton and Mora (1976) and Burton, Rolls and Mora (1976) found that pallidal units which altered their activity during feeding in the monkey had activity related to particular movements of for example the mouth or arm. Thus, for example some pallidal neurones altered their firing rates whenever the monkey drank different fluids such as isotonic glucose or saline, or water, independently of the taste of the solution, provided only that mouth movements were made. Similar responses were obtained when mouth movements were made to a syringe which was empty instead of containing fruit juice. It was also found that the responses of such pallidal neurones occurred whenever glucose or sucrose solutions were ingested, even after the monkey was fed to satiety (Burton, Rolls and Mora, 1976). This provides further evidence that the neuronal responses were movement-related, rather than reward-related, in that the responses occurred independently of whether the monkey would work for the sweet solution, and whenever the syringe from which the monkey was fed was introduced into the mouth for feeding. During the investigations it often seemed possible that sensory, e.g. tactile, inputs from the mouth were involved in the neuronal responses obtained, but this was usually difficult to test in the alert monkey. Lidsky, Buchwald, Hull and Levine (1975) have found that the activity of units in the globus pallidus of the cat occurring during ingestion was often phasically related to pulses of milk delivered to the mouth. At least some of these cells were also influenced by tactile stimulation of the mouth, and sensory (i.e. tactile) rather than purely motor aspects of the ingestion were held to account for the changes in firing rate. Lidsky et al. also found that altering the reward value of the fluid by changing it to water, or by testing the cats when satiated, did not influence the responses obtained, in agreement with the view that these were sensory or motor responses and were not related to motivational state.

Sparks and Travis (1973) have suggested that the activity of some pallidal neurones in the squirrel monkey is related to movements made to obtain food, and does not occur when the same movements are made to obtain non-food objects (see also Travis and Sparks, 1967, 1968; Travis, Hooten and Sparks, 1968; Sparks and Travis, 1968; Travis, Sparks and Hooten, 1968). For example, 31 of 97 pallidal units decreased their firing rate during food searching movements in a food hopper, during food grasping and transporting movements, or during actual consummatory (eating) behaviour (Travis, Hooten and Sparks, 1968). These units were found not to be responsive to elicited or spontaneous motor patterns involved in obtaining objects other than food. De Long (1971) in contrast found movement-related units in the globus pallidus, and suggested that the units which responded in association with food reward were aberrant cells of the substantia innominata lying within the internal or external medullary laminae, or bordering the globus pallidus. Sparks and Travis (1973) agreed that many of their reward-related cells were border cells, but held that others were within the globus pallidus. In our studies, we have typically found units associated with the sight of food reward in the region shown in Fig. 8, which includes the substantia innominata, and is ventral to the globus pallidus (see Rolls, Sanghera and Roper-Hall, 1978). Usually the food-related neurones are encountered immediately below the rapidly firing, closely packed cells of the globus pallidus (Rolls, Burton and Mora, 1976). Nevertheless, reward-related neurones are sometimes found higher than this (see one example in Fig. 8). Whether these neurones are pallidal neurones, or are aberrant cells of the substantia innominata, we do not know, but suggest that use of intracellular dye injection, perhaps in combination with an acetylcholinesterase technique to indicate neurones of the substantia innominata, could resolve the point.

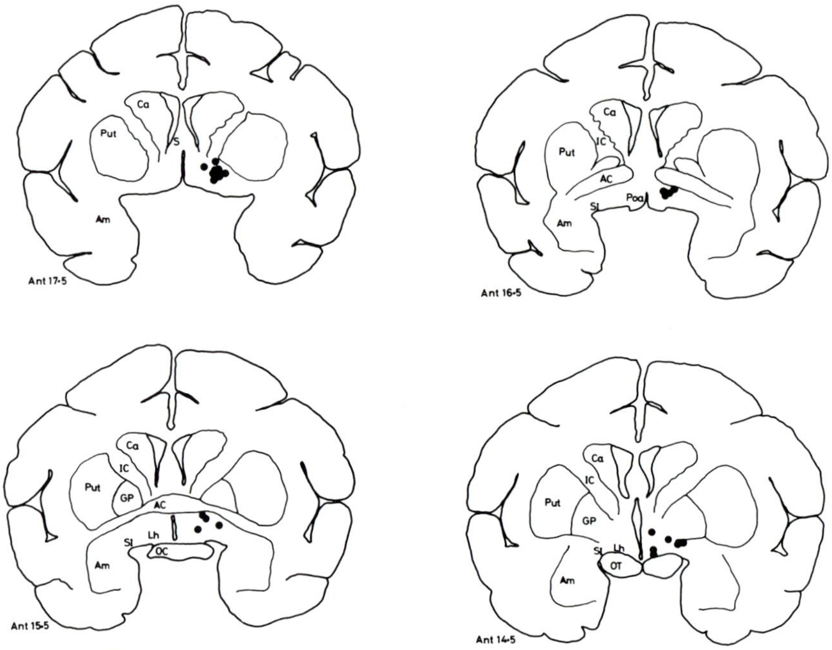

Fig. 8

Sites of feeding-related neurones recorded mainly in the hypothalamus and substantia innominata.

3. SUBSTANTIA NIGRA

The globus pallidus and substantia nigra together receive the main efferent projections of the caudate nucleus and putamen. Therefore the activity of neurones in the substantia nigra of the alert monkey is also considered here, to provide information on outputs from the neostriatum, and thus on its function. Neuronal activity in the substantia nigra is also of interest, because there are neurones in the substantia nigra which project to the neostriatum, for example those of the nigro-striatal dopaminergic pathway, which can thus influence neostriatal function.

Recordings have been made from neurones in the region of the substantia nigra in the alert monkey (Mora, Mogenson and Rolls, 1977). Forty-one of 104 neurones had responses which occurred during feeding. Clear responses often occurred when the monkey opened his mouth and drank fruit juice from a syringe. These responses depended on touch to the mouth or movements made by the monkey in that they could also be obtained by moving the mouth of the monkey with an empty syringe or other non-food object. The maximal response of these units was usually obtained when a syringe was held against the side of the mouth and the monkey moved his mouth laterally to accommodate it, or when the monkey protruded his lips to make contact with the syringe as it approached.

The relationship of the activity of these neurones to feeding was further investigated as follows. First, similar neuronal responses were obtained when solutions with different tastes were ingested, as illustrated in Fig. 9. Second, the

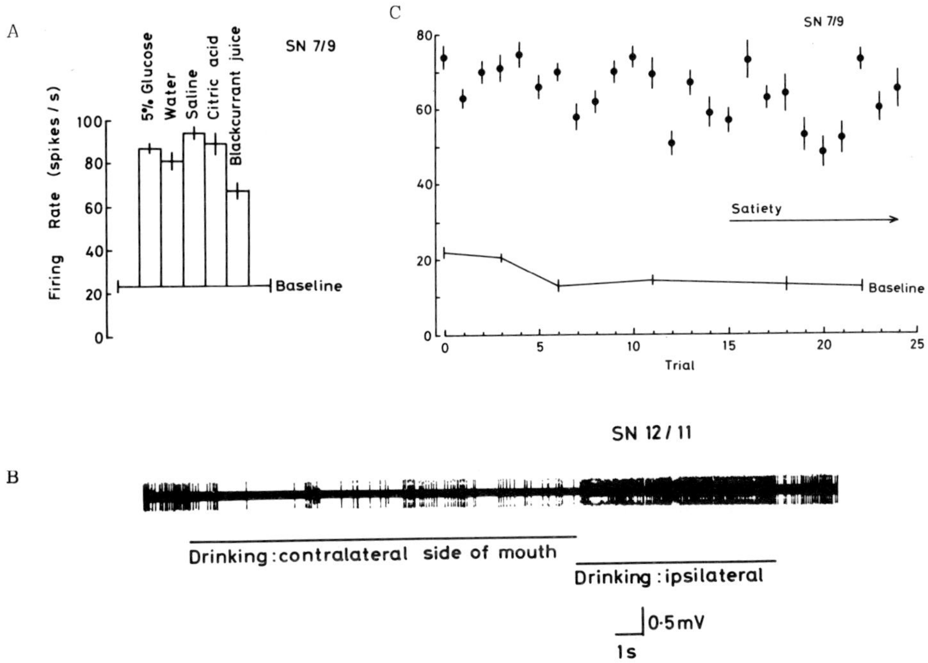

Fig. 9

A. The firing rate of a neurone in the region of the substantia nigra while the monkey drank different fluids. Similar responses (shown by the mean and the S.E. of the firing rate above the baseline spontaneous rate of 22 spikes/sec) were obtained to 5% glucose, water, 0.9% saline, 0.3% citric acid and blackcurrant juice.

B. Firing rate of a neurone in the region of the substantia nigra while the monkey drank from a syringe placed in the side of the mouth ipsilateral to or contralateral to the recording microelectrode. The control period with no drinking is shown before and after the drinking to indicate the spontaneous baseline firing rate of the neurone.

C. Effect of feeding the monkey to satiety on the responses of a neurone recorded in the region of the substantia nigra. The neurone responded (with an increase of firing rate to 50-80 spikes/sec) when the syringe was placed in the monkey's mouth (the mean and S.E.M. are shown). On each trial the monkey was fed 5 ml of the 25% glucose solution, until by trial 15 satiety, measured by attempted rejection of the syringe by hand and failure to swallow the glucose, started to develop. The unit continued to respond after this when the glucose was placed in the monkey's mouth.

(Reprinted from Mora, F., Mogenson, G.J., and Rolls, E.T., 1977, Brain Research, 133, 267-276)

neuronal response obtained depended on the mouth movements made rather than on the solution consumed, so that different effects on the firing rate of the neurone were sometimes obtained depending on whether the monkey drank from a syringe placed in the side of his mouth ipsilateral or contralateral to the recording electrode (see Fig. 9). Third, the neuronal responses were not directly related to the motivational state of the animal, in that they occurred whenever the monkey made a particular movement, irrespective of whether the animal was hungry or satiated. For example, the unit shown in Fig. 9 increased its firing rate to 60-70 spikes/sec from the baseline of 10-20 spikes/sec when a syringe containing 20% glucose was placed in the mouth, and the response did not diminish after the monkey had drunk 75 ml of the glucose and became satiated. Some of the neurones in this region responded both in relation to mouth movements and in relation to arm movements such as reaching. Although histological localisation showed that neuronal responses of this type were found within the substantia nigra and in a region of the tegmentum 2-3 mm dorsal to it, identification of the particular neurones recorded and for instance whether they were afferent or efferent to the neostriatum is not available at present (Mora et al., 1977). Thus the finding that the activity of a number of these neurones was related to identifiable movements made by the alert monkey provides support for the view that a function of neostriatal neurones is in the control of movement.

SYNTHESIS

Most neurophysiological studies of the neostriatum in alert animals have been aimed at determining how neostriatal neurones respond during movement. This emphasis stems from clinical studies which have shown that damage to the basal ganglia is associated with movement disorders, and from the efferent projections of the neostriatum to the globus pallidus and substantia nigra, and thus via the thalamus to the motor cortex. The neurophysiological evidence that the main output areas of the neostriatum, namely the globus pallidus and substantia nigra, contain neurones whose activity is closely related to particular movements such as arm extension or mouth opening, supports the view that the neostriatum is involved in the control of movement. This view also receives support from the observations that some movement-related neurones are found in the caudate nucleus and putamen, although the firing is not as specifically related to a particular movement, nor are such major changes in firing rate observed as in the globus pallidus and substantia nigra.

However, the neostriatum also contains a large number of neurones with other properties, which suggest that the neostriatum performs some considerably more sophisticated functions than might be suggested by simply viewing it as a "motor" structure. Further, the properties of these other neurones shed light on the deficits resulting from disruption of neostriatal function.

Firstly, the neostriatum receives a great deal of sensory information, as might be expected from its inputs from sensory association areas of neocortex, and this is clearly shown by the sensory responses recorded from neostriatal units in both the anaesthetised (Krauthamer, this volume) and the alert animal (see section 1b and Table 1). It may be expected that when recordings are made in the tail of the caudate nucleus, which has a direct input from temporal association cortex (Kemp and Powell, 1970), more effects of this influence will become evident.

Secondly, we have found that many neostriatal neurones respond particularly to events which are cues for the animal that a behavioural response may be required. Examples are the task-related neurones described above, which fire to cues such as a tone or the opening of a shutter which indicate when a visual discrimination will be required (see section 1c and Table 1). In some cases these neurones only respond to such events when they are important as cues. Other neurones respond to significant events in the environment such as the opening of the laboratory door, or when objects are presented to the monkey (see "clinical" responses described in section 1c and Table 1). Even more striking are the neurones with pre-presentation responses, which responded to particular movements of the experimenter's arms which provided a cue that an object which was sometimes food, was about to be shown to

TABLE 1 Classification of the units in the present study*

	Category	Number	Percentage
A.	MOTOR (i.e. activity related to particular movements)	21	6.8%
B.	SENSORY (i.e. activity related to sensory events independently of context)	16	5.2%
C.	TASK-RELATED (i.e. activity related to performance of visual discrimination task, but with no associated responses during 'clinical' testing)	54	17.6%
D.	'CLINICAL' RESPONSE (i.e. activity related to events during 'clinical' testing, but with no associated activity during performance of the visual discrimination task)	41	13.4%
E.	'PRE-PRESENTATION' RESPONSE (i.e. special category of units responsive to pre-presentation arm-movements made by the experimenter - see text)	34	11.1%
	TOTAL RESPONSIVE UNITS	166	54.1%
	TOTAL RECORDED UNITS	307	100%

* The recordings were made in four hemispheres of two rhesus monkeys using glass-coated tungsten microelectrodes. In the study, the recordings were probably from cell-bodies in that, (i) no units were recorded if the electrode was in white matter, (ii) the action potentials were biphasic, and (iii) if similar types of electrode are used during micro-iontophoresis, glutamate ions consistently excited such units. Recording sites were localised using X-ray photographs taken at the end of each track and with subsequent histological reconstruction, and were shown (in the case of all units described) to be largely in the head and body of the caudate nucleus, with the remainder (approximately 20% of the sample) in the rostral putamen (within the planes 22.5 - 16.0 mm anterior in the atlas of Snider and Lee, 1961).

the monkey, but stopped firing as soon as the monkey saw what the object was. Between them, neurones of this type could provide information relevant to the preparation and initiation of behavioural responses.

Thirdly, we found a population of neostriatal neurones with responses which were maintained throughout the period in which the animal was actively performing the task. For example, such units were excited from the first cue for the start of a trial until the animal had completed the behavioural response on that trial (these neurones are described in section 1c and included in Table 1; see e.g. Figs 4 and 6). These neurones could be involved in "enabling" the animal to make behavioural responses.

Finally, there are neurones in the neostriatum with activity related to movements, and these could be involved in the initiation and execution of particular responses (see below).

Thus, the neostriatum contains assemblies of neurones with properties which suggest that they are involved in different stages by which environmental cues prepare the animal and enable him to make behavioural responses. Evidence that these neurones are necessary for this type of function, and do not simply reflect responses being made by the animal, comes from the deficits which are produced by the dysfunction of the neostriatum. The deficits appear as a negative image of the characteristics of the neostriatal neurones described. For example, dysfunction of the neostriatum arising from damage to the nigro-striatal dopaminergic projection produces a sensori-motor deficit in which rats are deficient in orienting to environmental stimuli, and in initiating behavioural responses such as feeding and drinking (Marshall, Richardson and Teitelbaum, 1974; Ljungberg and Ungerstedt, 1976). The failure to initiate behaviour may be so extreme that catalepsy is produced by such lesions, as it is by blockade of dopamine receptors with neuroleptic drugs such as haloperidol. In man, also, dysfunction of this system is associated with an akinesia or failure to initiate movement in Parkinson's disease (see Hornykiewicz, 1973). Disorders in the control of movements are also produced by such damage. Damage directly to the basal ganglia also produces complex sensori-motor deficits (Teuber, 1976; Bowen, 1976) as well as movement disorders, and Denny-Brown and Yanagisawa (1976) have discussed deficits produced by neostriatal damage in terms of the absence of the activating "set" preparatory to motor performance oriented to the environment. These disorders then can be considered as the opposite of or a negative image of the response characteristics of the neostriatal neurones described here, and do provide evidence that these neostriatal neurones are involved in the functions described. A possible clinical implication of the neurophysiological results described here is that enhancement of environmental stimuli relevant to the initiation and maintenance of movements could help to minimize the impairments due to neostriatal dysfunction.

Although the exact way in which the neostriatum could perform these functions must be speculative at present, the following hypothesis which arises from the neuronal characteristics described is proposed. Neurones at a first stage of processing would receive sensory information from many cortical areas concerning events in the environment. Second, there would be a population of neurones which would detect those events which are relevant to the animal's performance. Whether this selection would be intrinsic to the neostriatum or would be performed elsewhere is not clear, but it will be interesting in future to consider whether the nigro-striatal dopaminergic projection might be involved in this plasticity. In either case, the neostriatum does indeed contain a large number of neurones which are sensitive to particular aspects of the environment, and some of these only respond when the event is a cue for the animal's performance. Third, there would be neurones with the function of "enabling" the animal and preparing him to make responses, which would be activated via the preceding neurones by those events which are cues for the performance of the particular task in which the animal was to become engaged, and would be deactivated when the task is complete, in our situation when the animal had made a response on each trial. Finally, these neurones, as well as others, could provide inputs to a fourth set of neurones, those with movement-related responses, which may be concerned with the initiation and execution of motor responses. Although this hypothesis is speculative at present, neostriatal neurones described here do have properties which suggest that they are involved in this type of function, and this suggestion is supported by the anatomical connections of the neostriatum, as well as by the deficits in behaviour produced by dysfunction of the neostriatum.

REFERENCES

(1) M. E. Anderson, Discharge patterns of Basal Ganglia neurons during active maintenance of postural stability and adjustment to chair tilt, Brain Res. 143, 325-338 (1978)

(2) F. P. Bowen, Behavioral alterations in patients with basal ganglia lesions. In: M. D. Yahr (Ed.), The Basal Ganglia, Raven, New York, 169-177 (1976)

(3) M. J. Burton, E. T. Rolls and F. Mora, Effects of hunger on the responses of neurones in the lateral hypothalamus to the sight and taste of food, Expl. Neurol. 51, 668-677 (1976)

(4) P. Buser, G. Pouderoux and J. Mereaux, Single unit recording in the caudate nucleus during sessions with elaborate movement in the awake monkey, Brain Res. 71, 337-344 (1974)

(5) P. Buser, G. Pouderoux and J. Mereaux, Single unit recording from the caudate nucleus in the awake monkey performing movements of the upper limb. In: T. Frigyesi (Ed.), Subcortical Mechanisms and Sensorimotor Activities, Hans Huber Publ., Bern Stuttgart Vienna, 177-190 (1975)

(6) M. R. DeLong, Activity of pallidal neurones during movements, J. Neurophysiol. 34, 414-427 (1971)

(7) M. R. DeLong, Activity of basal ganglia neurons during movement, Brain Res. 40, 127-135 (1972)

(8) M. R. DeLong, Putamen: Activity of single units during slow and rapid arm movements, Science 179, 1040-1042 (1973)

(9) M. R. DeLong and P. L. Strick, Relation of basal ganglia, cerebellum, and motor cortex units to ramp and ballistic limb movements, Brain Res. 71, 327-335 (1974)

(10) D. Denny-Brown and N. Yanagisawa, The role of the basal ganglia in the initiation of movement. In: M. D. Yahr (Ed.), The Basal Ganglia, Raven, New York (1976)

(11) I. Divac, Neostriatum and functions of prefrontal cortex, Acta Neurobiol. Exp. 32, 461-477 (1972)

(12) E. Dolbakyan, N. Hernandez-Mesa and J. Bures, Skilled forelimb movements and unit activity in motor cortex and caudate nucleus in rats, Neuroscience 2, 73-80 (1977)

(13) O. Hornykiewicz, Dopamine in the basal ganglia: Its role and therapeutic implications, Brit. Med. Bull. 29, 172-178 (1973)

(14) J. M. Kemp and T.P.S. Powell, The cortico-striate projection in the monkey, Brain, 93, 525-546 (1970)

(15) H. H. Kornhuber, Motor functions of cerebellum and basal ganglia: the cerebellocortical saccadic (ballistic) clock, the cerebellonuclear hold regulator, and the basal ganglia ramp (voluntary speed smooth movement) generator, Kybernetik 8, 157-162 (1971)

(16) T. I. Lidsky, N. A. Buchwald, C.D. Hull and M. S. Levine, Pallidal and entopeduncular single unit activity in cats during drinking, EEG clin. Neurophysiol. 39, 79-84 (1975)

(17) T. Ljungberg and O. Ungerstedt, Sensory inattention produced by 6-hydroxydopamine induced degeneration and ascending dopamine neurons in the brain, Exp. Neurol. 53, 585-600 (1976)

(18) J. P. Marshall, J. S. Richardson and P. Teitelbaum, Nigrostriatal bundle damage and the lateral hypothalamic syndrome, J. comp. physiol. Psychol. 87, 808-830 (1974)

(19) K. Matsunami and B. Cohen, Afferent modulation of unit activity in globus pallidus and caudate nucleus: changes induced by vestibular nucleus and pyramidal tract stimulation, Brain Res. 91, 140-146 (1975)

(20) F. Mora, G. J. Mogenson and E. T. Rolls, Activity of neurones in the region of the substantia nigra during feeding. Brain Res. 133, 267-276 (1977)

(21) H. Niki, M. Sakai and K. Kubota, Delayed alternation performance and unit activity of the caudate head and medial orbitofrontal gyrus in the monkey. Brain Res. 38, 343-353 (1972)

(22) E. T. Rolls, Neurophysiology of Feeding, Trends in Neurosciences 1, 1-3 (1978b)

(23) E. T. Rolls, M. J. Burton and F. Mora, Hypothalamic neuronal responses associated with the sight of food, Brain Res. 111, 53-66 (1976)

(24) E. T. Rolls, M. K. Sanghera and A. Roper-Hall, The latency of activation of neurons in the lateral hypothalamus and substantia innominata during feeding in the monkey, Brain Res. in press (1978)

(25) R. S. Snider and J. C. Lee, A Stereotaxic Atlas of the Monkey Brain, University of Chicago Press (1961)

(26) S. Soltysik, C. D. Hull, N. A. Buchwald and T. Fekete, Single unit activity in basal ganglia of monkeys during performance of a delayed response task, EEG Clin. Neurophysiol. 39, 65-78 (1975)

(27) D. L. Sparks and R. P. Travis Jun., Patterns of reticular unit activity observed during the performance of discriminative task, Physiol. Behav. 3, 961-968 (1968)

(28) D. L. Sparks and R. P. Travis, Firing patterns of extrapyramidal and reticular neurons of the alert monkey. In: M. I. Phillips (Ed.), Brain Unit Activity during Behavior, Thomas, Springfield, Ill., 288-300 (1973)

(29) H. L. Teuber, Complex functions of basal ganglia. In: M. D. Yahr (Ed.), The Basal Ganglia, Raven, New York, 151-168 (1976)

(30) R. P. Travis Jr and D. L. Sparks, Changes in unit activity during stimuli associated with food and shock reinforcement, Physiol. Behav. 2, 171-177 (1967)

(31) R. P. Travis Jr and D. L. Sparks, Unitary responses and discrimination learning in the squirrel monkey: The globus pallidus, Physiol. Behav. 3, 187-196 (1968)

(32) R. P. Travis Jr, T. F. Hooten and D. L. Sparks, Single unit activity related to behavior motivated by food reward, Physiol. Behav. 4, 309-318 (1968)

(33) R. P. Travis Jr, D. L. Sparks and T. F. Hooten, Single unit responses related to sequences of food motivated behavior, Brain Res. 7, 454-457 (1968)

(34) O. S. Vinogradova and E. N. Sokolov, Activation and habituation in neurons of the caudate nucleus. In: E. N. Sokolov and O. S. Vinogradova (Eds), Neuronal mechanisms of the orienting reflex, Erlbaum, 155-169 (1975).

Electrical Stimulation of the Neostriatum in Behaving Animals

Anthony G. Phillips

Department of Physiology, University of British Columbia, Vancouver, B.C., Canada V6T 1W5

Motor Effects of Electrical Stimulation

Neuroscientists have been probing the neostriatum (NS) with electrical stimulation for over 100 years in an attempt to decipher its role in the control of behavior. In 1873, Ferrier (10) observed strong contralateral pleurothotonus during stimulation of the caudate nucleus in dogs and concluded that this region was involved in contralateral turning of the body. A portent of future controversies was provided by Wilson's failure to elicit motor activity with electrical stimulation in his classic study of the corpus striatum (68). In time, Ferrier was vindicated and there now have been many reports of contraversive turning of the head and body in a wide variety of species, following unilateral stimulation of the caudate nucleus (9, 20, 33, 38, 39, 48, 52, 60, 66). It is noteworthy that these postural changes also occurred in animals in which the motor cortex had been extirpated and furthermore were not elicited from control placements in regions adjacent to the head of the caudate nucleus.

In addition to postural asymmetries and circling, caudate stimulation gives rise to stereotypied movements on the contralateral side of the head, and choreo-athetosis on the contralateral forelimb (9, 20, 33). A recent frequency/intensity analysis of the current parameters which elicit these motor responses has confirmed that only frequencies between 20 Hz and 30 Hz are effective with higher current intensity necessary at the lower frequency (9). In this same analysis of caudate stimulation in freely moving cats, Cools also has noted the striking similarities between the behaviors elicited by electrical stimulation at these frequencies, and those evoked by microinjection of dopamine (10μg) into the head of the caudate nucleus (11). Complete suppression of the electrical stimulation effects following inhibition of dopamine synthesis by pretreatment with a-methyl-para-tyrosine or blockade of dopamine receptors with the neuroleptic drug haloperidol, confirmed an important role for dopamine in these elicited motor responses. This observation is reminiscent of earlier reports of contraversive turning during stimulation of the dopaminergic nigrostriatal bundle (3) and the subsequent inhibition of this circling response by dopamine antagonists (4). Cools (10) has hypothesized that these motor effects are mediated in part by excitatory effects of dopamine. This hypothesis of excitatory dopaminergic synapses in the striatum has received support from recent neurophysiological experiments (53) and must be given careful consideration in the functional analysis of the NS.

It is important to emphasize that the motor responses described above are dependent on critical frequencies of stimulation (9, 38). When the frequency of electrical stimulation is lowered, a very different picture of neostriatal function emerges. Inhibitory effects including sleep (24) are seen at the lowest frequencies (0.2-10 Hz) and this effect gives way to the well known phenomenon of 'behavioral arrest' which occurs at certain current intensities with frequencies of 10-20 Hz (14, 38, 58). This effect is characterized by an abrupt cessation of ongoing behavior throughout the duration of caudate stimulation. Disruption of feeding behavior has been reported (32, 50), although procedural variables have precluded unanimous agreement on this point (8, 31). Similar reports of 'behavioral arrest' have accompanied stimulation at other sites including the septum and anteroventral region of the thalamus but the dramatic (if somewhat contrived) experiments of Delgado with charging bulls and ferocious monkeys have served to focus attention on the caudate (14, 15).

The experiments reviewed so far all seem to point to an important, if not exclusive, role for the NS in the control of motor activity (28). However, the picture is still far from complete as many are convinced that the neostriatum is involved in more complex aspects of behavior (54) which may include sensory motor integration and perhaps learning and memory (27, 30). Further evidence for the role of the neostriatum in more complex functions comes from the demonstration that animals will learn to engage in intracranial self-stimulation in this region of the brain.

Self-Stimulation Effects of Electrical Stimulation

As already indicated, much of the research on the behavioral correlates of electrical stimulation in the NS is characterized by conflicting reports from different laboratories. This pattern repeats itself again in the study of brain-stimulation reward at electrode loci in the caudate-putamen. Positive sites were identified in one of the earliest studies of self-stimulation (59) and this has been confirmed subsequently in rat (35, 48, 55), cat (29, 45), monkey (42, 50, 51) and man (25). However, these findings must be set against reports that caudate stimulation is either neutral (44, 58) or highly aversive (70). In one instance in which self-stimulation was observed in cats (29) with caudate electrodes, the effect was attributed to increased motor activity. Prado-Alcalá et al. (52) recently mapped the NS and were unable to obtain self-stimulation from any of the 37 electrodes located throughout this structure.

The reported failure to obtain self-stimulation from neostriatal electrodes is best attributed to procedural differences; in particular to the selection of current intensities and the number of training days. Some of the previous mapping studies (44, 55) have employed a fixed current intensity for all animals. Prado-Alcalá et al. (52) varied the current intensity from 30 to 90 μA, but the animals were tested for only 10 min/day at a given intensity. Furthermore, the experiment was terminated after only 5 consecutive training days. In contrast to these negative findings, Phillips et al. (48) have shown recently that self-stimulation can be elicited from all aspects of the NS in rat, provided that optimal current intensities are selected individually for each animal and that training is continued for up to 14 days.

A summary of the results of this mapping study is shown in Fig. 1. Fifty-nine of the 66 electrode placements tested in this study were positive for self-stimulation and 48 of these placements were located in the caudate-putamen. The mean number of test days to spontaneous self-stimulation was 8 days. Self-stimulation rates varied from a low rate of 55/15 min to a high of 1180/15 min. Fifty-six percent of the placements in the NS supported self-stimulation at rates greater than 250/15 min. Four negative placements were located in the NS. Although one of the most striking features of these data is the homogeneous distribution of self-stimulation sites throughout this region, electrodes located in the head of the caudate supported the highest rates ($\bar{X}$ = 1136/15 min). Of possible relevance to this finding

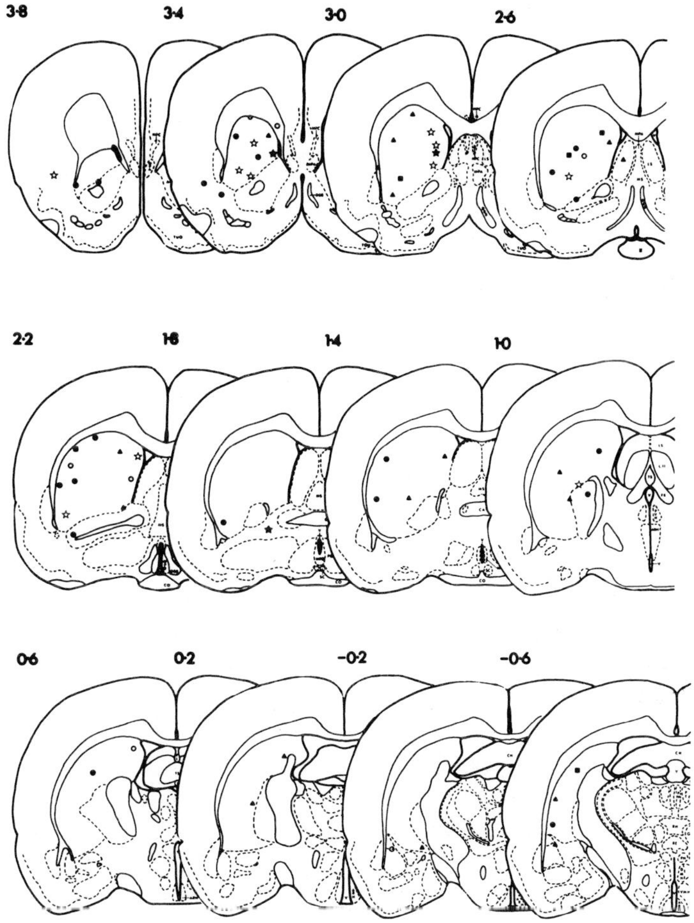

Fig. 1. Placements of self-stimulation electrode tips shown on coronal sections anterior and posterior to bregma. Symbols indicate self-stimulation rate/15 min. i.e., ★, 1000-1500; ☆ 500-999; ▲ 250-499; ● 100-249; △ 50-99; ○ <50.

is the observation that the highest concentrations of dopamine are found in the rostral aspect of the NS (32).

Self-stimulation in the NS was accompanied by distinct motor responses, as might be expected from previous studies. The majority of placements (i.e., 52%) elicited a 'clonus-rearing' response characterized by rapid rearing with superimposed forelimb clonus and facial twitches. A second class of motor response consisted of head turning which in more severe cases manifested itself as a strong contralateral body turn. These two classes of motor response are very similar to those seen in other studies (10, 33, 52). In the remaining 8% of the animals stimulation elicited only forward locomotor behavior reminiscent of the response seen during self-stimulation in the lateral hypothalamus.

The confirmation of brain-stimulation reward in the NS gives rise to two important questions. First, can we identify the neuroanatomical and neurochemical correlates, and secondly, can a careful analysis of this phenomenon yield any insights into the functional significance of the NS?

Neurochemical correlates of self-stimulation in the NS. Biochemical (32, 65) and histochemical (21, 62) studies have provided evidence for high concentrations of dopamine in the NS. The cell bodies of this dopamine system reside in the substantia nigra pars compacta and the projection of the nigrostriatal pathways has been traced by histochemical fluorescence (36, 62). Of relevance to the present discussion is the high incidence of self-stimulation sites along the trajectory of the dopamine pathways from its origins in the substantia nigra through the dorsolateral hypothalamus and entopenduncular nucleus to its terminal region in the NS (19, 22). This dopamine pathways appears to play an important role in self-stimulation of the NS as 6-OHDA lesions of the dopamine cell bodies in the substantia nigra result in a severe disruption of self-stimulation in the ipsilateral caudate-putamen (48). During the initial phases of post-lesion testing, self-stimulation was disrupted by both ipsilateral and contralateral lesions, indicating that such lesions can produce a general disruption in bar-pressing behavior. In subsequent tests, however, responding in the ipsilateral lesion group remained depressed, whereas the contralateral group exhibited a significant recovery of ICS rate to 72% of the pre-lesion scores. An interesting incidental observation was the enhancement of stimulation-induced motor responses following destruction of the contralateral nigrostriatal bundle. This behavioral effect may be related to the apparent interdependence of the two nigrostriatal dopamine systems as revealed by a lesion-induced increase in dopamine release in the contralateral NS (43).

Pharmacological data also suggest a role for dopamine in self-stimulation of the NS. Self-stimulation in rat is facilitated to an equal degree by both isomers of amphetamine (Phillips & Carter, unpublished observations). Similar effects also are observed in the monkey after treatment with the dopamine receptor agonist apomorphine (50), although this drug produced both facilitation and inhibition of caudate self-stimulation in rats (42). The caudate nucleus contains relatively high concentrations of opiate receptor (47), and it has been suggested that enkephalin may subserve brain-stimulation reward (5). However, recent observations from our laboratory suggest an independence of the opiate reinforcement and brain-stimulation reinforcement system at the level of the NS. Naloxone, the highly specific opiate receptor blocker, at a wide range of doses (5-40 mg/kg), had no significant effect on self-stimulation (64).

These pharmacological data, along with the disruption of self-stimulation following depletion of dopamine in NS, clearly suggest an important role for dopamine but it would be an oversimplification to argue for an exclusive role. The NS contains moderate concentrations of serotonin (6), which arises in part from the dorsal raphe nucleus (40). There is now evidence (49) that a serotonergic mechanism contributes to brain stimulation reward in the NS as self-stimulation is suppressed after inhibition of serotonin biosynthesis with injections of para-chlorophenylalanine (PCPA). Maximal effects are observed 3 to 4 days after treatment at a time when serotonin levels are maximally depleted (63). Furthermore, the disruption of self-stimulation in the NS cannot be attributed to some specific drug effects, as self-stimulation in the lateral hypothalamus is enhanced after treatment with PCPA (49). Taken together the data from the nigrostriatal lesion study (48) and the PCPA experiment (49) suggest that self-stimulation in the NS may be subserved by an interaction between dopaminergic and serotonergic mechanisms. This suggestion is similar to Cools' (10) explanation of motor stereotypy. According to this model, a cholinergic mechanism provides a connection between the dopaminergic and serotonergic systems in the NS with the serotonergic mechanism forming an essential link for processes initiated by activation of the dopamine system. Further support for such a model awaits an analysis of the effects of cholinergic drugs on brain-stimulation reward in the NS.

Functional correlates of self-stimulation in the NS. The absence of a complete neurochemical analysis of brain-stimulation reward in the NS does not preclude speculation as to why an animal should choose to repeatedly deliver electrical stimulation to this region of its brain. An explanation based on stimulation-induced motor automatism can be dismissed because at the beginning of the session the animals go directly to the bar to initiate responding. In addition, the motor responses which accompany self-stimulation tend to interfere with the barpress response, often causing the animal to turn away from the bar. Despite these disruptive effects, the subjects are capable of devising strategies to maintain regular responding that can reach a level of over 4000 responses per hour. Given the compelling suggestion that normal activity of the NS is involved in the initiation and execution of learned sequences of motor activity, perhaps we can seek an explanation of neostriatal self-stimulation in terms of this function. Glickman and Schiff (23) have argued that the occurrence of species-specific responses per se, is reinforcing. If this theoretical perspective is extended to include important learned responses such as those that permit the animal to avoid danger and obtain nutrients, then perhaps it is not surprising that an animal will take the opportunity to artificially activate a system that controls movement when the movement is one essential for survival.

The Effect of Electrical Stimulation on Learning and Memory

Another possible function for the neural substrates of brain stimulation reward in the NS is their involvement in learning and memory formation. The finding that prolonged stimulation at self-stimulation sites in various regions of the brain can cause apparent retrograde amnesia has been taken as evidence for the role of these sites in memory consolidation (56). Further support for this hypothesis comes from the recent report of memory facilitation under certain conditions following self-stimulation (68). Other studies analyzing the behavioral effects of low frequency stimulation of the caudate have shown that such stimulation can produce selective inhibition of learned motor movements (8, 12, 13, 37, 67). For example, Buchwald et al. (8) reported a complete cessation of an operant response for milk reward, without any attendant disruption of grooming or orienting to visual and auditory stimuli. Others have confirmed this observation (31, 66) and extended it to show that caudate stimulation also can disrupt the execution ot complex learned motor movements.

There is an extensive literature showing memory disruption following electrical stimulation of the caudate (27, 30). For example, in rats, posttrial caudate stimulation impaired passive avoidance of drinking at a spout paired with shock (71, 73) but stimulation preceding the footshock had no effect (72). Food rewarded maze learning (46), one-trial appetitive learning (74) and extinction of approach learning (26) are also all impaired. Taken together, these findings suggest a major involvement of the NS in the formation of memory (27, 30, 46).

Memory disruption also can be obtained in rats following stimulation of some of the major striatal afferent projections. Routtenberg and Holzman (57) have demonstrated that unilateral low intensity (10-25 μA) electrical stimulation of the zona compacta of the substantia nigra during or immediately after the acquisition of a passive avoidance step-down task impaired the retention of that task 24 hr later. It was suggested that the dopaminergic nigrostriatal bundle was involved critically in memory consolidation and storage. Fibiger and Phillips (18) have confirmed the importance of the nigrostriatal pathway in mediating the disruption of memory. However, in view of the fact that lesions of this projection a) did not by themselves, affect acquisition or retention of passive avoidance and b) failed to influence the memory deficit observed after electrical stimulation of the NS, it was suggested that the nigrostriatal system is not involved critically in learning and memory. Rather, it was hypothesized that nigral stimulation resulted in an excess release of dopamine in the NS which in turn produced a disruption of memory.

A serotonergic projectionfrom the dorsal raphe nucleus to the striatum has recently been described (40). It is therefore of interest to note that electircal stimulation of the dorsal raphe also disrupts retention of a step-down passive avoidance response (17). Inhibition of serotonin biosynthesis with PCPA blocked the disruptive effects of dorsal raphe stimulation. Pretreatment with PCPA had no effect on memory in implanted and unimplanted controls. Furthermore, these ameliorative effects of PCPA on memory disruption were locus specific as PCPA injections failed to attenuate the disruptive effect of amygdala stimulation. These data do not support the hypothesis that amnesic effects of localized brain-stimulation are mediated by interference with storage processes in cells directly excited by the current. Rather they indicate that memory disruption may result from effects in other regions of the brain far removed from the locus of stimulation; in this instance, the NS.

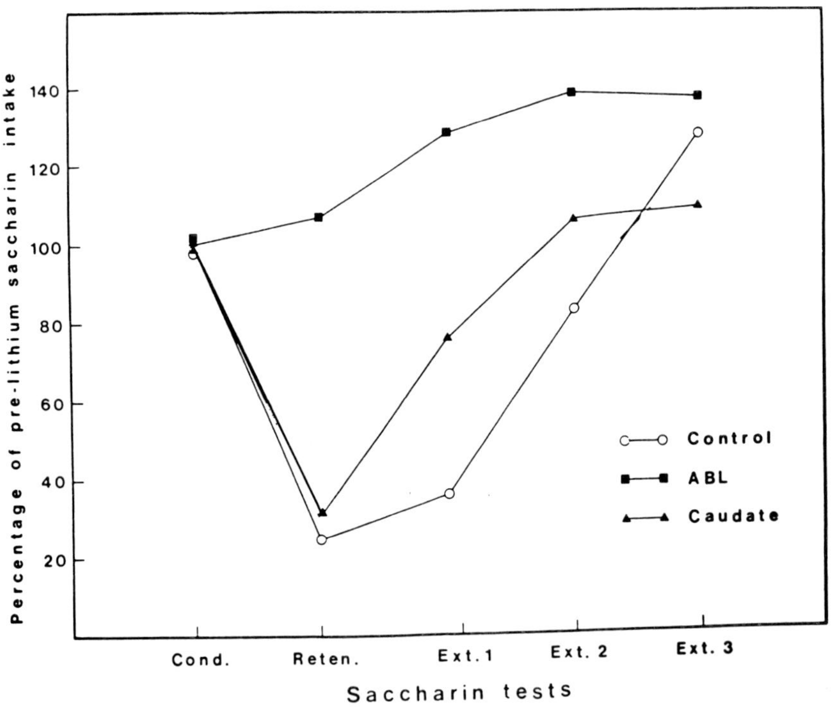

Fig. 2. Effect of electrical stimulation in the caudate nucleus or basolateral nucleus of the amygdala (ABL) on retention of a conditioned taste aversion to saccharin.

At this juncture it should be noted that the analysis of neostriatal function in terms of memory processing is not necessarily at odds with the previous suggestion that this structure is concerned with the initiation and execution of learned motor responses. Acceptance of the latter hypothesis would however limit the categories of memory processing to those related to certain learned motor activities such as those elicited by exteroceptive sensory stimuli. Preliminary evidence suggests that such a relationship may indeed prevail (34). This experiment compared the disruptive effects of low-intensity electrical stimulation of the amygdala with similar stimulation in the caudate, on the memory of two distinctly different avoidance responses. Namely, step-down shock avoidance and conditioned taste aversion. In the latter paradigm, a novel taste (saccharin) is paired with toxicosis induced by lithium chloride and the rat normally avoids the saccharin on the next encounter. Stimulation of the amygdala prior to the onset of illness already has been shown to disrupt retention of the taste aversion (4) and as may be seen in Fig. 2, intermittent stimulation of the amygdala during the 10 min. saccharin test produced a similar effect. The location of the electrode placements used in this study are shown in Fig. 3. Interestingly, comparable stimulation of the caudate-putamen had no disruptive effect on the retention of the conditioned taste aversion, although similar parameters of caudate stimulation did interfere with retention of step-down passive avoidance response (Fig. 4).

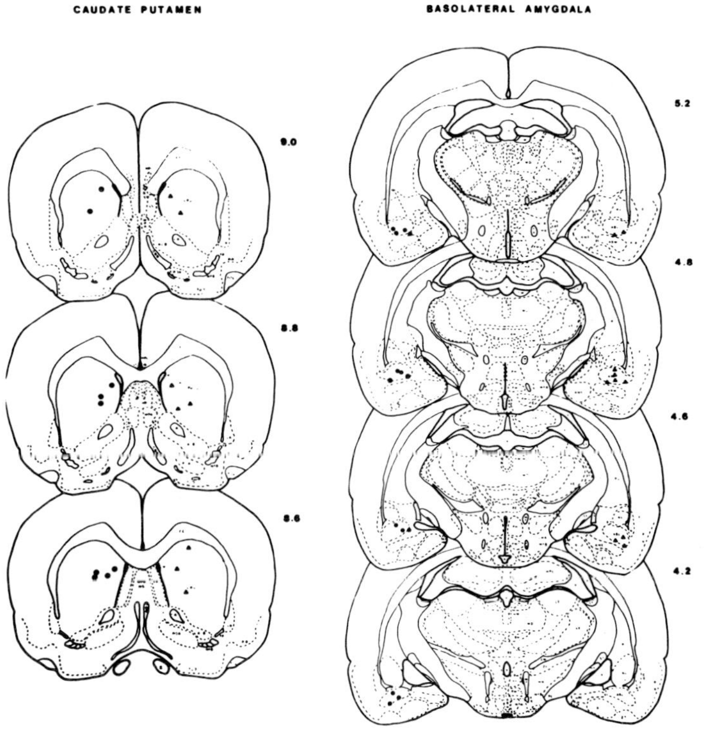

Fig. 3. Electrode placements in caudate putamen and basolateral nucleus of the amygdala of subjects in the conditioned taste aversion study (●) and step-down passive avoidance test (▲).

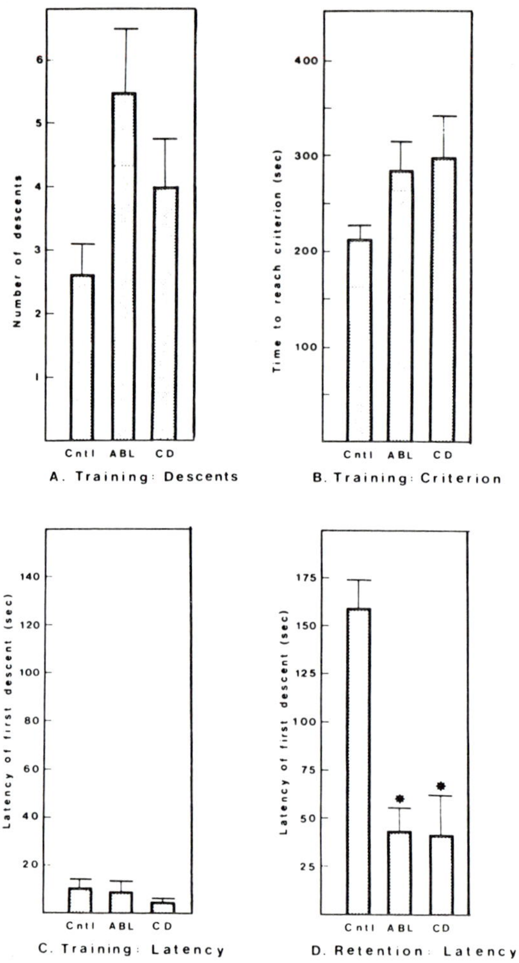

Fig. 4. Effect of electrical stimulation in the caudate nucleus (CD) or basolateral nucleus of the amygdala (ABL) on acquisition and retention of a step-down passive avoidance response. Stars indicate retention scores significantly different ($p<.05$) from implanted controls (Cntl).

On the basis of these data, it appears as though the NS indeed may be selectively involved in memory formation. However, as attractive as this hypothesis may seem, it is important to remember that there is as yet, no direct proof that neural events in the striatum are essential for even a limited facet of memory. The role of this structure in memory formation is <u>inferred</u> from the absence of the expected behavior during a retest, by animals that received brain stimulation during or shortly after the learning trial. The need for a more cautious interpretation of these data is underscored further by the realization that none of the experiments

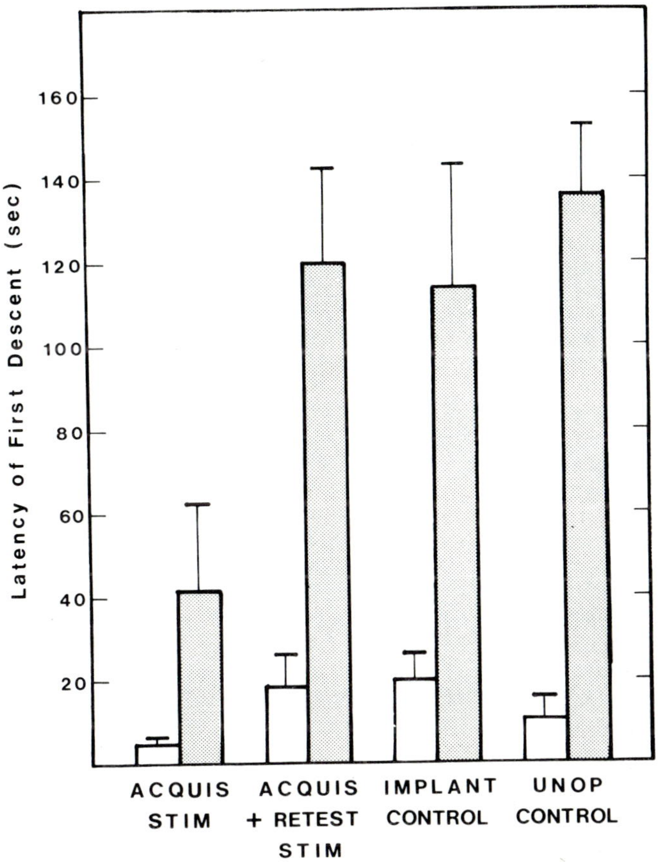

Fig. 5. Comparison of effects of electrical stimulation of caudate putamen during acquisition test (ACQUIS-STIM) or during acquisition <u>and</u> retention tests (ACQUIS + RETEST-STIM) on retention of step-down passive avoidance response. Data represent mean scores ($\pm$S.E.M.) of latency to first descent from platform on a 24 hr. retention test.

described so far had a control for either state dependency or the stimulus properties of brain-stimulation.

If brain-stimulation, administered during or shortly after learning, has interfered with some aspect of memory formation, then delivering brain-stimulation in an identical manner during the retention trial should have no effect on behavior. On the other hand, if brain-stimulation produced a unique state under which the memory was stored, the memory may be irretreivable under normal conditions but retrievable if the original brain-stimulation condition was reinstated. Therefore it is necessary to include a group of animals that receive brain-stimulation during both acquisition and retention. When such a group is included in the study of the effect of caudate stimulation on retention of a passive avoidance response, a very different picture emerges. As may be seen in Fig. 5, the group receiving intermittent caudate stimulation only during acquisition showed a marked impairment in retention of the avoidance response. In contrast, the group that was stimulated during both acquisition and retention retained the passive avoidance response as well as implanted and unoperated controls. It remains to be seen whether similar effects can be shown with post-training stimulation, especially with the brief intense stimulation used by Wyers and his co-workers (27, 71, 72, 73). Nevertheless, the need for such a control is obvious and until such data are provided, we must reserve judgment as to whether experiments with brain-stimulation in behaving animals can be used to provide evidence for the role of the NS in memory formation.

Conclusions

1. Electrical stimulation of the NS gives rise to a number of motor effects including postural asymmetries and stereotyped behavior. These effects are related to the frequency of the electrical current as stimulation at lower frequencies (i.e., 0.2-20 Hz) is accompanied by inhibitory effects, including 'behavioral arrest'.

2. The NS may also be involved in motivation as electrical stimulation at high frequencies (60 Hz) can maintain self-stimulation behavior. Pharmacological studies and lesion experiments indicate important roles for both dopamine and serotonin in this phenomenon. It is suggested that the reinforcing property of NS stimulation may be related to activation of systems that initiate patterns of learned motor responses.

3. Studies of the disruptive effects of NS stimulation on learning and memory suggest a role for this structure in cognition. Experiments comparing the disruptive effect of NS or amygdaloid stimulation on retention of conditioned taste aversion or step-down passive avoidance in rats, suggest that NS stimulation will only disrupt tasks involving the association of exteroceptive stimuli with specific motor responses.

4. Despite the extensive literature on disruption of memory by localized electrical stimulation of the NS, recent experiments indicate that other factors may be responsible for these disruptive effects. Specifically, it is suggested that NS stimulation can serve as a conditioned stimulus which can be associated with footshock. Failure to provide this CS in subsequent retests may give rise to apparent memory deficits which really reflect conditioning effects.

Acknowledgements

This research was supported by grants from the National Research Council and Medical Research Council of Canada. The author is a Killam Senior Research Scholar. The original research described in this review is the product of close collaboration with my colleagues H.C. Fibiger, F.G. LePiane, D.A. Carter and D. van der Kooy.

References

1. K. AKERT and B. ANDERSSON, Acta physiol. Scand., 22, 281-298 (1951).
2. G.W. ARBUTHNOTT, and T.J. CROW, Brain Res., 24, 471-483 (1970).
3. G.W. ARBUTHNOTT and U. UNGERSTEDT. Exp. Neurol., 47, 162-172 (1975).
4. J.B. ARTHUR, Behav. Biol., 13, 369-376 (1975).
5. J.D. BELLUZZI, and L. STEIN. Nature, 266, 557-558 (1977).
6. D.F. BOGDANSKI, H. WEISSBACH, and S. UDENFRIEND. J. Neurochem. 1, 272-278, (1957).
7. N.A. BUCHWALD, and C.D. HULL, Brain Res., 6, 1-11 (1967).
8. N.A. BUCHWALD, E.J. WYERS, T. OKUMA, and G. HEUSER. Electroenceph. clin. Neurophysiol., 13, 509-518 (1961).
9. A.R. COOLS, Brain Res., 58, 437-451 (1973).
10. A.R. COOLS, Psychopharmacol. Berl. 36, 17-28 (1974).
11. A.R. COOLS, and J.M. van ROSSEUM, Arch. int. Pharmacodyn., 187, 163-173 (1970).
12. S.A. DEADWYLER,and E.J. WYERS. Behav. Biol., 7, 55-64 (1972).
13. S.A. DEADWYLER, N. HIRASUNA, and E.J. WYERS. Brain Res., 75, 59-70 (1974).
14. J.M.R. DELGADO. Int. Rev. Neurobiol. 6, 349-449 (1964).
15. J.M.R. DELGADO. Physical Control of the Mind. New York, Harper & Row (1969).
16. D. FERRIER, West Riding Lunatic Asymlum Med. Rep., 3, 30-46 (1873).
17. H.C. FIBIGER, F.G. LePIANE, and A.G. PHILLIPS, Brain Res., in press (1978).
18. H.C. FIBIGER, and A.G. PHILLIPS. Brain Res., 116, 23-33 (1976).
19. H.C. FIBIGER and A.G. PHILLIPS. The Neurobiology of Dopamine. In: A.S. Horn, B. Westerink and J. Korf, (Eds.), Academic Press, London, in press (1978).
20. D. FORMAN and J.W. WARD. J. Neurophysiol., 20, 230-244 (1957).
21. K. FUXE, Acta physiol. Scand., 64, Suppl. 247, 39-85 (1965).
22. D.C. GERMAN and D.M. BOWDEN. Brain Res., 73, 381-419 (1974).
23. S.E. GLICKMAN and B.B. SCHIFF, Psychol. Rev., 74, 81-109 (1967).
24. R.G. HEATH and R. HODES. Trans-Amer. Neurol. Ass., 204-210 (1952).
25. R.G. HEATH, S.B. JOHN and C.J. FONTANA. In: Computers and Electronic Devices in Psychiatry, N. Kline and E. Laska (Eds.), Grune and Stratton, New York, 214-248 (1968).
26. M.J. HERZ, and H.V.S. PEEKE, Brain Res., 33, 510-522 (1971).
27. M.J. HERZ, E.J. WYERS and H.V.S. PEEKE. In: Subcortical mechanisms and sensorimotor activities,T.L. Frigyesi (Ed.), Huber, Bern, 191-215 (1975).
28. R. JUNG and R. HASSLER, In: Handbook of Physiology Section I. Neurophysiology, Vol. 2. J. Field (Ed.), The American Physiological Society, Washington, 863-927 (1960).
29. D.R. JUSTESEN, J.C. SHARP and P.B. PORTER, J. comp. physiol. Psychol., 56, 371-374 (1963).
30. R.P. KESNER and M.W. WILBURN. Behav. Biol., 10, 259-293 (1974).
31. A. KITSIKIS, and A. ROUGEUL, Physiol. Behav., 3, 831-837 (1968).
32. S.H. KOSLOW, G. RAGAGNI, and E. COSTA. Neuropharmacol., 13, 1123-1130 (1974).
33. A.M. LAURSEN, Acta physiol. Scand., 59, Suppl. 211 (1963).
34. F.G. LePIANE and A.G. PHILLIPS. Physiol. Behav., submitted (1978).
35. M.S. LEVINE, N. FERGUSON, C.S. KREINICK, J.W. GUSTAFSON and J.S. SCHWARTZBAUM, J. Comp. physiol. Psychol., 77, 282-293 (1971).
36. O. LINDVALL and A. BJORKLUND. Acta phsyiol. Scand., Suppl. 412, 1-48 (1974).
37. C.G. LINEBERRY and J. SIEGEL. Brain Res. 34, 143-161 (1971).
38. H. McLENNAN, P.R. EMMONS, and P.M. PLUMMER. Can. J. Physiol. Pharmacol., 42, 329-339 (1964).
39. F.A. METTLER, Res. Publ. Ass. Nerv. Ment. Dis., 21, 150-227 (1942).
40. J.J. MILLER, T.L. RICHARDSON, H.C. FIBIGER and H. McLENNAN. Brain Res., 97, 133-138 (1975).
41. F. MORA, A.G. PHILLIPS, J.M. KOOLHAAS and E.T. ROLLS. Brain Res. Bull., 1, 421-424 (1976).
42. F. MORA, E.T. ROLLS, M.J. BURTON and S.G. SHAW. Pharmacol. Biochem. Behav. 4, 211-216 (1976).

43. A. NIEOÚLLON, A. CHÉRAMY and J. GLOWINSKI. Science, 198, 416-418 (1977).
44. M.E. OLDS and J. OLDS. J. comp. Neurol., 120, 259-295 (1963).
45. N.E. O'DONOHUE and W.D. HAGAMEN. Brain Res., 5, 289-305 (1967).
46. H.V.S. PEEKE and M.J. HERZ. Science, 173, 80-82 (1971).
47. C.G. PERT, M.J. KUHAR and S.H. SNYDER. Life Sci., 16, 1849-1854 (1975).
48. A.G. PHILLIPS, D.A. CARTER, and H.C. FIBIGER. Brain Res., 104, 221-232 (1976).
49. A.G. PHILLIPS, D.A. CARTER and H.C. FIBIGER. Psychopharmacol., 49, 23-27 (1976).
50. A.G. PHILLIPS, F. MORA, and E.T. ROLLS. Psychopharmacol., in press (1978).
51. R. PLOTNIK, D. MIR and J.M.R. DELGADO. Int. J. Psychobiol., 2, 1-21 (1972).
52. R.A. PRADO-ALCALÁ, E.W. KENT, and L.D. REID. Brain Res., 84, 531-540 (1975).
53. T.L. RICHARDSON, J.J. MILLER and H. McLENNAN. Brain Res., 127, 219 (1977).
54. H.E. ROSVOLD, In: Mind as a tissue. C. Rupp (Ed.) Hoeber, New York, 21-38 (1968).
55. A. ROUTTENBERG, J. comp. physiol. Psychol., 75, 264-276 (1971).
56. A. ROUTTENBERG, Nebr. Symp. Motiv., 27, 161-182 (1975).
57. A. ROUTTENBERG, and N. HOLZMAN, Science, 181, 83-86 (1973).
58. E.H. RUBINSTEIN, and J.M.R. DELGADO. Am. J. Physiol., 205, 941-948 (1963).
59. M. SIDMAN, J.V. BRADY, J.J. BOREN, D. CONRAD and A. SCHULMAN. Science, 122, 830-831 (1956).
60. J.R. STEVENS, C. KIM, and P. MACLEAN. Arch. Neurol. (Chic)., 4, 733-741. (1971).
61. H.L. TEUBER. In: Basal Ganglia, M.D. Yahr (Ed.), Raven Press, New York, 151-168 (1976).
62. U. UNGERSTEDT. Acta Physiol. Scand., Suppl. 367, 1-48 (1971).
63. D. van der KOOY, H.C. FIBIGER and A.G. PHILLIPS, Brain Res., 136, 119-130, (1977).
64. D. van der KOOY, F.G. LePIANE and A.G. PHILLIPS, Life Sci., 20, 981-986 (1977).
65. P.F. VON VOIGTLANDER and K.E. MOORE. Neuropharmacol., 10, 733-741 (1971).
66. R.P. WHITE and H.E. HIMWICH. J. Neurophysiol., 20, 81-90 (1957).
67. M.W. WILBURN and R.P. KESNER. Exp. Neurol., 45, 61-71 (1974).
68. S.A.K. WILSON, Brain, 36, 427-492 (1914).
69. N. WHITE and R. MAJOR. Can. J. Psychol., in press (1978).
70. R.H. WURTZ and J. OLDS, J. comp. physiol. Psychol., 56, 941-949 (1963).
71. E.J. WYERS and S.A. DEADWYLER. Physiol. Behav., 6, 97-103 (1971).
72. E.J. WYERS, S.A. DEADWYLER, N. HIRASUNA and D. MONTGOMERY. Physiol. Behav., 11, 809-819 (1973).
73. E.J. WYERS, H.V.S. PEEKE, J.S. WILLISTON and M.J. HERZ. Exp. Neurol., 22, 350-366 (1968).
74. S.F. ZORNETZER and R.B. CHRONISTER. Physiol. Behav., 10, 747-750 (1973).

Behaviour After Neostriatal Lesions in Animals

Susan D. Iversen

Department of Experimental Psychology, Downing Street, Cambridge CB2 3EB, England

INTRODUCTION

Over the last 40 years a number of attempts have been made to understand the function of the neostriatum. These efforts have been marked by periods of intense activity interspersed by long periods of neglect. Traditionally any discussion of the striatum begins with a statement defining it as a motor structure. Typical of such comment would be that of Jung and Hassler (39) stating that "the caudate nucleus and putamen, jointly forming the striatum, are the highest subcortical centres in the extrapyramidal motor system". Some of the earlier lesion studies are to be found in the neurological (82) literature (see Laursen (48) and Jung & Hassler (39) for reviews) where, not surprisingly, efforts were made to define neostriatal function in terms of motor control. This emphasis on motor function generated the first era of experimental studies in animals epitomised in the work of Denny Brown and Mettler (13,53). Brain lesion experiments performed on monkeys and cats, helped to define the role of the various sectors of the pyramidal and extrapyramidal motor systems in a delicately balanced motor apparatus. Throughout the 1950's, despite changing views on the details of the anatomical connections between the various components of the motor system (see Nieuwenhuys, (62) for review) and the introduction of new techniques, such as brain stimulation in behaving animals, the consolidated knowledge of the striatum was still viewed within an essentially motor framework. Achievements in this area were reviewed by Jung and Hassler (39) in their classic contribution to the Handbook of Neurophysiology. However, they include in that review some short but interesting paragraphs (pp 914-916) in which they reflect on the role of the striatum and related structures in the broad context of behaviour, concluding that "very little is known about the role of the extrapyramidal motor system in conditioned behaviour". Even enthusiasts may have been discouraged from pursuing this line of investigation by an earlier categorical statement of Lashley's (46) that "the evidence seems conclusive that in mammals the basal ganglia are not an essential link in the patterning of learned activities".

However, since 1960 the functions of the striatum have received broader attention and the use of experimental lesions has played an important role in changing concepts of striatal* function. Anatomical findings were largely responsible for opening up this new era of studies. It was shown in the rat (77), rabbit (9), cat (78) and monkey (55,15,11,40) that the striatum receives a highly organized projection apparently from all areas of the neocortex. Immediately it was asked why an essentially motor structure should receive projections from areas of cortex with no apparent involvement in motor performance. These observations added a new dimension to theories of striatal function and the hypothesis that the forebrain components of the extrapyramidal motor system might indeed serve cognitive, rather than strictly motor functions, was entertained for the first time. Laursen (48) was one of the first to grasp the importance of this development and reviewing all the previous experimental studies on the striatum in 1963, concluded with the comment "the function of the corpus striatum is on a high level of integration". A prophetic statement considering knowledge at the time.

The Organization of the Cortico-striatal Projections

Interest in anatomical connections between the neocortex and the striatum was encouraged by positive findings obtained using strychnine neuronography (24). Using both the Marchi and the then novel Nauta silver staining method for degenerating axons, Webster (77) examined the claim of Dusser de Barenne et al (24) that the striatum received input from areas of cortex other than the motor regions. In agreement with earlier workers Webster failed to find cortico-striatal connections with the Marchi technique but found a topographical projection of cortex onto the striatum using the Nauta technique. The summary diagram from that study is shown in Fig. 1 which combines Lashley's (45) differentiation of the various motor areas with Webster's observations on the areas of striatum receiving input from topographically defined cortical areas. Similar anatomical studies were undertaken in the monkey and in 1964 Nauta (55) reviewed the progress. There were a number of conflicting results in the literature (see Webster, (77)) but Nauta presented convincing evidence that striatal degeneration in the head of the caudate nucleus followed lesions to the prefrontal cortex. Furthermore, he made a preliminary observation that lesions confined to the dorsolateral frontal cortex produced degeneration in the dorsolateral region of the head of caudate, whereas lesions to the orbital (ventral) surface of the frontal lobe were followed by degeneration in a ventrolateral part of the head of caudate. This proved to be a particularly important observation (made possible by Nauta's judicious use of small, well defined cortical lesions) as it was already becoming apparent that the dorsolateral and orbitofrontal cortical lesions in the monkey resulted in different behavioural deficits (54). Further evidence for the existence of topographically organized cortico-striatal connections in the monkey (15,11) was soon obtained.

---

*The term striatum includes the caudate and putamen. There have been few studies on the putamen. In the manuscript, as is common practice, the term 'striatal' is often used when lesions involve only the caudate nucleus.

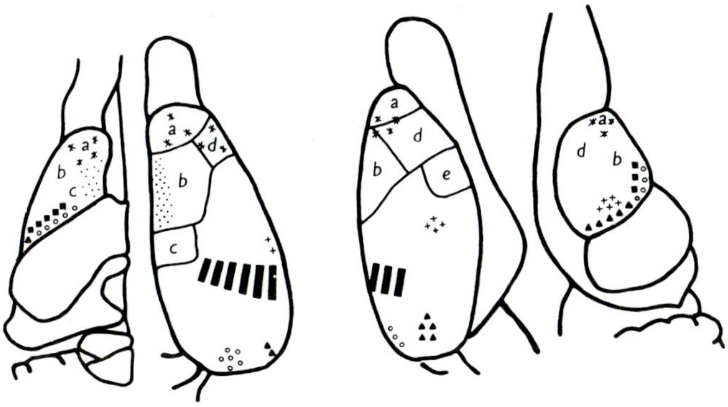

Plan of cortico-striate projections in the rat, to summarize the topographical arrangement. Explanation given with Text-fig. 1. The motor regions are modified from Lashley (1941): *a* = neck; *b* = forelimb; *c* = hindlimb; *d* = snout; *e* = mouth and tongue.

Fig. 1. Plan of cortico-striatal projections in the rat, to summarize the topographical arrangement. The motor regions are modified from Lashley (1941): a = neck; b = forelimb; c = hindlimb; d = snout; e = mouth and tongue. The cortical lesions are viewed from above (centre, left) and from the side (centre, right). On either side are diagrams of a decorticate rat brain viewed from above (far left) and from the side (far right), forming mirror images of the cortical diagrams. (Reproduced from Webster, 1961).

In these latter studies attention had focussed on the projections from the frontal lobe to striatum. However, in studies on the rabbit (9) and cat (78), it became clear that the whole of the cortical mantle including the occipital and temporal lobes projects topographically to the striatum. This brought into focus an earlier observation in the monkey that the inferotemporal cortex projected to the tail of the caudate nucleus and ventral putamen (79).

In recent years anatomical results have played a large role in guiding functional studies of striatum and thus some reference to such work was deemed appropriate in a behavioural contribution. However, Grofova's chapter in this book provides a more detailed appraisal of striatal anatomy.

The Comparison of Cortical and Striatal Lesions on Cognitive Tasks in the Monkey

Rosvold and Szwarcbart (70) originally suggested that if the caudate nucleus received projections from frontal cortex, then it was reasonable to ask if lesions to these areas of caudate produced the same deficits as lesions to the cortical origins of the afferent caudate projections. A number of studies have now investigated the effect on behaviour of lesions to discrete areas of the caudate which receive projections from dissociable areas of cortex. It was known that frontal lesions in the monkey produced deficits on a range of cognitive tasks including delayed response, delayed alternation, spatial alternation, object learning set, object reversal, go, no-go discrimination in all modalities, bar pressing during extinction; conditional position discrimination, adaption to prism misreaching (32,68, for reviews). Over the years in a number of independent studies, monkeys with bilateral electrolytic lesions to the head of caudate have been tested on most of these tasks and found deficient, amply fulfilling Rosvold and Szwarcbart's prediction. Some of the particular tasks and studies are cited in Table 1.

TABLE 1 Studies Demonstrating Certain Behavioural Deficits After Caudate Lesions in the Monkey

Delayed response	Battig et al, 1960	(3)
	Rosvold and Delgado	(69)
Delayed alternation	Rosvold et al, 1958	(71)
Go, no-go discimination	Battig et al, 1962	(4)
Conditional position discrimination	Rosvold, 1968	(68)
Adaption to prism misreaching	Bossom, 1965	(6)
Bar pressing extinction	Butters and Rosvold, 1963	(8)

In the course of further behavioural investigation of the frontal lesion syndrome results had been obtained suggesting that lesions confined to the orbitofrontal and dorsolateral cortex produced their greatest effect on different tasks from the "frontal repetoire" (54). Comparing performance on object discrimination reversal and go, no-go discrimination, it was found that the orbitofrontal lesion produced the more severe deficit on these tasks whereas on delayed alternation and conditional position discrimination the dorsolateral lesion resulted in the more severe impairment. Knowing that the dorsolateral and orbitofrontal cortex projected to different sectors of the head of caudate and that the inferotemporal cortex (involved in visual discrimination learning) projected to the tail of caudate, Divac et al (22) undertook the first double dissociation lesion study of monkey caudate (CN). The study is summarized in Table 2.

TABLE 2 Cortical Focus, Projection Area in Caudate and Behavioural Deficit After Caudate Lesion

Cortical focus	Projection Area in CN	Behaviour Impaired
Dorsolateral-frontal	Anterodorsal head	Delayed alternation (DA)
Orbitofrontal	Ventrolateral head	Object reversal
Inferotemporal	Tail	Visual pattern discrimination

A posteroventral lesion in the head produced the most severe DA impairment which was surprising, in view of claims that this area of caudate just dorsal to the anterior commissure does not receive a frontal projection. In the rabbit and cat the comparable area of the CN does receive input from an area of frontal cortex and lesions to this part of frontal cortex do indeed produce deficits on a delayed response task. This area of monkey caudate should therefore be re-examined both from anatomical and functional points of view. The inferotemporal cortex also projects to the putamen (79) and more recently, Buerger et al (7) have found visual pattern discrimination loss in the monkey after ventral putamen lesions.

Since the time of Divac's study a number of experiments have been done which make it necessary to look again at the fronto-caudate relationships. Firstly, the border between the dorsolateral and orbital cortex on the lateral surface of the frontal lobe has been redefined experimentally. Tissue on the convexity of the frontal lobe (Walker area 12), which was previously included in dorsolateral lesions is now known to be functionally part of the orbital cortex. Restriction of the dorsolateral removal to exclude area 12 tissue, results in a more clearly defined behavioural syndrome (49). Furthermore, the more restricted dorsolateral lesion was further divided by Goldman and Rosvold (29) into a sulcus principalis lesion and an arcuate sulcus lesion. The former lesion severely impaired delayed alternation performance and the latter conditional position discrimination. Secondly the orbitofrontal lesion, including area 12, defined and studied by Lawicka et al (49) has been the subject of further dissociation lesion studies. It was found that lesions to area 12 produce the typical orbitofrontal syndrome involving response disinhibition on tasks which require withholding of a response, (notably, go, no-go discrimination; and object reversal). The remaining part of the orbital lesion (i.e. the medial orbital cortex) did not produce such deficits (35). Aware of this finer behavioural parcellation of frontal cortex, Johnson et al (37) re-investigated in the monkey frontal-striatal projections using cortical lesions carefully confined to the behaviourally defined sectors (Fig. 2). Tissue dorsal

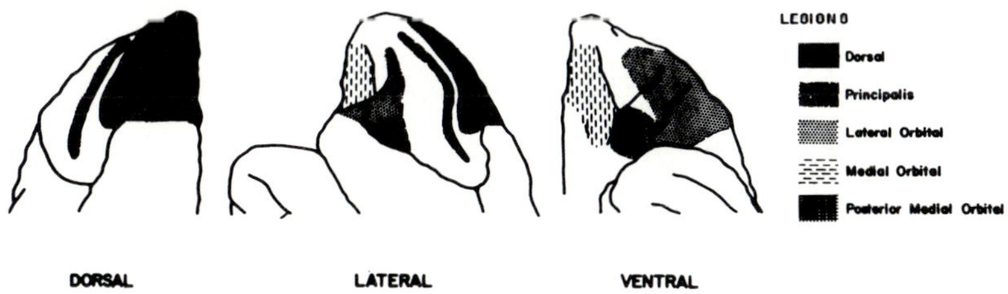

Diagrams of the different lesions based on photographs taken at autopsy.

Fig. 2. Diagrams of the different frontal cortex lesions based on photographs taken at autopsy. (Reproduced from Johnson et al, 1968).

to the sulcus principalis produced degeneration in the most dorsal rostral part of the head of caudate, while principalis lesions produced more widespread degeneration in this area; after lateral orbital lesions degenerating fibres were seen in more lateral, ventral and caudal areas of CN and this direction of degeneration was more pronounced after medial orbital damage (Fig. 3). Thus at least four areas of influence have been delineated in the head of caudate and it would be interesting to study selective lesion to these areas of CN on a wider range of the more recently developed behavioural tests.

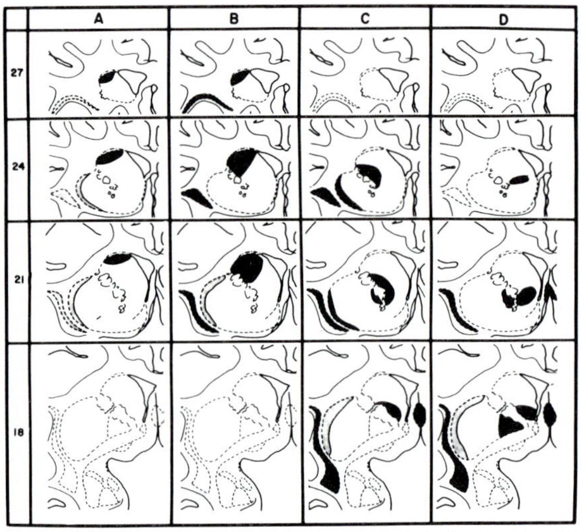

Schematic representations of the locus and extent of preterminal fiber degeneration in the basal ganglia and septum after dorsal (A), principalis (B), lateral orbital (C), and medial orbital (D) lesions. Numbers refer to stereotactic coordinates of frontal sections of a normal brain matched to those of Olszewski (14). Moderate degeneration is shown in black and light degeneration in gray.

Fig. 3. Schematic representations of the locus and extent of preterminal fibre degeneration in the basal ganglia and septum after dorsal (A), principalis (B), lateral orbital (C), and medial orbital (D) lesions. Numbers refer to stereotatic coordinates of frontal sections of a normal brain matched to those of Olszewski. Moderate degeneration is shown in black and light degeneration in grey. (Reproduced from Johnson et al, 1968).

Comparative Lesion Studies of the Striatum

Divac has demonstrated both in the cat and the rat deficits comparable to those seen in the monkey after selective lesions to frontal cortex and to the sectors of CN they innervate. For example in the cat, bilateral prefrontal or anterior caudate lesions were found to impair delayed response performance (16). The effect of the anterior caudate lesion was greater than that of the frontal lesion and as great as that produced by a combined frontal and caudate lesion. Similar results were obtained on the delayed alternation task (18). However, the caudate nucleus in this species appears to be involved in performance on these delay tasks only when the frontal cortex is intact. If cats relearn delayed response or delayed alternation after frontal lesions, a subsequent caudate lesion is without effect on performance (19,80), despite the fact that such a lesion would produce a severe impairment if made in an intact animal. In the rat the picture is less clear because (i) there is uncertainty about homologies (50,81) and (ii) the cortico-subcortical connections differ from those in higher forms (50). However, it is clear that both frontal cortex lesions and lesions in the corresponding neostriatal region in the rat result in behavioural deficits which are similar to those seen in monkeys with comparable damage. For example, lesions to either the anteromedial cortex (which receives input from the n.medialis dorsalis of the thalamus) or the anteromedial sector of the neostriatum result in deficits on spatial alternation (81), go, no-go alternation (2) and spatial reversal (17). Retention of spatial alternation in a T-maze was impaired to a greater degree after caudate than after frontal cortical lesions (81) suggesting that the caudate deficit cannot be ascribed to undercutting of the frontal cortex. Rats with caudate lesions were also impaired on alternation tasks tested in two lever operant conditioning chambers (10,30). In the latter study changes in activity level, response perseveration, motivation and memory were excluded as explanations of the alternation deficit. After caudate lesions, deficits have also been reported on discrimination reversal paradigms (41) and over-responding observed on schedules of reinforcement demanding slow, paced responding, such as those with differential reinforcement of low rates of responding (73,60). Few experiments in the rat have compared the effect of lesions confined to different areas of the caudate. Studies by Potegal (65) using tests of spatial orientation and Winocur (83) using 1 and 2 way active avoidance and passive avoidance are notable exceptions and both conclude that the caudate is functionally heterogeneous as regards involvement in these behaviours.

It is clear that the topographical division in the monkey between the anterodorsal and ventrolateral caudate is maintained in the output from the CN to other subcortical sites. Johnson and Rosvold (36) placed electrolytic lesions in these behaviourally defined sectors of the caudate and compared the pattern of degeneration. The dissociation in the CN was maintained in the projections to the globus pallidus and substantia nigra, representing at least on the anatomical level, the continuation to subcortical sites of the two functional systems which originate in the dorsolateral and orbitofrontal cortex. A hypothesis is proposed that the dorsolateral frontal system is concerned with the solution of spatial problems and the orbitofrontal with the control of response tendencies and that the efferent circuits concerned with these aspects of behavioural control remain distinct, not only in the striatum, but also in the globus pallidus, thalamus and substantia nigra. A further suggestion is made that deficits in avoidance behaviour which have been reported in cats, rats and monkeys after frontal, caudate and globus pallidus lesions (41,72,63, 47) are most likely related to interference with orbitofrontal projections. In conclusion, Johnson and Rosvold (36) state that "taken together, these studies lead to the conclusion that the head of the caudate nucleus is a composite of subdivisions". They go on to say "However, it is not generally evident that the subdivisions which subserve the different behavioural functions are the same as those defined by anatomists and electrophysiologists". However, their observations on the efferent projections of anterodorsal and ventrolateral sectors of caudate strengthens the view that behavioural heterogeneity correlates at least with anatomical heterogeneity.

Attempts to Relate Other Aspects of Striatal Neural Organization with Behaviour

The striatum is heterogeneous in terms of its neuronal structure (Pasik et al, this volume), neurochemical substrates (Fonnum, this volume) and electrophysiological responses (Feger, this volume) as well as in terms of its afferent and efferent connections. It remains a challenge to relate the behavioural functions of the striatum to these other features of its organization. Neill and Grossman (58) have approached the problem pharmacologically, showing that the blockade of cholinergic neurotransmission in different areas of caudate (achieved by local injections of an anti-cholinergic drug, scopolamine) resulted in different behavioural effects. In the dorsal caudate scopolamine produced an impairment of shock avoidance behaviour, whereas the behaviour was facilitated after scopolamine injections into the ventral striatum. Similar facilitatory effects on bar pressing tasks requiring slow, paced responding have been obtained from the ventral striatum after injection of scopolamine, dopamine and amphetamine (57). The cholinergic substrate in the ventral striatum appears to interact with dopamine terminals in the mediation of avoidance behaviour. Electrolytic lesions (59) to ventral striatum, which resulted in dopamine loss from this structure, impaired avoidance performance, blocked the behavioural effects of amphetamine and the severity of the impairments correlated with the degree of dopamine depletion. By contrast, lesions in the dorsal area produced poor dopamine depletion, gave a small effect on avoidance, and did not alter amphetamine responses. There are many drugs which interfere with the neurotransmitters found in the striatum and their use in local injection studies offers a new way of interfering with and studying striatal functions. Lesion techniques, which destroy one class of neurone or one neurotransmitter system are particularly valuable in studying structures as heterogeneous as the striatum. Kainic acid is one such chemical toxin, which destroys neuronal cell bodies rather than axons or terminals and particularly neurones which receive terminals of glutamate containing neurones. One such pathway has been identified projecting from frontal cortex to the striatum (20) and kainic acid injected into this area of striatum results in substantial cell loss. Divac et al (21) have found that if kainic acid is injected into the anteromedial caudate of the rat, a severe deficit on delayed alternation is seen. Electrolytic lesions to the same area of caudate produced a similar impairment.

The role of the Afferent Dopamine Tract in Striatal Function

Developing knowledge of the functions of the nigro-striatal dopamine pathway forms a basis for the hypothesis that the striatum is involved in sensorimotor organization. In the mid 1960's fluorescent histochemical techniques made it possible to stain *in situ* dopamine containing pathways in the rat brain. The major group of dopamine neurones was found in the substantia nigra with ascending axons innervating the striatum and globus pallidus. Hornykiewicz (26) reported loss of dopamine in the striatum of patients with Parkinson's disease. The loss of movement in this disease was considered consistent with degeneration of the dopaminergic input to the striatum. However, experimental work on the function of the nigro-striatal DA pathway indicates that loss of the pathway is associated with a more general failure of sensorimotor co-ordination than the clinical term akinesia implies. Rats with bilateral nigro-striatal lesions do not move, feed or drink; they simply do not respond to the external world and die rapidly. Using bilateral cryogenic probes to reversibly disrupt caudate function in the monkey, Proctor-Bowen (66) found that tracking a moving light around a screen was severely impaired during cooling. Of course we do not know how the cryogenic procedure disrupts striatal organization but the study does provide further evidence for a striatal role in sensorimotor co-ordination. The sensory aspect of Parkinson's disease has never been emphasized but in fact it is clear that patients are not under the normal control of salient cues for sensorimotor co-ordination. Indeed, if these cues are made more dramatic the initiation of appropriate motor responses improves. For example, the shuffling walking pattern is improved if clear parallel lines are painted on the floor to provide guide lines for stepping (52).

In some ways unilateral NS lesions have proved more illuminating for understanding nigro-striatal organization than models involving bilateral interference with the dopamine pathways. Ungerstedt (76) emphasized the multimodal contralateral sensory neglect associated with a unilateral neurotoxin lesion to the substantia nigra, a deficit described many years before by Teitelbaum and Epstein (75) after what we now assume was a transection of the nigro-striatal dopamine tract at the level of the hypothalamus. These results support the view that DA release in the striatum is essential for sensorimotor co-ordination, or to put it in other words, is necessary if the animal is to select and come under the control of the most important and relevant environmental stimuli at a particular time. Results consistent with this view have come from electrical stimulation studies aimed at activating the nigro-striatal tract. Bandler and Flynn (1) for example, found that unilateral hypothalamic stimulation in the cat induced lunging towards a mouse in the contralateral but not the ipsilateral visual field. In a conditioning paradigm Beagley and Holley (5) used lateral hypothalamic stimulation to induce spontaneous eating and lever pressing for food. The latter response was trained in the presence of a light stimulus. Subsequently it was found that electrical stimulation of the LH induced lever pressing for food only when the light was presented to the eye contralateral to the brain stimulation. These stimulation experiments probably activate the NS pathway and suggest that its activation sensitises the animal to a particular stimulus modality in one part of space. In animals with unilateral damage to the nigro-striatal dopamine system orientation is towards the half of space programmed by the intact striatum. General stimulation activated the intact side of the brain and produced rotational behaviour towards this half of space. Drugs which release dopamine from the intact dopamine terminals, activate the post synaptic dopamine receptors, or block these receptors, produce predictable alterations in the direction and rate of rotational behaviour. The "nigro-striatal model" has played a major role in identifying and quantifying the potency of drugs which interact with dopamine neurones.

Recently interest has focussed on the efferent systems mediating rotational behaviour. Marshall and Ungerstedt (51) abolished turning with cuts of the striato-nigral pathways. More recent observations of Crossman et al (12) that motor decortication has only transitory disruptive effects on nigro-striatal turning suggest that thalamo-cortical pathways are not essential for the behaviour. York and

Faber (85) suggest that projections from substantia nigra to the midbrain tectum represent an important further stage in this efferent pathway.

Glick has proved that in rats there are endogenous differences in dopamine levels in the two striata which may account for the spontaneous asymmetries observed; the stable preferences, for example, for the right or left lever in conditioning experiments in Skinner boxes (27). When a preference has developed it is assumed to be controlled by the contralateral striatum. The non-dominant striatum can then be activated with implanted electrodes and this is followed by a shift of preference to the other lever (86). These endogenous asymmetries may have a biological advantage. Zimmerberg et al (87) have screened rats for rotation to determine their degree of asymmetry and then trained them in a T-maze in which they could escape from the shocked arm to either left or right. Rats with strong endogenous asymmetry learned and remembered better than non-rotators. Furthermore, learning was most rapid if the safe side in the shock box coincided with their natural preference.

Bilateral nigro-striatal lesions in rats and Parkinson's disease in man illustrate the deficiency of sensorimotor integration seen after loss of forebrain dopamine. By contrast, treatment with amphetamine, which acts to release brain dopamine from synaptic storage sites, may throw light on the normal role of dopamine. Amphetamine is a classical stimulant drug and in low doses increases the output of behaviour whether it be unconditioned or conditioned. In man amphetamine is used to increase mental agility. The stimulation is dose dependent but if excessive doses are used disruption of ongoing behaviour is observed. In all species studied, including man, high doses of amphetamine induce an abnormal repetitive motor behaviour described as stereotyped (67). The animal appears trapped in one small element of its world and endlessly repeats a particular act. Rats tend to sniff and lick a particular spot, monkeys pick the skin in one part of the body, people repeat complex mechanical manipulations. The act selected for repetition can be influenced by training, suggesting that behaviours with a high probability of occurrence under normal conditions are those most likely to be enhanced by amphetamine (33). Perhaps then, the normal physiological release of dopamine in the striatum is responsible for selective access of stimuli to appropriate action patterns. Overstimulation of this dopamine mechanism prevents the normal tight coupling between intention and action. This is easy enough to understand when one considers simple stimulus-response correlations. But returning to an earlier section it must be emphasized again that any theory of striatal function must account for the fact that conditioned behaviour as well as the simpler stimulus-response behaviour is mediated by striatum. Could it be that the term "cognitive" is misleading? What do we mean by cognitive? Very often that the behaviour observed cannot be easily accounted for in terms of simple stimulus-response elements. What goes on, for example, when a rat reliably alternates between left and right positions? What and where are the stimuli guiding this behaviour? If they were known we should be much happier to talk of the behaviour in terms of sensorimotor co-ordination. It may be proposed, albeit tentatively, that dopamine in the striatum is essential for all forms of behaviour which involve information processing in that substrate. There are few studies relating striatal dopamine levels to cognitive performance. Kitsikis et al (43) noted in the cat that low levels of striatal dopamine are associated with poor performance on a delayed-response task. In "poor performers", L-DOPA therapy improved delayed response performance.

Interaction of Striatal Afferent Influences and Behaviour

What then is the nature of intrinsic striatal information processing? We do not have an answer to this question. The concept which has received the most general acceptance, views the striatum as a tonic inhibitory structure. Release from its influence leads to general or more specific loss of behavioural control. But does the striatum merely act as a switch permitting the access of cortex to lower motor control centres or is information processed in transit through the structure. The intricate neuronal architecture suggests that processing does occur. The evidence we have already on the anatomical and functional properties of striatal afferent systems suggests a selective rheostat mechanism, which sets up areas of preferential information flow in specific areas of the striatum and thus increases the probability of certain types of sensorimotor integration occurring.

There is apparently a greater degree of selective anatomical coding in the striatal connections then previously realised. Goldman and Nauta (28) reinvestigating in the monkey fronto-striatal connections with modern tracing methods and Jones et al (38) focussing on the sensory-motor cortex both conclude that :-

(i) specific cortical areas do project to a particular part of caudate or putamen

(ii) but within that area the terminals form islands of innervation with the space in between occupied, presumably, by other afferent systems to that part of striatum e.g. thalamo-striatal or nigro-striatal projections.

(iii) there is a degree of overlap in the projection fields of different areas of cortex.

In a more recent study Yeterian and van Hoesen (84) have added an important new dimension to these anatomical considerations. The fact that certain cortical areas have reciprocal trans-cortical connections (64) (e.g. parietal lobe and arcuate sulcus) is reiterated and developed with the observation that the projections from such reciprocally connected cortical areas to the striatum overlap in that structure. The "island" organization (28,38) would be compatible with such an arrangement. Thus cortical interactions continue at the level of the striatum. It would be interesting to know how the afferent dopamine terminals fit into this overlapping anatomical arrangement. What we do know is that the nigro-striatal DA input to caudate is, itself, topographically organized both in terms of the input to the striatum from the zona compacta cells of the substantia nigra and in terms of the striato-nigral feedback pathways onto the dopamine cell bodies (23,25). It is possible that the nigro-striatal dopamine pathway is capable not only of striatal arousal but striatal arousal in a highly localized area which is processing a particular pattern of information. The selective activation of such a focus would be maintained by the topographically organized feedback to substantia nigra.

Mason and Fibiger (personal communication) have performed lesion experiments with kainic acid which suggest that striato-nigral feedback is essential for balanced striatal function. The dorsolateral striatum was injected with kainic acid to produce severe loss of striatal neurones. These rats were not spontaneously hyperactive but when treated with low doses of the stimulant drug, d-amphetamine (0.5 mg/kg) showed a greatly enhanced motor response. They suggest that damage to the descending modulatory limb of the dopamine circuit, leaves the striatum unbalanced. It was mentioned earlier that Divac's animals with similar bilateral lesions were unable to perform on a cognitive task. Kainic acid may therefore provide an interesting new tool for lesioning striatal neurones and studies should be extended to areas of striatum receiving topographically organized input from posterior cortical regions. The neurotoxin 6-hydroxydopamine is also available for selective lesioning of dopamine terminals in the striatum. Now that the topography of the dopamine terminals has been worked out more precisely it would be possible to compare kainic acid and 6-hydroxydopamine lesions to striatal foci with identified behavioural properties. Why, it may be asked has such emphasis been placed on the role of dopamine in the striatum when it is known that ACh, serotonin, glutamate, GABA enkephalin and perhaps other neurotransmitters are present at striatal synapses. The answer is that firstly we do not know as much about these transmitters but secondly, and much more importantly, it appears that these other transmitter systems do not innervate the striatum in the highly organized manner seen with the dopamine neurones. This suggests to Fallon and Moore (25) that dopamine in the striatum plays an important organizational role. In speculations on circuitry one is left with the question of the origin of the stimuli which set the nigro-striatal system in operation. It is known that sensory stimuli activate the nigro-striatal dopamine pathway (61) and it is tempting to suggest that such influences reach the substantia nigra via the cortico-striatal, striato-nigral route and thereby invoke the dopamine arousal system to facilitate their processing in the striatum.

The nigro-striatal dopamine system is paralleled by an ascending non-striatal dopamine system (76) which according to Heimer and Wilson (31) innervates the components of a "ventral striatum" system (e.g. nucleus accumbens, tuberculum olfactorium) rather than the dorsal striatum. Just as processing in the dorsal striatum allows sensorimotor co-ordination to the external world, so the ventral striatum receives input from limbic areas and provides the means by which affective stimuli gain access to response patterns (34). The nucleus accumbens projects to the globus pallidus and hence to the motor systems (74) but also to the substantia nigra. The latter connection is the focus of much interest as it could provide a basis for "forebrain crosstalk" between motivation and sensorimotor arousal an interface between affect, cognition and action (56).

CONCLUSIONS

Despite impressive advances in our knowledge of striatal structure and function it remains difficult to formulate a theory of striatal function. At least older concepts are being disproved. The striatum appears more than a general inhibitory centre on the output system of brain. Nor is it merely modulating efferent motor messages from cortex. The striatum is involved in action and Kornhuber believes in an important way. External and internal stimuli goad animals to action; stimuli processed by sensory-associative cortex and by its limbic equivalent. Both these processing areas impinge on striatum. Movements have been classified as ramp or ballistic i.e. fast or slow. Kornhuber (44) proposes that when an organism is moved to act in a certain way programmes of ramp and ballistic movements are generated; the former by the striatum and the latter by the cerebellum. These instructions activate muscles via thalamo-cortical and subcortical routes. What we do not understand is how the subroutine of actions appropriate to a particular set of circumstances is selected. There is no evidence that striatal stimulation induces co-ordinated motor acts, indeed, Jung and Hassler (39) have commented that only in the tegmentum can purposeful acts be elicited by electrical stimulation. Despite this fact, it has been shown in the monkey that neuronal activity in the basal ganglia is correlated with the performance of learned responses, particularly those involving slow, ramp movements (14). But how does the striatum gain access to such coded motor acts? Attention to the efferent rather than the afferent inputs of the striatal complex may prove to be a useful approach in pursuing this difficult question.

REFERENCES

(1) R. Bandler and J.P. Flynn, Science, 171, 817-818 (1971).

(2) D.J. Barker and G.J. Thomas, Physiol. Behav. 1, 313-317 (1966).

(3) K. Battig, H.E. Rosvold and M. Mishkin, J. comp. physiol. Psychol. 53, 400-404 (1960).

(4) K. Battig, H.E. Rosvold and M. Mishkin, J. comp. physiol. Psychol. 55, 458-463 (1962).

(5) W.K. Beagley and T.L. Holley, Science, 196, 321-322 (1976).

(6) J. Bossom, Psycho. Sci., 2, 45-46 (1965).

(7) A.A. Buerger, C.G. Gross and C.E. Rocha-Miranda, J. comp. physiol. Psychol., 86, 440-446 (1974).

(8) N. Butters and H.E. Rosvold, J. comp. physiol. Psychol., 65, 397-403 (1968).

(9) W.M. Carmen, J.B. Cowan and T.P.S. Powell, Brain, 86, 525-562, (1963).

(10) S.L. Chorover and C.G. Gross, Science, 141, 826-827 (1963).

(11) W.M. Cowan and T.P.S. Powell, J. Neurol. Neurosurg. Psychiat. 29, 426-439 (1966).

(12) A.R. Crossman, M.A. Sam Brook, S.W. Gergies and P. Slater, J. Neurol. Sci., 34, 407-414 (1977).

(13) D. Denny-Brown. In: M.D. Yahr and D.P. Purpura (Eds.), Neurophysiological Basis of Normal and Abnormal Motor Activites Raven Press, New York, 415-444 (1967).

(14) M.R. De Long. In: F.O. Schmitt and F.G. Worden (eds.), The Neurosciences: Third Study Program MIT Press, Cambridge Mass, 319-325.

(15) J.L. De Vito and O.A. Smith, J. comp. Neurol. 123, 413-424 (1964).

(16) I. Divac, Acta. Biol. Exp. 28, 149-167 (1968).

(17) I. Divac, Neuropsychologia, 9, 175-183 (1971).

(18) I. Divac, Physiol. & Behav. 8, 519-522 (1972).

(19) I. Divac. Physiol. Psychol. 2, 104-106 (1974).

(20) I. Divac, F. Fonnum and J. Storm-Mathisen, Nature, 266, 377-378 (1977).

(21) I. Divac, H.J. Mar Kowitsch and M. Pritzel, Brain Res. 151, 523-532 (1978).

(22) I. Divac, H.E. Rosvold and M.K. Szwarcbart, J. comp. physiol. Psychol. 63, 184-190 (1967).

(23) V.B. Domesick, R.M. Beckstead and W.J.H. Nauta, Neurosciences Abstracts, II, 61 (1976).

(24) J.G. Dusser de Barenne, H.W. Garol and E.W.S. McCulloch. In: Diseases of the Basal Ganglia, Res. Publ. Ass. Nerv. Ment. Dis. 21, 246 (1942).

(25) J.H. Fallon and R.Y. Moore, J. comp. Neurol.180, 545-580 (1978).

(26) I.J. Farely, K.S. Price and O. Hornykiewicz. In: E. Costa and G.L. Gessa (Eds.), Advances in Biochemical Psychopharmacology Vol. 16, Raven Press, New York 57-64 (1977).

(27) S.D. Glick and T.P. Jerussi, J. Pharmacol. exp. Ther. 188, 714-725 (1974).

(28) P.S. Goldman and W.J.H. Nauta, J. comp. Neurol. 171, 369-386 (1977).

(29) P.S. Goldman and H.E. Rosvold, Exp. Neur. 27, 291-304 (1970).

(30) C.G. Gross, L. Chorover and S.M. Cohen, Neuropsychologia, 3, 53-68 (1965).

(31) L. Heimer and R.D. Wilson. In: M.M. Santini (Ed.), The Golgi Centennial Symposium: Perspectives in Neurobiology, Raven Press, New York (1975).

(32) S.D. Iversen. In: A. Deutch (Ed.), The Physiological Basis of Memory, Academic Press, 305-364 (1973).

(33) S.D. Iversen. In: A.R. Cools, A.H.M. Lohman and J.H.L. van den Bercken (EDS.), Psychobiology of the Striatum,Elsevier, North-Holland, 99-118 (1977).

(34) S.D. Iversen. In: L.L. Iversen, S.D. Iversen and S.H. Snyder (Eds.), Handbook of Psychopharmacology, Vol. 8, Plenum Press, New York, 333-384 (1977).

(35) S.D. Iversen and M. Mishkin, Exp. Brain Res. 11, 376-386 (1976).

(36) T.N. Johnson and H.E. Rosvold. Exp. Neurol. 33, 584-596 (1971).

(37) T.N. Johnson, H.E. Rosvold and M. Mishkin, Exp. Neurol. 21, 20-34 (1968).

(38) E.G. Jones, J.D. Coulter, H. Burton and R. Porter, J. comp. Neur. 173, 53-80 (1977).

(39) R. Jung and R. Hassler. In: J. Field (Ed.), Handbook of Neurophysiology Vol. II, Field American Physiol. Soc. Washington, 863-928 (1960).

(40) J.M. Kemp and T.P.S. Powell, Brain, 93, 525-546 (1970).

(41) R.J. Kirkby and D.P. Kimble, Exp. Neurol. 20, 215-227 (1968).

(42) R.J. Kirkby, Physiol. Behav. 4, 451-545 (1969).

(43) A. Kitsikis, A.G. Roberge and G. Frenette, Exp. Brain Res. 15, 305-317 (1972).

(44) H.H. Kornhuber. In: F.O. Schmitt and F.G. Worden (Eds.), The Neurosciences: Third Study Program, MIT Press, Cambridge, Mass. 267-280 (1974).

(45) K.S. Lashley, Brain, 44, 256-286 (1941).

(46) K.S. Lashley, S.E.B. Symposium IV, 454-482 (1950).

(47) A.M. Laursen, Acta Physiol. Scand. 57, 81-89 (1963).

(48) A.M. Laursen, Acta Physiol. Scand. (suppl.) 211, (1963).

(49) W. Lawicka, M. Mishkin and H.E. Rosvold, Acta Neurobiol. 35, 581-607 (1975).

(50) C.M. Leonard, Brain Behav. Evol. 6, 524-541 (1972).

(51) J.F. Marshall and U. Ungerstedt, Science, 198, 62-64 (1977).

(52) J.P. Martin, The Basal Ganglia and Posture Lippincott, Philadelphia (1967).

(53) F. Mettler. In: M.D. Yahr and D.P. Purpura (Eds.), Neurophysiological Basis of Normal and Abnormal Motor Activities, Raven Press, New York, 445-497 (1967).

(54) M. Mishkin. In: J.M. Warren and K. Akert (Eds.), The Frontal Granular Cortex and Behaviour, McGraw Hill, New York, 219-241, (1964).

(55) W.J. Nauta. In: J.M. Warren and K. Akert (Eds.), The Frontal Granular and Cortex and Behaviour, McGraw Hill, New York, 397-407 (1964).

(56) W.J. Nauta and V.B. Domesick. In: K.E. Livingston and O. Hornykiewicz(Eds.), Limbic Mechanisms, Plenum Publishing Co. 75-93 (1978).

(57) D.B. Neill, Brain Res, 105, 89-103 (1976).

(58) D.B. Neill and S.P. Grossman, J. comp. physiol. Psychol. 71, 311-317 (1970).

(59) D.B. Neill, W.O. Boggan and S.P. Grossman, J. comp. physiol. Psychol. 86, 1019-1030 (1974).

(60) D.B. Neill, J.F. Ross and S.P. Grossman, Physiol. Behav. 13, 297-305 (1974).

(61) A. Nieoullon, A. Cheramy and J. Glowinski, Nature, 269, 340-342.

(62) R. Nieuwenhuys. In: A.R. Cools, A.H.M. Lohman and J.H.L. van den Bercken (Eds.), Psychobiology of the Striatum, Elsevier, North-Holland, 1-19 (1977).

(63) M.E. Olds and J. Olds, J. comp. Neurol. 120, 259-295 (1963).

(64) D.N. Pandya and H.G.J. Kuypers, Brain Res. 13, 13-36 (1969).

(65) M. Potegal, J. comp. Physiol. Psychol. 69, 756-764 (1969).

(66) F. Proctor-Bowen, Neuropsychologia, 7, 59-65 (1969).

(67) A. Randrup and I. Munkvad. In: E. Costa and S. Garattini, (Eds.), Amphetamines and Related Compounds Raven Press, New York, 695-713 (1970).

(68) H.E. Rosvold. In: C. Rupp (Ed.), Mind as a Tissue, Harper and Row, New York, 21-38 (1968).

(69) H.E. Rosvold and J.M.R. Delgado, J. comp. Physiol. 49, 365-372 (1956).

(70) H.E. Rosvold and M.K. Szwarcbart. In: J.M. Warren and K. Akert (Eds.), The Frontal Granular Cortex and Behaviour McGraw Hill 1-15 (1964).

(71) H.E. Rosvold, M. Mishkin and M.K. Szwarcbart, J. comp. physiol. Psychol. 51, 437-444 (1958).

(72) E.H. Rubinstein and J.M.R. Delgado, Am. J. Physiol. 205, 941-948 (1963).

(73) L.W. Schmaltz and R.L. Isaacson, J. comp. physiol. Psychol. 65, 343-348 (1968).

(74) L.W. Swanson and W.M. Cowan, Brain Res. 92, 324-330 (1975).

(75) P. Teitelbaum and A.N. Epstein, Psychol. Rev. 69, 74-90 (1962).

(76) U. Ungerstedt, Acta physiol. Scand. Suppl. 367, 95-122 (1971).

(77) K.E. Webster, J. Anat. 95, 522-545 (1961).

(78) K.E. Webster, J. Anat. 99, 329-337 (1965).

(79) D.G. Whitlock and W.J.H. Nauta, Neurol, 106, 183-212 (1956).

(80) R.G.E. Wikmark and I. Divac, Israel J. Med. Sci. 9 (Suppl.) 92-97 (1973).

(81) R.G.E. Wikmark, I. Divac and R. Weiss, Brain Behav. Evol. 8, 329-339 (1973).

(82) K. Wilson, Brain, 36, 427-492 (1914).

(83) G. Winocur, J. comp. physiol. Psychol. 86, 432-439 (1974).

(84) E.H. Yeterian and G.W. van Hoesen, Brain Res. 139, 43-63 (1978).

(85) D.H. York and J.E. Faber, Brain Res, 130, 383-386 (1977).

(86) B. Zimmerberg and S.D. Glick, Brain Res, 86, 335-338 (1975).

(87) B. Zimmerberg, A.J. Strumpf and S.D. Glick, Brain Res, 140, 194-196 (1978).

Functional Concepts

Introduction

I. Divac and R. G. E. Öberg

In this part of the book, four general concepts about NS functions are being considered, namely, the dominating notions of motor and inhibitory functions as well as the much less widely accepted concepts of sensory and "cognitive" functions. These four categories were not chosen because we think they represent particularly meaningful entities or clearcut alternatives (we do not), but rather because they reflect distinct traditions of research and interpretation. (1) It may validly be argued e.g. that initiation of movement and behavioural inhibition are two sides of the same function, that some sensory processes and certain inhibitory phenomena ought to be sorted under "cognitive" mechanisms, or, more radically, that the routes leading from the senses via the intermediate net to the final common path form continuous systems where exact borders and subclassificaitons are artificial and useless. However, current research on NS originated in times with other ways of thinking, and the main conceptual frameworks to be discussed here cannot be annihilated by general considerations about the organisation of the nervous system and behaviour. They can, on the other hand, gradually be compared, related and integrated through exchange of thoughts and findings. The initiation of such a rapprochement was the modest but exciting aim of the fourth session of the meeting.

The first chapter below presents a short history of the four concepts of NS functioning as well as a few reflections on their current status. The subsequent chapters review the available evidence for each class of funtions. The book concludes with a joint contribution by Drs Albe-Fessard and Buser in which the two last sections are discussed.

---

(1) In retrospect we regret having restricted our attention to four concepts of the functions of NS. In the last few years, a number of studies have indicated that this formation may be of importance also for motivational processes. This notion is currently even less widely accepted than the concepts of cognitive and sensory processes, but some of the evidence in its favor is worthy of serious attention.

Current Conceptions of Neostriatal Functions History and an Evaluation

Ivan Divac* and R. Gunilla E. Öberg**

**Institute of Neurophysiology, University of Copenhagen, Juliane Maries Vej 36 Copenhagen, Denmark*
***Laboratory of Psychology, University of Copenhagen, Njalsgade 92, Copenhagen, Denmark*

They (concepts) are too broad and (or) ill defined to be meaningful and they are too likely to lead to an ad hoc interpretation of symptoms. The attractiveness or plausibility of such formulations is part of their danger; it is too easy to think of all the facts they account for and not to ask whether there are others which they do not explain or whether symptoms are absent which they require.

D.O. Hebb, 1954[43]

In contemporary textbooks, the neostriatum (NS) is almost invariably ascribed motor and inhibitory functions. Alternative or supplementary views about sensory and "cognitive"* functions are maintained by few, and the general impact of these notions is presently slight. All four conceptions have, however, a long history, and the favored doctrine has changed more than once. In this chapter we shall give a mini-review of the history of each of these ways of looking at NS. We shall also make a brief assessment of their current status, in which we attempt to challenge the privileged position presently granted to concepts of motor and inhibitory functions.

HISTORY**

Neostriatal functions have been the subject of speculation for many centuries. The earliest theorists were impressed by NS' position next to the cerebral ventricles, which were then thought to be the seat of "psychic spirits" such as imagination and memory[9]. By virtue of its topographic proximity to the ventricles, the NS was also ascribed highly sophisticated functions. In the influential book by Mundinus, written at the very beginning of the fourteenth century, the NS was pictured as

---

*This term is used only as a provisional label for various "integrative" or "intermediate" functions which do not seem to be usefully described by any other current notion. We are loosely referring to processes such as learning, memory, or perception, but current ignorance precludes definite formulations. See also Öberg and Divac, this volume.

**This historical sketch is largely based on several reviews[9,65,68,87,88,139], to which the reader is referred for more comprehensive accounts.

controlling cognitive processes in the following fashion: "If, then, the individual wishes to suspend cogitation, he elongates his worms (caudate nuclei), and this prevents memory and imagination from uniting and forming 'common sense' and eventually 'ratio'" (cited in ref. 68).

More than three centuries later, Willis published the first serious attempt at an anatomical description of NS[138]. Willis believed that all incoming sensation is received by the NS, which was therefore called "sensorium commune". Swedenborg further promoted the significance of the NS when he declared, in 1780, that "the royal road of the sensation of the body to the soul is through the corpora striata, and all determinations of the will also descend by that route" (cited in ref. 68).

A new era in the exploration of NS functions was entered in the nineteenth century when empirical research was winning ground and speculations found certain restrictions in experimental results. It is inherent in this shift of method that functional theories changed content: when animal behavior is observed, results are generally interpreted in terms of motor mechanisms rather than as manifestations of ratio or will. Hence, the early phase of experimental brain research quite naturally saw an excessive emphasis on motor functions.

The empirical analysis of NS' behavioral roles was initiated in 1820 by Flourens, but he failed to elicit movements by stimulation of the NS, and his studies were of little influence[139]. Two decades later, however, Magendie found that rabbits with large lesions in the NS lept forward as if under an irresistible impulse. Magendie reached the conclusion that corpus striatum controls backwards locomotion, and that it normally acts in balance with the cerebellum, which controls forward locomotion[139]. According to Mettler[88], the notion of motor functions of NS was related to Bell-Magendie's law: the NS, localized anteriorly in the hemisphere, was thought to be the site of origin of pes pedunculi and the anterior funiculi, and thus directly related to the motor roots of the spinal cord. The thalamus, placed posteriorly, was thought to be the target of the dorsal columns.

Magendie's description of "nodus cursorius" in the anterior part of the NS stimulated an endless series of experiments in which caudate nuclei were pinched, pierced, injected and electrified, all in attempts to find out whether or not such treatments would produce movements[139]. Sometimes they did, sometimes they did not, and as a consequence the belief in motor functions was waxing and waning, but it never vanished again. The manner in which the motor roles of NS were conceived did, however, undergo numerous transformations in the process.

According to Wilson[139], the NS was considered in the early days of scientific neurology to be an "organ as high in the cerebral hierarchy as the cortex itself, endowed with motor functions as elaborate and as detailed. But the change took place as it was realized that many of the functions assigned to it were the property of the cortico-spinal paths, and almost at once it seemed to fall from its high estate and depreciate in physiological significance". One of the most influential formulations of NS' presumed motor functions can be traced via its promotor, Ariëns Kappers, to Rezek (1887, cited in ref. 87), who suggested that NS mediates movements which, "after being called into existence by psychic processes, are continued automatically". The view of NS as a motor structure received significant impetus from the clinico-pathological findings of Wilson[139] and many successors (e.g. ref. 133), and from the birth of the "extrapyramidal motor system", of which NS became an important constituent[52].

Today, many authorities continue to regard NS involvement in regulation of posture and movement as the best documented notion of its functions[2,18,53,81], serious criticisms notwithstanding[65,92].

The notion of inhibitory functions of NS was first advanced in 1875 by the Russian physiologist Danilewski[16], who was probably influenced by Sechenov's ideas about inhibitory processes in the brain. In a now forgotten experiment Danilewski showed that limb and body movements produced by stimulation of the peripheral nerves could be prevented by electrical stimulation aimed at the caudate nucleus[16]. In the 1920's the view of NS as an inhibitory structure won new adherents: Tilney and Riley[122] postulated that the role of the paleostriatum in fish is to produce oscillatory movements, and that the neostriatum evolved in higher vertebrates in order to inhibit this mechanism which was unnecesary for life on the ground. The Vogts[134] suggested that the NS inhibits, and the globus pallidus facilitates movements.

In more recent times, two groups of investigators contributed significantly to the acceptance of the idea that NS has inhibitory functions. In 1938, Dusser de Barenne and McCulloch[25] reported that the inhibitory cortical strips (later shown to be artefacts) project to the caudate nucleus; these authors introduced the notion of a particular neuronal circuit mediating inhibition. One year later, Mettler *et al.* found that stimulation aimed at the caudate nucleus can inhibit movements[89]. These two reports dove-tailed so well that many were convinced, and the inhibitory functions of NS were sufficiently established to stay to our days.

In spite of the heavy emphasis on NS' motor and inhibitory roles during the past century, some authors ventured to ascribe more "complex" functions to this formation. Hering proposed, in 1870, that the NS deals with memory for organised movements[87], a formulation which is akin to some modern conceptions[59]. As late as 1946, the well-known neurosurgeon Dandy dared suggest that the NS is the seat of consciousness[14]. This view was dismissed by Meyers[91] — and for good reasons. More recently, Mettler and Crandell proposed that the NS is able to free the organism from "proprioceptive dominance", thereby permitting it to make choices[90]; this hypothesis has not been followed up either in theory or by experimentation.

A close relation between the cerebral cortex and the NS was suggested by several authors in the last century[92], but it was rejected in the influential paper by Wilson[140], who favored the position that the cortex and the NS have independent descending routes. In the middle of the present century, however, new knowledge and current streams of thought converged to revitalize the interest in cortico-striatal associations and to initiate exploration of their importance for "cognitive" processes. Anatomical work had at that time confirmed the projections from neocortex to NS[87] and Rioch's ablation studies suggested to Fulton[36] and others that NS functions depend on the presence of the neocortex. A recurrent issue of the time was whether or not the effects of frontal lobotomy could be attributed to damage in the caudate nucleus. In an important experiment performed in Fulton's laboratory by Rosvold and Delgado[110], it was for the first time demonstrated that stimulation or lesion in the head of the caudate nucleus produce the same effects as ablation of the prefrontal cortex in certain tasks with a memory component. The impairment in these learning tasks could be produced at stimulus-intensities which were too low to cause any "motor" signs. This study, although insufficiently recognized even today, gave the impetus for a slowly growing interest in cortico-striatal "cognitive" mechanisms.

In spite of Willis' early emphasis on the inflow of sensory information to the NS[138], the participation of this structure in sensory processing has been almost totally disregarded for centuries. Only during the last few decades have neurophysiologists started to pay attention to this neglected aspect; Albé-Fessard, Buser and their collaborators were the pioneers in this field (Krauthamer, this volume). Denny-Brown has long stressed the importance of sensory, particularly visual, processes for NS' motor functions[18].

This sketchy review illustrates that notions about NS functions have often emerged on the basis of scanty evidence under the influence of philosophical and scientific fashions of the time. It is interesting to note that predominant doctrines are sometimes easily conquered, but at other times remain resistant to contradictive evidence. Thus, the NS fell from his "high estate" of physiological significance in the nineteenth century as it was realised that many of the "elaborate and detailed" functions earlier ascribed to it may instead be mediated by the adjacent pathways. When the same kind of artefacts were demonstrated in this century with respect to various motor and inhibitory phenomena (e.g. refs. 32, 65), the impact on contemporary conceptions was minimal. On the other hand, the NS was for centuries considered to be involved in the highest functions known to man on no better evidence than its vicinity to the cerebral ventricles, a view which completely vanished in the dawn of the experimental era. When reliable evidence for "cognitive" functions was finally advanced in 1956[110], the prevalent notions about NS functions were remarkably resistant to modification.

CURRENT STATUS

The beginning student of NS is likely to grow increasingly bewildered while consulting contemporary handbooks, monographs, and reviews on the functions of this formation. Current theories are remarkably diverse and often vaguely stated. Some hypotheses seem untestable; others are founded on, at best, weak evidence. Certain formulations could be applied to almost any part of the brain. A selection of views from the last two decades will serve to illustrate the newcomer's predicament and the rudimentary state of our knowledge.

Authoritative reviews have suggested that the NS takes part in: "regulation of intentional movements by suppressing and screening additional motor mechanisms in some functional association with the nonspecific thalamoreticular system ..."[53], as well as in "the initiation and control of locomotion by virtue of activities which are postural"[81], and in regulation of the "set function" in motility, the "filter mechanism" for muscle-tone control, and elaboration of the "strategy of learning"[2]. A much-cited hypothesis states that the basal ganglia "serve as a ramp generator for slow voluntary smooth movements of different speeds"[61,60]. The caudate nucleus has been claimed to be "concerned in the integrative balance of cortical responses to environmental contact stimuli with those of visual context"[18]; the same author conceives of the putamen as a "selective resonator for cortical activity"[19]. Quite obviously, there is an "involvement of striatal function in sensorimotor integration"[125], and the "caudate nucleus contains or is contained in a basic system which enables the organism to perform biologically adaptive responses"[11]. The NS has also been said to belong to a system which mediates "egocentric localization"[102], or "corollary discharges"[119], or which "figuratively speaking ... translates cognition into action"[23]. An exhaustive review simply concluded that the NS functions on a "high level of integration"[65].

If our novice is intrigued by diversity of opinion and proceeds to search the literature for relevant data, he/she will soon discover that many of the experiments which were taken to support well-known concepts of motor and inhibitory roles of the NS (e.g. in refs. 18, 53, 81, 88, 100) must now be re-evaluated. The dramatic behavioral alterations which were obtained with large lesions involving not only NS, but also non-striatal formations and fibers of passage (e.g. refs. 19, 121), as well as with presumably wide-spreading electrical currents in interaction with anaesthesas (e.g. ref. 89; for review and criticism, see ref. 65), differ very drastically from the intricate syndromes which are caused by well localised treatments. The many subtle effects of well controlled manipulations of the NS raise neglected questions about the usefulness of broad functional concepts and unitary notions in the description of NS, and of the possibility to integrate diverse findings. The remainder of this chapter will touch on some of these problems.

Much of what we have to say might seem too evident to require discussion, yet, to paraphrase Justice Frankfurter: at this time we may need education in the obvious as much as investigation of the obscure.

Motor functions

The influence of NS upon posture and movements is sometimes referred to as the "simple" function of this structure (as opposed to the presumably more "complex" mediation of tasks requiring learning and memory). However, in the face of now available evidence it seems impossible to maintain that the motor functions of NS are either elementary or direct. Experimental treatments or pathological conditions of the NS do not always give rise to motor symptoms, and when such symptoms do occur, they are commonly context-dependent to a significant degree.

It has not been demonstrated that severe motor disabilities can be produced by destruction of NS tissue alone. Lesions restricted to different subregions of NS in rats, cats, and monkeys do not cause any "clinical" motor signs such as tremor, ataxia, rigidity, or akinesia after a brief postoperative recovery period (personal observation on large numbers of animals with destruction of approximately 20% of the NS or less). Much more extensive NS lesions in cats were followed by "surprisingly" mild disorders of movement[132]; large unilateral ablations caused neglect of the contralateral forelimb in a conditioned bar-pressing task[99], but the animals apparently used the same leg proficiently in, for example, running and jumping[132]. Similarly slight and situation-dependent defects were observed in rats[42,66,67]. In monkeys, cooling of the globus pallidus interfered selectively with certain patterns of movement in a situation-contingent way[48].

Electrical stimulation restricted to the NS does not give more substantial, unequivocally "motor" signs than does the lesion technique. Unilateral stimulation of the NS in freely moving cats was found to elicit turning to the contralateral side, but no other "motor" effects[65]; Laursen noted that the cats seemed as if they were looking at something[65], and Ursin *et al.* have found that this turning response habituates with repeated stimulation[126]. In a recent review of behavioral consequences of NS stimulation, Krauthamer concluded that the effects depend on "the type of movement and effort required, degree of motivation, intensity of stimulation, and site of stimulation within NS"[62].

It may be suggested that lesions and stimulation fail to produce massive motor syndromes in animals because only the human NS plays an important role for motor mechanisms. (The possibility of such species differences has been mentioned both by Kennard[56] and by Jung and Hassler[53]). But the available data on stimulation of the human NS[82,108,128] and on effects of lesions restricted to this formation[93,105] are in good agreement with the results of animal experiments: the motor effects are negligible. Martin has written about this enigma of the clinic thus: "The clinicopathological method has encountered peculiar difficulties in dealing with the basal ganglia ... almost all the diseases which have symptoms of the kind that we attributed to the function of the basal ganglia are diffuse pathological conditions ...; *when local lesions do occur* from disease, they are often not helpful for investigational purposes ... *we do not recognize any symptoms resulting from them*"[81] (italics added).

The only severe motor impairments for which NS can probably be deemed responsible are obtained through biochemical dysbalance affecting large portions of the NS or the whole formation[2,22,124] (see also Steg and Johnels, this volume; the more intricate effects produced by regionally restricted transmitter dysbalance are nicely illustrated in ref. 37). Interestingly, even the profound akinesias which result from dopamine depletion in NS (and elsewhere) may vary considerably with the requirements of the situation. Thus, akinetic rats can escape aversive stimulation[27,125] and Parkinsonian patients who have been sitting motionless for years may suddenly catch something which is thrown at them, or run in emergency situations[113]. A patient who cannot walk may dance or march to music[20]. The staggering gait, a motor symptom par excellence, may be improved when the patient walks on a floor which is painted with conspicuous stripes[81].

The intricacy of the movement disorders in which NS dysfunctioning plays a role suggests that NS has neither exclusive nor very direct influence on many motor functions. These results are consistent with the anatomical picture according to which NS is part of one of several systems converging on the final common path neurons, and which shows that at least four synapses separate the NS from these effector neurons. (It is instructive to recall Evarts' finding that pyramidal tract neurons and peripheral motor neurons discharge independently under certain conditions, yet these neurons are only one or two synapses apart[26]).

Physiological evidence indicates that movements may be "initiated" not only from traditional "motor" structures such as NS[17], globus pallidus and substantia nigra (Rolls, this volume; De Long, personal communication), motor cortex[26], cerebellum[120] and nucleus ruber[38], but also from many formations which are not normally included into any "motor system". Thus, neurons with movement-related response characteristics were found in the following structures: the prefrontal[64] and posterior parietal[51,96] cortical areas, the basal forebrain and mesencephalon[79], the pontine reticular formation[114], and the pontine gigantocellular neurons[131]. In the hippocampus, both gross[142] and single unit[103] recordings revealed activity associated with movements.

Similarly, movements can be elicited by stimulation in the following "non-motor" formations: all parts of the cerebral convexity[72,73,123], the hypothalamus[14], thalamic nuclei which do not project to the motor cortex[45,136], the hippocampus[55,77], and all regions from which self-stimulation can be obtained[127]. Both the amygdala[40] and the hypothalamus[47] appear to influence fusimotor activity. Lesions of the medial thalamus (sparing the ventrolateral nucleus) cause deficient "initiation of movement"[129].

This survey of movement-related activities in the brain shows that many parts of the nervous system may be implicated in motor functions. All mentioned formations obviously have anatomical access to the final common path, and many of them are closer to effector neurons than is the NS. Interestingly, however, movements which occur during stimulation of structures which are not traditionally labelled "motor" are often considered to be "side effects" (see, for example, ref. 106) — a rather curious notion which cements established views on the limits of "motor systems".

To conclude our discussion of NS motor functions, we do not doubt that this formation takes part in regulation of movements under certain conditions, as do numerous other regions of the nervous system. The role of NS in motor control is undoubtedly "complex". It may also be "important", as claimed by most textbooks, but its importance is not understood. No convincing formulation of what is unique about NS contribution to motor mechanisms is yet available. Of many contemporary attempts to define this contribution (see above), none seems to be founded on firm evidence. The hypothesis of Kornhuber[60] is presently most influential. It is, however, unsupported by clinical as well as experimental evidence[4,12,28,29,41,46,94,104,116,137].

Inhibitory functions

Since the original observations of Danilewski[16], electrical stimulation of the NS has been found to cause a multitude of "inhibitory" phenomena (Delgado, this volume).

These effects, as well as some "disinhibitory" symptoms produced by NS lesions (Delgado, this volume), have established NS as a structure with special inhibitory powers on both neural and behavioral levels. Many of the stimulation effects may have been due to spread of current to non-striatal tissue, or are obtainable only under the influence of certain anaesthetic agents[32,65,69,70,89,92,98,107], but others can indeed be ascribed to the activation of NS neurons (e.g. refs. 58, 98; see also ref. 62 and Delgado, this volume). There is no reason to disbelieve NS' involvement in inhibitory processes. Whether such processes are particularly characteristic of NS' way of working, whether they form a meaningful functional category, and whether they are more "centered" in NS than in other parts of the brain can, however, be debated.

As mentioned by Delgado in this volume, electrical stimulation of NS is often followed by inhibition of one and simultaneous facilitation of another event. Thus, for example, stimulation of the caudate nucleus was found to induce IPSP:s in pallidal neurons, EPSP:s in the putamen[78], and IPSP-EPSP sequences in substantia nigra[35] (for details see Feger *et al.*, this volume). Both excitatory and inhibitory effects can be obtained in the same formation by stimulating one NS site with different frequencies[6]. Sensory stimulation may affect the occurence of inhibitory phenomena elicited from the NS[34,49]. Movements can be either inhibited or evoked by NS stimulation[8,62], and NS lesions can impair both inhibition and initiation of responses in learning tasks (Öberg and Divac, this volume). These findings are only mentioned to illustrate the obvious but occasionally forgotten fact that NS cannot be assigned any generalized inhibitory role without specifying the dependent and independent variables.

A more fundamental question is, however, brought forward by the heterogeneous spectrum of inhibitory consequences of NS stimulation. First, it seems essential to consider whether all these phenomena are likely to have anything in common other than the label "inhibitory". It is, for example, by no means obvious that one basic mechanism is responsible for inhibiting single cell activity and evoked potentials in many regions of the brain (Feger *et al.*, this volume; Krauthamer, this volume), and a diverse set of behaviors, "learned", "spontaneous", and "general" (e.g. refs. 7, 8, 58, 75, 101, 112; see also Delgado, this volume). Some of these effects have been dissociated by virtue of differences in distribution of effective stimulation sites within the NS (see, for example, refs. 63, 98, 107; for a review of work up till 1968, see ref. 21). The identification of inhibitory points depends on what is to be inhibited (Delgado, this volume), and there is thus precious little evidence both for a particular inhibitory subregion of NS[70] and, more importantly, for any fundamental kinship amongst all the different inhibitory phenomena. And if it is true that "inhibition" does not refer to any unitary process or kind of operation in NS, it seems reasonable to question the use of this term as a descriptive and explanatory concept.

The notion of inhibition, albeit properly used at the level of the synapse, is deceptive when applied to more complex biological functions. Since the consequences of different manipulations of the nervous system are usually assessed by quantitative measurements, the results can often be described as increases or decreases in certain parameters. Sometimes, then, "less" is translated into "inhibited" and "more" is transcribed into either "excited" or "disinhibited", or even referred to as "inhibition of inhibition". These translations make more or less explicit reference to underlying mechanisms, and they often attain "explanatory" status even in cases where the presumed basic mechanism was not studied or even theoretically specified. The lure of the concept of inhibition is amply illustrated by the use of such pseudoexplanations in the literature. About the NS we have been told, for example, that "... dopamine inhibits a mechanism which is behaviorally inhibitory"[97], or "... the main function of the dopamine is to modulate and antagonise the inhibitory influences of the neostriatum on psychomotor activity"[100].

By this manner of reasoning one could obviously embrace any effect as supportive evidence for the inhibitory functions of NS. (One may even succeed in explaining the peculiar fact that profound inactivity, akinesia, is the most outstanding symptom of severe dopamine deficiency in this presumably inhibitory formation).

Various inhibitory and disinhibitory phenomena have been obtained by stimulation or ablation in numerous parts of the brain, including the following formations: the hippocampus[24,30,57,85,130], septum[33,54,83], hypothalamus[1], the basal forebrain region[44,115,118], amygdala[31,39,54], thalamus[50,111], cerebellum[95], the reticular formation[74], and frontal cortical areas[5,10,115]. Inhibition of sensory evoked potentials in the intralaminar thalamic nuclei is obtained not only with caudate stimulation[63], but also with stimulation of the olfactory bulb[3]. Most of these structures have at some time been referred to as "inhibitory centres". They are not interconnected so as to form any plausible "inhibitory circuit".

The very number of suspected centres for inhibition suggests that the brain contains no particular foci for multi-level inhibitory processes, but rather that inhibition, broadly conceived, is a generalized aspect of neural functioning in all parts of the nervous system[135]. The great variety of inhibitory phenomena and the wide dispersion of "inhibitory loci" may in part be understood by the assumption that any functional system in the brain is able to inhibit other systems which are not needed for the moment, or to preempt their access to, for example, the senses and the final common path. This very obvious and common-sense assumption is in accord with psychological knowledge, but it has only sporadically been demonstrated at the level of neuronal organization. If it has any virtue, it suggests that "inhibition" should not be regarded as a "function" (any more than "excitation" is so regarded), but as a set of heterogeneous mechanisms which are characteristic of the way in which the brain subserves its various functions.

In conclusion, we feel that the research on inhibitory influences of the NS has given us much useful information both about neural relations and about selective behavioral impairments (Delgado, this volume). We do not think, however, that the results of this research should be interpreted to mean that inhibition is a useful functional category and that NS is a seat for such a function.

"Cognitive" functions

The notion of cognitive functions of NS is extensively discussed in a later chapter of this book. Here, we only wish to point out that many of the problems which are encountered in assigning motor and inhibitory functions to NS are met also by the attempts to study the cognitive roles of this formation. Thus, we have tried to argue above that broadly conceived motor and inhibitory mechanisms are not more localized in the NS than in many other structures of the brain, and that the particular contribution of NS to such mechanisms is far from understood. But this argument is not meant to imply that any other concept can define a behavioral role which is unique for the NS. The cognitive functions which can be ascribed to this structure are shared by other formations which are anatomically related to the NS, and which can therefore be regarded as parts of the same functional system (see, for example, refs. 27, 109; also Öberg and Divac, this volume). The way in which NS contributes to the implicated cognitive processes is not better defined than its role in initiation, execution, or inhibition of movements.

We have also claimed above that the manifold inhibitory influences of the NS are not related by common mechanisms and that they thus form no meaningful functional category. The same can be said for the various classes of cognitive functions in which NS seems to be implicated.

Unitary concepts and functional heterogeneity

The NS looks homogeneous under the microscope, and unitary functional concepts, combined with more or less explicit assumptions about functional homogeneity, have always dominated the main-stream conceptions of this formation. Today, neither of these views can be upheld. As indicated above, and as amply illustrated in the rest of this book, the NS is involved in mediation of different kinds of processes and functions, and different sectors of NS are important for different categories of behavioral tasks (Öberg and Divac, this volume). This functional heterogeneity and regional specialization is unequivocally demonstrated but poorly understood. It is not known, for instance, whether different NS subregions are functionally equivalent with regard to the physiology of the local circuitry and thus are differentiated only by virtue of their different sources of information (Grofova, this volume), or whether these subregions also process information in different ways (a possibility which is worthy of attention after recent descriptions of regional variations in the neurochemical substrate, Fonnum and Walaas, this volume; for discussion see also ref. 23). More pertinent to the present discussion is, however, the question of whether all behavioural functions of NS can be shown to be "focussed" in particular parts of the structure, or whether there are also functions which depend on all subregions equally.

Steg and Johnels suggest, in this volume, that the NS is relatively equipotential with respect to motor functions. The possibility that this formation is regionally specialized for cognitive processes but not for motor control is supported by the fact that properly chosen learning tasks are sensitive detectors of all manners of regionally restricted interference with NS processing (Öberg and Divac, this volume), while motor deficits are rarely seen unless large portions of the NS are malfunctional. But the many indications of redundant organization of motor functions within NS must be contrasted with anatomical and physiological findings which suggest that one part of this formation may be more directly involved in motor control than are other regions: one sector of the NS receives abundant projections from the motor cortex (Grofova, this volume), and De Long has found that neurons with movement-related response characteristics are concentrated in this very region (personal communication). How the apparent equipotentiality of NS for motor control is to be reconciled with the hodological and electrophysiological evidence for localization of motor function is not known. We are, however, reminded of the times when it could be maintained that the hodologically differentiated neocortex is equipotential for learning. This claim was reinterpreted as it was realized that the employed behavioral tasks (mazes) could be solved by different basic mechanisms and were therefore too unspecific for the purpose of functional analysis. It seems possible that the "motor" syndromes which result from dysfunctioning of large parts of NS should be understood as composite behavioral expressions of a breakdown of motor as well as of sensory and cognitive processes in the NS. The disorders of movement may in other words result from disconnection (or, more effectively, from dysfunctioning connections) between cortical processes and motor mechanisms[23]. The global "motor" (behavioral) function of the NS as a whole may thus be possible to dissociate from the specific role played by the region which is associated with the motor cortex.

Broad concepts and equivocal syndromes

The distinction between motor functions and sensory or cognitive mechanisms is, as we have just seen, occasionally far from obvious. The NS is situated in the core of the "intermediate net" of the forebrain, several synapses distant from senses as well as muscles, and by virtue of its projections from all parts of the cortex associated with different functional systems. It is no wonder, then, that NS malfunctioning leads to syndromes which have sensory, cognitive, and motor aspects, and which can be differently interpreted depending on the preconceived notions of the interpreter. Examples abound.

An animal which turns during unilateral stimulation of NS[65] may be doing so because of direct activation of the motor commands or because of indirect motor activation via sensory/attentional mechanisms[126]. The famous turning syndrome of animals with unilateral dopamine depletion in the NS has been regarded as a manifestation of severe malfunctioning of elementary motor mechanisms, but it has been successfully reinterpreted as secondary to profound sensory neglect[125]. The stereotyped behavior patterns which can be induced by dopamine agonists have been understood in terms of disturbed motor mechanisms, but this syndrome is strongly influenced by sensory stimulation[117], and it can be elegantly explained by non-motor mechanisms[76]. (Incidentally, stereotypy does not even seem to depend crucially on NS mediation; see, for example, refs. 13, 22, 80, 84, 141). The pacing paw movements which are sometimes observed in cats with large lesions in the caudate nucleus have been regarded as a clinically relevant "motor" sign and hence labelled "choreo-athetoid"[71], but the same movements have also been seen as normally executed "kneading" and understood as one of several expressions of the amiable emotional state of these preparations[132]. Rats with disruption of dopamine transmission in one region of NS show topographically (clinically) intact movements, but they interupt behavioral sequences too soon[37]; this interesting syndrome could be "explained" by impairments of, for example, arousal or motivation as well as of motor patterning. Some variants of "cognitive" deficits may alternatively be described as disturbances in sensory or inhibitory mechanisms (Öberg and Divac, this volume). Many such equivocal syndromes are ambiguous because they have not been thoroughly studied, but others will remain equivocal as long as our concepts are insufficiently specified and therefore considerably overlap with one another.

The problem of choosing labels for functions and dysfunctions of NS may in one sense be regarded as trivial. It is rather obvious that sensory and cognitive as well as motor and inhibitory notions can be so vaguely stated or so widely conceived that they are capable of covering almost any kind of behavioral aberration. This may not much matter as long as a label is not regarded as an "explanation", and the search continues to delineate carefully in just what way this structure is involved in the respective functional category. Unfortunately, however, the study of NS to date shows that each research tradition tends to focus on those very syndromes and anatomical and physiological facts which are most clearly in agreement with one's own view, and to formulate hypotheses which give the appearance of exhaustive accounts while in fact covering only a fraction of the available evidence.

From this gloomy perspective, it is most encouraging to note that all of the authors of the succeeding chapters take a liberal stand with respect to the range of behavioral roles which may be ascribed to the NS. While evidence is mustered in support of NS involvement in motor, inhibitory, sensory and cognitive processes in turn, no author claims that the NS is exclusively – or even predominantly – concerned with the respective class of functions. This tolerant atmosphere suggests that the NS may be entering an era of well deserved conceptual freedom after decades of confinement in motor and inhibitory schemes.

REFERENCES

1. Akert, K., in: *Electrical Stimulation of the Brain*, D.E. Sheer (ed.), University of Texas Press, Austin, pp. 288-310 (1961).

2. Barbeau, A., in: *Biology of Brain Dysfunction*, G.E. Gall (ed.), Plenum, New York, pp. 333-350 (1973).

Bernstein, S., Lamarche, M. and Buser, P., *Arch. Sci. Biol.*, 53, 73-88 (1969).

Brooks, V.B., *J. Canad. Sci. Neurol.*, August, 265-277 (1975).

Brutkowski, S., *Physiol. Rev.*, 45, 721-746 (1965).

Buchwald, N.A., Wyers, E.J., Carlin, J. and Farley, R.E., *Exp. Neurol.*, 4, 23-26 (1961).

Buchwald, N.A., Horvath, F.E., Soltysik, S. and Romero-Sierra, C., *Bol. Inst. Estud. Med. Biol. (Mex.)*, 22, 363-377 (1964).

Buchwald, N.A. and Hull, C.D., *Brain Res.*, 6, 1-11 (1967).

Clarke, E. and Dewhurst, K., *An Illustrated History of Brain Function*, Sandford Publications, Oxford (1972).

Clemente, C.D., Chase, M.H., Knauss, T.K., Sauerland, E.K. and Sterman, M.B., *Experientia*, 22, 844-845 (1966).

Cools, A.R., *The Caudate Nucleus and Neurochemical Control of Behavior*, Thesis. Nijmegen, The Netherlands (1973).

Corin, M.S., Elizan, T.S. and Bender, M.B., *J. Neurol. Sci.*, 15, 251-265 (1972).

Costall, B. and Naylor, R.J., *Eur. J. Pharmacol.*, 24, 8-24 (1973).

Cox, V.C. and Valenstein, E.S., *J. Comp. Physiol. Psychol.*, 69, 730-733 (1969).

Dandy, W.E., *Bull. Johns Hopkins Hosp.*, 79, 34-57 (1946).

Danilewski, V.Y., *Arch. ges. Physiol. Pflüger*, 11, 128-138 (1875).

De Long, M.R. and Strick, P.L., *Brain Res.*, 71, 327-335 (1974).

Denny-Brown, D., *The Basal Ganglia and their Relation to Disorders of Movement*, Oxford University Press, Oxford (1962).

Denny-Brown, D. and Yanagisawa, N., in: *The Basal Ganglia*, M.D. Yahr (ed.), Raven Press, New York, pp. 115-149 (1976).

Dinnerstein, A.J., Frigyesi, T. and Lowenthal, M., *Percept. & Mot. Skills*, 15, 667-680 (1962).

Divac, I., *Acta Biol. Exp. (Warsaw)*, 28, 107-120 (1968).

Divac, I., *Psychopharmacol.*, 27, 171-178 (1972).

Divac, I., in: *Psychobiology of the Striatum*, A.R. Cools, A.H.M. Lohman, and J.H.L. van den Bercken (eds.), Elsevier/North Holland, Amsterdam, pp. 21-30 (1977).

Douglas, R.J., *Psychol. Bull.*, 67, 416-442 (1967).

Dusser de Barenne, J.G. and McCulloch, W.S., *J. Neurophysiol.*, 1, 364-377 (1938).

Evarts, E.V., *Life Sci.*, 15, 1393-1399 (1974).

27. Fibiger, H.C., in: *Psychobiology of the Striatum*, A.R. Cools, A.H.M. Lohman and E.H.L. van den Bercken (eds.), Elsevier/North Holland, Amsterdam, pp. 73-84 (1977).

28. Flowers, K., *Neurology*, 25, 413-421 (1975).

29. Flowers, K., *Brain*, 101, 19-34 (1978).

30. Flynn, J.P. and Wasman, M., *Science*, 131, 1607-1608 (1960).

31. Fonberg, E. and Delgado, J.M.R., *Fed. Proc.*, 20, 1 (1961).

32. Forman, D. and Ward, J.W., *J. Neurophysiol.*, 20, 230-244 (1957).

33. Fox, S.S., Kimble, D.P. and Lickey, M.F., *J. Comp. Physiol. Psychol.*, 58, 380-386 (1964).

34. Fox, S. and O'Brian, J.H., *Science*, 137, 423-425 (1962).

35. Frigyesi, T.L. and Purpura, D.P., *Brain Res.*, 6, 440-456 (1967).

36. Fulton, J.F., *Physiology of the Nervous System*, 3rd ed., Oxford University Press, New York, p. 491 (1949).

37. Gaddy, J.R. and Neill, D.B., *Brain Res.*, 119, 439-446 (1977).

38. Ghez, C. and Kubota, K., *Brain Res.*, 129, 383-388 (1977).

39. Gloor, P., in: *Handbook of Physiology*, J. Field, H.W. Magoun and V.E. Hall (eds.), Amer. Physiol. Soc., Washington D.C., pp. 1395-1420 (1960).

40. Granit, R. and Kaada, B.R., *Acta Physiol. Scand.*, 27, 130-160 (1953).

41. Hallett, M., Shahani, B.T., and Young, R.R., *J. Neurol. Neurosurg. Psychiat.*, 40, 1129-1135 (1977).

42. Hansing, R.A., Schwartzbaum, J.S. and Thompson, J.B., *J. Comp. Physiol. Psychol.*, 66, 378-388 (1968).

43. Hebb, D.O., *Arch. Neurol. Psychiat.*, 54, 421-438 (1954).

44. Hernandez-Peon, R., O'Flaherty, J.J. and Mazzuchelli-O'Flaherty, A.L., *Brain Res.*, 4, 243-267 (1967).

45. Hess, W.R., *Diencephalon. Autonomic and Extrapyramidal Functions.* Grune and Stratton, New York (1959).

46. Highstein, S., Cohen, B. and Mones, R., *Trans. Amer. Neurol. Ass.*, 94, 277-279 (1969).

47. Hongo, T., Kubota, K., and Shimazu, H., *J. Neurophysiol.*, 26, 568-580 (1963).

48. Hore, J., Meyer-Lohmann, J. and Brooks, V.B., *Science*, 195, 584-586 (1977).

49. Hull, C.D., Buchwald, N.A., and Vieth, J., *Brain Res.*, 6, 12-21 (1967).

50. Hunter, J. and Jasper, H.H., *Electroenceph. Clin. Neurophysiol.*, 1, 305-324 (1949).

51. Hyvärinen, J. and Poranen, A., *Brain*, 97, 673-692 (1974).

52. Jacob, A., *Arch. Neurol. Psychiat. (Chic.)*, 13, 596-620 (1925).

53. Jung, R. and Hassler, R., in: *Handbook of Physiology*, J. Field, H.W. Magoun and V.E. Hall (eds.), Amer. Physiol. Soc., Washington, D.C., pp. 863-927 (1960).

54. Kaada, B.R., *Acta Physiol. Scand.* (Suppl. 23), 24, 1-285 (1951).

55. Kaada, B.R., Jansen, J. and Andersen, P., *Neurology (Minneap.)*, 3, 844-857 (1953).

56. Kennard, M.A., *J. Neurophysiol.*, 7, 127-148 (1944).

57. Kimble, D.P., *Psychol. Bull.*, 70, 285-295 (1968).

58. Kitsikis, A. and Rougeul, A., *Physiol. Behav.*, 3, 831-837 (1968).

59. Konorski, J., *Integrative Activity of the Brain: An Interdisciplinary Approach*, University of Chicago Press, Chicago (1967).

60. Kornhuber, H.H., *Kybernetik*, 8, 157-162 (1971).

61. Kornhuber, H.H., in: *The Neurosciences. Third Study Program*, F.O. Schmitt and F.G. Worden (eds.), M.I.T. Press, Cambridge, Mass., pp. 267-280 (1974).

62. Krauthamer, G., in: *Catecholamines and Behavior, Vol. 1*, A.J. Friedhoff (ed.), Plenum, New York, pp. 59-87 (1975).

63. Krauthamer, G. and Albé-Fessard, D., *J. Neurophysiol.*, 28, 100-124 (1965).

64. Kubota, K. and Niki, H., *J. Neurophysiol.*, 34, 337-347 (1971).

65. Laursen, A.M., *Acta Physiol Scand.* (Suppl. 211), 59, 1-106 (1963).

66. Levine, M.S., Ferguson, N., Kreinick, C.J., Gustafson, J.W. and Schwartzbaum, J.S., *J. Comp. Physiol. Psychol.*, 77, 282-293 (1971).

67. Levine, M. and Schwartzbaum, J.S., *J. Comp. Physiol. Psychol.*, 85, 615-635 (1973).

68. Lewy, F.H., in: *The Diseases of the Basal Ganglia.*, *Res. Publ. Ass. Nerv. Ment. Dis.*, 21, 1-20 (1942).

69. Liles, S.L. and Davis, G.D., *J. Neurophysiol.*, 32, 564-573 (1969).

70. Liles, S.L. and Davis, G.D., *J. Neurophysiol.*, 32, 574-582 (1969).

71. Liles, S.L. and Davis, G.D., *Science*, 164, 195-197 (1969).

72. Lilly, J.C., *Science*, 124, 937 (1956).

73. Lilly, J.C., Hughes, J.R. and Galkin, T.W., *Fed. Proc.*, 15, 119 (1956).

74. Lindsley, D.B., in: *Handbook of Neurophysiology*, J. Field, H.W. Magoun and V.E. Hall (eds.), American Physiological Society, Washington D.C., pp. 1553-1593 (1960).

75. Lineberry, C.G. and Siegel, J., *Brain Res.*, 34, 143-161 (1971).

76. Lyon, M. and Robbins, T., in: *Current Developments in Psychopharmacology, Vol. 2*, W. Essman and L. Valzelli (eds.), Spectrum, New York, pp. 81-163 (1975).

77. MacLean, P.D., *Arch. Neurol. Psychiat. (Chic.)*, 78, 128-142 (1957).

78. Malliani, A. and Purpura, D.P., *Brain Res.*, 6, 341-354 (1967).

79. Malmo, H.P. and Malmo, R.B., *Electroenceph. Clin. Neurophysiol.*, 42, 501-509 (1977).

80. Marcus, R.J. and Villablanca, J., *Proc. West. Pharmacol. Soc.*, 17, 219-222 (1974).

81. Martin, J.P., *The Basal Ganglia and Posture*, Pitman Medical, London (1967).

82. Matsui, I., Yoshioka, H., Ishikawa, O. and Ihara, K., *Folia Psychiat. Neurol. Jap.*, 11, 150-156 (1957).

83. McCleary, R.A., in: *Progress in Physiological Psychology*, E. Stellar and J.M. Sprague (eds.), Academic Press, New York, pp. 209-272 (1966).

84. McKenzie, G.M., *Psychopharmacol. (Berl.)*, 23, 212-219 (1972).

85. McKenzie, J.S. and Rogers, D.K., *Brain Res.*, 64, 1-15 (1973).

86. McNair, J.L., Sutin, J. and Tsubokawa, T., *Exp. Neurol.*, 37, 395-411 (1972).

87. Mettler, F.A., in: *The Diseases of the Basal Ganglia, Res. Publ. Ass. Nerv. Ment. Dis.*, 21, 150-227 (1942).

88. Mettler, F.A., in: *Handbook of Clinical Neurology, Vol. 6*, P.J. Vinken and G.W. Bruyn (eds.), North Holland, Amsterdam, pp. 1-55 (1968).

89. Mettler, F.A., Ades, H.W., Lipman, E. and Culler, E.A., *Arch. Neurol. Psychiat. (Chicago)*, 41, 984-995 (1939).

90. Mettler, F.A. and Crandell, A., *J. Nerv. Ment. Dis.*, 129, 551-563 (1959).

91. Meyers, R., *Arch. Neurol. Psychiat. (Chicago)*, 65, 659-671 (1951).

92. Meyers, R., *Neurology (Minneap.)*, 3, 627-655 (1953).

93. Meyers, R. Knott, J.R., Skultety, F.M. and Imler, R., *J. Neurosurg.*, 11, 7-23 (1954).

94. Miles, F.A. and Fuller, J.H., *Science*, 189, 1000-1001 (1975).

95. Moruzzi, G., *Problems in Cerebellar Physiology*, Thomas, Springfield, Ill. (1950).

96. Mountcastle, V.B., Lynch, J.C., Georgopoulos, A., Sakata, H. and Acuna, A., *J. Neurophysiol.*, 38, 871-908 (1975).

97. Neill, D.B., *Brain Res.*, 105, 89-103 (1976).

98. Newton, R.A. and Price, D.P., *Brain Res.*, 85, 403-422 (1975).

99. Olmstead, C.E., Villablanca, J.R., Marcus, R.J. and Avery, D.L., *Exp. Neurol.*, 53, 670-693 (1976).

100. Papeschi, R., *Psychiat. Neurol. Neurochir. (Amsterdam)*, 75, 13-48 (1972).

101. Plumer, S.I. and Siegel, J., *Physiol. Psychol.*, 1, 254-256 (1973).

102. Potegal, M., *Acta Neurobiol. Exp.*, 32, 479-494 (1972).

103. Rank, J.B. Jr., *Exp. Neurol.*, 41, 461-555 (1973).

104. Raphan, T. and Cohen, B., *Ann. Rev. Physiol.*, 40, 527-552 (1978).

105. Richardson, E.P. Jr., in: *Handbook of Clinical Neurology, Vol. 6*, P.J. Vinken and G.W. Bruyn (eds.), North Holland, Amsterdam, pp. 497-505 (1968).

106. Roberts, W.W., in: *The Neural Control of Behavior*, R.E. Whalen, R.F. Thompson, M. Verzeano and N.M. Weinberger (eds.), Academic Press, New York, pp. 175-208 (1970).

107. Rogers, D.K. and McKenzie, J.S., *Brain Res.*, 56, 345-349 (1973).

108. Rosner, B.S., Blankfein, R.J. and Davis, R.A., *Electroenceph. Clin. Neurophysiol.*, 20, 391-396 (1966).

109. Rosvold, H.E., in: *Mind as a Tissue*, C. Rupp (ed.), Harper & Row, New York, pp. 21-38 (1968).

110. Rosvold, H.E. and Delgado, J.M.R., *J. Comp. Physiol. Psychol.*, 49, 365-372 (1956).

111. Rougeul, A., Perret, C. and Buser, P., *Electroenceph. Clin. Neurophysiol.*, 6, 164-180 (1967).

112. Schlag, J.D. and Villablanca, J., *Brain Res.*, 6, 119-142 (1967).

113. Schwab, R.S. and Zieper, I., *Psychiat. Neurol.*, 150, 345-357 (1965).

114. Siegel, J.M. and McGinty, D.J., *Science*, 196, 678-680 (1977).

115. Siegel, J. and Wang, R.Y., *Exp. Neurol.*, 42, 28-50 (1974).

116. Starr, A., *Brain*, 90, 545-564 (1967).

117. Stevens, J., Livermore, A. and Cronan, J., *Physiol. Behav.*, 18, 809-812 (1977).

118. Sterman, M.B. and Clemente, C.D., *Exp. Neurol.*, 6, 91-102 (1962).

119. Teuber, H.-L., in: *The Basal Galnglia*, M.D. Yahr (ed.), Raven Press, New York, pp. 151-168 (1976).

120. Thach, W.T., *J. Neurophysiol.*, 31, 785-797 (1968).

121. Thompson, R.L. and Mettler, F.A., *Amer. J. Ment. Defic.*, 67, 126-134 (1963).

122. Tilney, F. and Riley, H.A., *The Form and Function of the Central Nervous System*, 2nd Ed., Lewis & Co., London (1923).

123. Towe, A.L., in: *Efferent Organization and the Integration of Behavior*, I.D. Maser (ed.), Academic Press, New York, pp. 67-97 (1973).

124. Ungerstedt, U., *Acta Physiol. Scand.*, 82, Suppl. 367 (1971).

125. Ungerstedt, U., Ljungberg, T. and Ranje, C., in: *Psychobiology of the Striatum*, A.R. Cools, A.H.M. Lohman and J.H.L. van den Bercken (eds.), Elsevier/North Holland, Amsterdam, pp. 85-97 (1977).

126. Ursin, H., Sundberg, H. and Menaker, S., *Neuropsychol.*, 7, 313-318 (1969).

127. Valenstein, E.S., *Psychol. Rev.*, 71, 416-437 (1964).

128. Van Buren, J.M., *J. Neurosurg.*, 24, 416-417 (1966).

129. Vanderwolf, C.H., *Canad. J. Psychol.*, 16, 318-330 (1962).

130. Vanegas, H. and Flynn, J.P., *Brain Res.*, 11, 489-506 (1968).

131. Vertes, R.P., *Brain Res.*, 128, 146-152 (1977).

132. Villablanca, J.R., Marcus, R.J. and Olmstead, C.E., *Exp. Neurol.*, 52, 389-420 (1976).

133. Vinken, P.J. and Bruyn, G.W., *Handbook of Clinical Neurology, Vol. 6*, North Holland, Amsterdam (1968).

134. Vogt, C. and Vogt, O., *J. Psychol. Neurol. (Leipzig)*, 25 (Suppl. 3), 627-846 (1920).

135. Weiskrantz, L., in: *Analysis of Behavioral Change*, L. Weiskrantz (ed.), Harper & Row, New York, pp. 400-414 (1968).

136. Wester, K., *Physiol. Behav.*, 9, 643-647 (1972).

137. Westheimer, G. and Blair, S.M., *Exp. Brain Res.*, 21, 463-472 (1974).

138. Willis, T., *Cerebri Anatome cui accessit Nervorum Descriptio et usus*, Roycroft, London (1664).

139. Wilson, S.A.K., *Brain*, 36, 427-492 (1914).

140. Wilson, S.A.K., *Arch. Neurol. Psychiat.*, 11, 385-404 (1924).

141. Wolfarth, S., Grabowska, M., Lacki, M., Dulska, E. and Antkiewicz, L., *Activ. Nerv. Super. (Praha)*, 15, 132-133 (1973).

142. Young. G.A., *J. Comp. Physiol. Psychol.*, 90, 78-90 (1976).

Motor Functions of the Striatum

Göran Steg and Bo Johnels

University of Göteborg, Department of Neurology, S-413 45 Göteborg, Sweden

"The royal road of the sensations of the body to the soul is through the corpora striata, and all determinations of the will also descend by that road".

Emanuel Swedenborg, 1740

The conditions for increasing our knowledge of motor functions of the striatum are favourable today due to recent advances. Neuroanatomy has provided us with a more complete picture of the connections between striatum and other parts of the central nervous system; histochemistry has contributed with new techniques to visualize some constituents of the nervous system and pharmacology has given us a variety of new drugs that can be used to alter brain functions and thus "manipulate" these in specific ways.

Several animal models of "extrapyramidal" diseases can link clinical research closer to the theoretical disciplines, giving further insight in the relations between clinical signs and symptoms and the motor functions of the striatum.

As being clinicians, we feel that our ideas of the motor functions must rest upon a combination of facts from clinical knowledge of extrapyramidal disease and the findings from anatomical and physiological research. We will restrict ourselves to the more "elementary" motor functions like locomotion, muscle tone and posture and do not take into judgment newer findings of more complex animal behavior after lesions in, and stimulation of the NS as these will be discussed by other authors in this volume.

We will present important findings from different fields of research or different techniques pointing to motor functions for the striatum and in the summary we will try to put these pieces of evidence together and to give our opinion on which motor functions may be associated with the striatum.

CLINICAL AND PATHOLOGICAL EVIDENCE FOR MOTOR FUNCTIONS OF THE STRIATUM

Parkinson's disease presents at least three major symptoms of disturbed motor control, namely hypokinesia, augmented muscle tone (rigidity) and a characteristic alternating tremor of the limbs and the trunk. Trétiakoff (42) in 1919 stated that the only common finding at autopsy of Parkinson patients was degeneration of nerve cells in the substantia nigra. This was later confirmed by Hassler (22). Today it is an established fact that this disease, at least in part, is due to degeneration of the nigro-striatal dopamine pathways (Ref. 15).

Several other diseases with involvement of the striatum, such as Wilsons disease and Strio-nigral degeneration, show similarities with Parkinson's disease.

Bilateral infarctions in the striata will often give rise to a Parkinson-like rigidity and a flexor posture (Ref. 13).

Chorea Huntington presents in part the opposite motor syndrome in comparison with Parkinsonism, with hyperkinesia and lowered muscle tone. Pathological changes have been found in the striatum and in the frontal cortex (Ref. 19).

In Parkinson's disease it is often possible to reverse the hypokinetic-hypertonic syndrome into a hyperkinetic-hypotonic counterpart by giving an overdose of levodopa or dopamine agonists like apomorphine. With levodopa it is, in fact, possible to modulate these motor functions between the two extremes by dose adjustments. It seems that at one endpoint of function we have the hypokinetic-hypertonic syndrome (severe Parkinsonism, the reserpine syndrome) associated with hypofunction of the dopamine pathways and at the other endpoint the hyperkinetic-hypotonic syndrome (Chorea Huntington, overdose of levodopa).
It is our experience that different profiles of these syndromes may exist at the same time in different parts of the body during levodopa treatment so that rigidity in the legs may be concomitant with e.g. hyperkinesias in the fingers or perioral region.

Major symptoms of Parkinsonism

Tremor. Parkinsonian tremor is thought to be a centrally induced phenomenon relying upon the integrity of certain loops in the brain including the efferent pathways of the striatum and the cortex cerebri as it can be relieved after lesions in the ventrolateral thalamic nuclei and as it is absent after removal of the motor cortex in man (Ref. 6, 11, 27). This type of tremor may be induced by tegmental lesions in the monkey and it persists after cutting the dorsal spinal roots (Ref. 51). Interesting studies by Dietz et al. (14) show that tremor in the Parkinson patient has its greatest amplitude and the lowest frequency at rest and that these parameters approach those of physiological tremor with increasing muscle load.

Rigidity. Clinical data indicate that the augmented muscle tone (rigidity) in Parkinson's disease depends on striatal dysfunction and that there is no primary malfunction at spinal or peripheral levels (Ref. 18, 36). There is however, still much debate whether the descending neuronal hyperactivity is acting directly on the alfa motoneurons or upon the gamma motoneurons and via the muscle spindles and monosynaptic reflex arc indirectly upon the alfa motoneurons. The terms alfa and gamma rigidity have been coined in this discussion.

Studies with the microneurographical techniques of Hagbarth et al. (21, 50) of the activity in the group Ia fibers on Parkinson patients indicate that there is an increase at both alfa and gamma motoneuron excitability. This may indicate multiple ways of influence from descending pathways on the spinal levels.

Rigidity in Parkinsonism persists after section of the dorsal spinal roots but may be relieved by stereotaxic lesions in the pallidum or the ventrolateral nuclei of the thalamus (Ref. 11, 32). Rigidity is also a common side effect after long term treatment with neuroleptic drugs and the risk seems to be greatest with those drugs known to be potent blockers of dopamine receptors in animals. It has been shown that the age distribution of patients with drug induced Parkinsonism is the same as for those with the idiopatic form of the disease (Ref. 5). Centrally acting anticholinergic drugs counteract rigidity and tremor in both types of Parkinsonism. Thus there is a probable common pathophysiological mechanism.

Hypokinesia (akinesia): This is the most disabling of the symptoms in the Parkinson syndrome and the one that is best overcome with levodopa (in the idiopatic form of the disease). Levodopa may even turn parkinsonian hypokinesia into hyperkinesia. The most important aspect of hypokinesia from the clinical point of view is the difficulty to initiate movements. When it is a debute symptom the patient often experiences that e.g. one arm has become useless to him although he has full strength, sensibility and normal tests for coordination. Clinical studies have not given much evidence about the site of origin of hypokinesia or whether the primary disturbance is one of neuronal hyper- or hypoactivity.

Martin (30) has studied postencephalitiv Parkinson patients and found that they seemed to lack the ability to adjust the body centre of gravity forward and to induce lateral, alternating movements necessary for walking. He thought, from clinical analogy, that these disturbances were due to impairment of postural reflexes which are concerned in normal locomotion. One of the defects behind akinesia may be an inability to process proprioceptive information in the brain. Exteroceptive stimuli, e.g. placing a ladder on the floor in front of the patient, may help him to initiate walking. Hypokinesia is not influenced by the stereotaxic thalamothomy used to treat Parkinsonian tremor and rigidity.

Hyperkinetic syndromes. The choreo-athetosis of Chorea Huntington, the akathisia and the hyperkinetic syndrome following overdose of levodopa are especially interesting in relation to the striatum. The hyperkinesias in Chorea Huntington consist of unintentional, unpurposeful, exaggerated movements of the limbs and trunk. They are worsened by drugs that enhance dopaminergic functions in the brain such as levodopa. The best treatment seems to be by those drugs that have a potent dopamine receptor blocking capacity like haloperidol. Prolonged therapy with neuroleptics, especially in elderly patients, can induce akathisia, a state of restless small movements of stereotyped character in the perioral region and in the extremities. Akathisia is thought to depend upon hypersensitivy of the functionally denervated dopamine receptors. On withdrawal of the neuroleptic drug these symptoms usually clear out (Ref. 26).

The hyperkinetic movements seen after an overdose of levodopa usually consist of small movements of stereotyped character in the perioral region and in the distal extremities.

There are several similarites between this syndrome and the apomorphine syndrome in animals described later in this essay. Levodopa-induced hyperkinesias have been reported not to occur on the side operated upon with thalamotomy in Parkinson patients (Ref. 38).

EVIDENCE FOR STRIATAL MOTOR FUNCTIONS FROM ANIMAL MODELS OF EXTRAPYRAMIDAL DISEASE

The reserpine syndrome. Carlsson (7) found that in the rat, like in man, high doses of reserpine caused a syndrome with similarities to Parkinson's disease. The reserpinized rat is akinetic and shows muscle rigidity and a fine tremor (Ref. 39). Reserpine is known to deplete monoamine stores in the brain and it gives a significant reduction of dopamine in the striatum (which normally holds about 80% of the brain dopamine). Levodopa or dopamine receptor stimulating drugs such as apomorphine will counteract the reserpine syndrome when given systemically (Ref. 25). It has been found that local application of apomorphine to either side or bilaterally in the corpus striatum of reserpinized rats will abolish the rigidity but not the akinesia. The latter can be counteracted by injections of apomorphine into the nucleus accumbens associated with the mesolimbic dopamine pathways. Injections in the nucleus accumbens, however, will not oppose the rigidity (Ref. 3, 35). Thus it seems that the rigidity and akinesia are mediated by different pathways in the brain.

It is possible to induce reserpine rigidity after bilateral decortication in the rat and this effect may be antagonized by apomorphine. This is, however, not the case after bilateral cortico-striatectomy, decerebration at the level of the anterior collicculi or spinal transection (Ref. 24). We have interpreted these findings to indicate that there must exist direct descending pathways that can influence muscle tone and that are independent of the thalamo-cortical loops.

Anticholinergic drugs with central effects such as atropine counteract reserpine rigidity which is not the case with those drugs such as methyl atropine that do not cross the blood-brain barrier (Ref. 24). Thus, it seems that in this animal model, like in Parkinsonism, there is a cholinerigic hyperactivity.

Investigations on the spinal level show that there is an enhanced alfa motoneuron, and a reduced gamma motoneuron discharge concomitant with the reserpine rigidity and that these changes are opposed by giving levodopa (Ref. 39).

Accordingly to our opinion the reserpine model is in many respects representative of the end stage of Parkinsonism with almost complete dopamine cell degeneration.

<u>6-hydroxy dopamine lesions</u>. Bilateral stereotaxic injections of 6-hydroxy dopamine (6-OH DA) into the substantia nigra or slightly in front thereof, where dopaminergic pathways assemble into bundle-like formations, may selectively destroy these pathways. Rats so treated become akinetic but there is little rigidity or tremor in the acute phase. The motor deficits may be counteracted by apomorphine which is thought to act upon the denervated dopamine receptors in the striatum and other locations (Ref. 43).

Unilateral destruction of the dopamine pathways with 6-OH DA does not produce any gross motor deficit apart from an inclination of the rat to turn its head and tail towards the side of the lesion. Given apomorphine the rat will start to circle towards the unoperated side. This effect is thought to be caused by an assymetry of striatal motor function due to hypersensitivity of the postsynaptic dopamine receptors on the side of the lesion. This rotational behavior of rats has been widely utilized to test, among other things, the dopamine receptor activating potency of drugs. The perhaps best known example is the "rotometer" of Ungerstedt (44).

<u>Ventromedial tegmental lesions</u>. Bilateral electrolytic lesions in the ventral tegmentum of monkeys, involving as well striopetal and striofugal fibres as a part of the red nucleus, have been shown to induce akinesia, rigidity and Parkinson-like tremor. This tremor is antagnonized by injections of apomorphine (Ref. 34).

<u>Unilateral destruction of the corpus striatum</u>. This operation will not produce any gross motor disturbance in the rat. When given levodopa, the animal will present an asymmetric posture turning its head and tail towards the operated side and also start to circle in this direction. Given neuroleptic drugs these rats will turn to the unoperated side. This effect can be counteracted by atropine (Ref. 1,2). These effects are the opposite to those obtained after 6-OH DA induced degeneration of the ascending dopamine system on one side. In this case the effects are thought to depend on stimulation and blockade of dopamine receptors in the intact side.

Unilateral striatectomy does not influence reserpine induced rigidity measured in the rat hindlegs as does the bilateral operation (Ref. 25). In monkeys and cats after similar lesions, there has been noted a slowness of movements and an unwillingness to use the contralateral arm (Ref. 8, 33). Bilateral destruction seems to induce poverty of movements and disturbances of more complex behavioral functions (Ref. 49).

<u>Stereotyped behavior after apomorphine or amphetamine</u>. Systemic treatment of

animals with dopamine agonists such as apomorphine or amphetamine induces compulsive stereotyped behavioral patterns which can be used in the analysis of more complex functions of the striatum. These topics are dealt with by other authors in this volume but some aspects may be taken up here.

The effects of the dopamine agonists may be blocked by pretreatment with dopamine receptor blocking agents such as haloperidol. The effects are potentiated after pretreatment with reserpine or after lesions of the ascending dopamine pathways. This type of potentiation is thought to depend upon hypersensitivity of the denervated dopamine receptors.

To some extent the apomorphine syndrome can be elicited also by local application of the drug to the corpus striatum (Ref. 16, 44). The syndrome has been analyzed in considerable detail by e.g. Ungerstedt (45). There is common belief that part of it (i.a. gnawing and licking and sniffing) depends upon acitivy in the striatum and we have noted that these phenomena disappear after decortication. The augmented locomotion is probably the result of dopamine receptor stimulation in the nucleus accumbens (Ref. 3. 23).

EVIDENCE FROM ELECTROPHYSIOLOGICAL STIMULATION AND RECORDING

Studies with electrical stimulation and recording in the striatum and adjacent structures have yielded different and sometimes conflicting results depending on the techniques used. We will therefore not go into any detail but only mention some of the findings that we find interesting in relation to what has been said above. The reader may find more extensive reviews in Ref. 10, 18, 29, 49. Among the effects that have been observed with unilateral stimulation in the striatum are the following:

a. contralateral turning of the head and flexion movements of the contralateral limbs in cats,
b. inactivation of the animal without sedation,
c. blocking of a number of learned behavioural responses (Ref. 10).

In many cases, the same head and limb movements can be elicited by chemical stimulation with e.g. dopamine or by non-specific stimulation such as the introduction of a steel cannula into the corpus striatum (Ref. 10, 24).

In humans low level electrical stimulation of the striatum is reported to have caused arrest of speech and voluntary movements and unresponsiveness to exteroceptive stimuli and a post-stimulation amnesia. Higher levels of stimulation caused a contralateral turning of the head and eyes as would be expected after stimulation of corticospinal pathways (Ref. 48).

In the curarized, awake, cat electrical stimulation in both the pallidum and the substantia nigra has been shown to facilitate alfa and to inhibit gamma type motoneurons (Ref. 40). There is no common agreement on whether the nigrostriatal dopamine pathways act inhibitory or exitatory or both on their target cells in the striatum (Ref. 9, 31, 37, 52).

It seems that there is a very wide range of motor activities that may be affected by stimulation in the striatum from simple motor functions to more complex behavior.

Recording in the striatum. There is very little spontaneous firing of cells in the putamen and nucleus caudatus in contrast to pallidal neurons (Ref. 49). Thus most investigation of the former had to be done with help of more or less physiological means of exciting these neurons. According to our opinion, the most interesting information has come from recordings in awake monkeys performing learned tasks. It has been shown that cells in the putamen, pallidum and ventrolateral nuclei of the thalamus shown an increased activity related to slow movements of the arms and that his activity precedes that in the cortex cerebri (Ref. 12, 28). Evarts (17) draws the conclusion that both the cerebellum and the basal ganglia are structures that send motor command to the cortex rather than to be relay stations or subordinate structures comparing proprioceptive information to cortical efference.

Recording at the spinal and peripheral levels. It is well known that some parts of elementary motor performance as stepping gait and supportive muscle tone are present in cats even after spinal transection. Thus it is probable that the supraspinal descending signals have the character more of command and release qualities than that they contain the whole "message" necessary for e.g. walking (Ref. 20). One of the ways of trying to evaluate the motor functions of the striatum would of course be to study the alterations on the spinal level after manipulation of striatal function. Some of this work has been reported earlier in this text in context with the Parkinson syndrome and evidence has been found for both alfa and gamma motoneuron hyperactivity in man (Ref. 4). In both the reserpinized rat and rats with bilateral destination of dopaminergic paths by means of 6-OH DA treatment it has been shown that there is an elevated alfa motoneuron activity concomitant with a lowered gamma motoneuron activity (Ref. 39, 46). These models may represent the "end stage" of severe Parkinsonism. In the future it will be of great interest to study the alterations in the levels of reciprocal inhibition and recurrent inhibition in the Parkinson patients and in the related animal models by means of developing techniques (Ref. 41, 47).

SUMMARY

Most of what we know about the less complex motor functions of the striatum comes from comparisons between the motor deficits and hyperfunction in the extrapyramidal diseases and studies on animal models of disturbed dopamine mechanisms. It is evident that the striatum takes part in the regulation of muscle tone and body posture and that interaction between the dopamine and the cholinergic pathways of the striatum play an essential role for these. The initiation of motor acts and normal locomotion seem to be dependent on the pathways corresponding to the mesolimbic dopamine tract in the animal and the loss of function here will give rise to a totally incapacitating motor deficit. The different effect of the dopamine agonists on the major symptoms of extrapyramidal disease and in the animal models may indicate different types of dopamine receptors in different parts of the brain. It takes a widespread lesion in the striatum or a profound metabolic disturbance to provoke overt motor symptoms. This indicates that compensatory mechanisms are powerful. In the progression of Parkinson's disease it becomes increasingly difficult to regulate the motor performance between hypo- and hyperfunction with dose adjustments of dopamine agonists and this development may be parallel to the continuous decrease in number of dopamine neurons in the substantia nigra and an increasing hypersensitivity in the remaining postsynaptic receptors.

There are several facts that point to a role for the striatum in processing proprioceptive information ranging from the disturbed balance and poverty of protective reflexes in Parkinson patients to the "sensory neglect" found with the unilaterally dopamine denervated rats.

REFERENCES

1. Andén, N.-E. and Bédard, P., Influences of cholinergic mechanisms on the function and turnover of brain dopamine. J. Pharm. Pharmacol. 23, 460-462 (1971).

2. Andén, N.-E., Dahlström, A. and Larsson, K., Functional role of the nigro-neostriatal dopamine neurons. Acta pharmacol. et toxicol. 24, 263-274 (1966).

3. Andén, N.-E. and Johnels, B., Effect of local application of apomorphine to the corpus striatum and to the nucleus accumbens on the reserpine-induced rigidity in rats. Brain Res., 133, 386-389 (1977).

4. Andrews, C.J., Burke, D. and Lance, J.W., The response to muscle stretch and shortening in Parkinsonian rigidity. Brain, 95, 795-812 (1972).

5. Ayd, Jr., F.J., Neuroleptics and extrapyramidal reactions in psychiatric patients. In Extrapyramidal system and Neuroleptics. J.-M.Bordelean (ed), Ed. Psychiatrique, Montreal, 355-363 (1961).

6. Bucy, P.C. and Case, T.J., Tremor, physiologic mechanisms and abolition by surgical means. Arch. Neurol. Psychiatr. 41, 721-746 (1949).

7. Carlsson, A., The occurence, distribution and physiological role of cathecolamines in the nervous system. Pharmacol. Rev. 11, 490-493, (1959).

8. Carpenter, M.B. and Whittier, J.R., Study of methods for producing experimental lesions of the central nervous system with special reference to stereotaxic techniques. J. comp. neurol. 102, 151-322 (1952).

9. Connor, J.D., Caudate nucleus neurones: correlation of the effects of substantia nigra stimulation with iontophoretic dopamine. J. Physiol., 208, 691 (1970).

10. Cools, A.R., The function of dopamine and serotonin in the caput caudati nuclei of cats. (Dissertation paper), Brakkenstein, Nijmegen, The Netherlands (1973).

11. Cooper, I.S., Results of 1000 consecutive basal ganglia operations for Parkinsonism. Ann. Int. Med. 52, 483-499 (1960).

12. DeLong, M.R. and Strick, P.L., Relation of basal ganglia, cerebellum and motor cortex units to ramp and ballastic limb movements. Brain Res. 71, 327-335 (1974).

13. Denny-Brown, D., The Basal Ganglia. Oxford Univ. Press 41 (1962).

14. Dietz, V. Hillesheimer, W. and Freund, H.-J.: Correlation between tremor, voluntary contraction and firing pattern of motor units in Parkinson's disease. J. Neurol. Neurosurg. Psychiat., 37, 927-937 (1974).

15. Ehringer, H. and Hornykiewicz, O., Verteilung von Noradrenalin und Dopamin im Gehirn des Menschen und ihr Verhalten bei Erkrankungen des extrapyramidalen Systems. Klin. Wschr. 38, 1236-1239 (1960).

16. Ernst, A.M., Mode of action of apomorphine and dexamphetamine on gnawing compulsion in rats. Psychopharmacologia (Berl.), 10, 316-323, (1967).

17. Evarts, E.V., Brain mechanisms in motor control. Life Sci., 15, 1393-1399 (1975).

18. Evarts, E.V., Neurophysiological mechanisms in Parkinson's disease. In Advances in Parkinsonism, Editories Roche, Basel, 37-54 (1976).

19. Forno, L.S. and Jose, C., In Advances in Neurology, 1, Raven Press, N.Y. 453-472 (1973).

20. Grillner, S., Locomotion in vertebrates. Central mechanisms and reflex interaction. Physiol. Rev. 55, 247-304 (1975).

21. Hagbarth, K.-E. and Vallbo, A.B., Discharge characteristics of human muscle afferents during muscle stretch and contraction. Exp. Neurol., 22, 674 (1968).

22. Hassler, R., Zur Pathologie der Paralysis agitans und des postenzephalitischen Parkinsonismus, J. Psychol. Neurol. 48, 387 (1938).

23. Jackson, D.M., Andén, N.-E. and Dahlström, A., A functional effect of dopamine in the nucleus accumbens and some other dopamine-rich parts of the rat brain. Psychopharmacologia, 45, 139-149 (1975).

24. Johnels, B. and Steg, G., Unpublished observations.

25. Johnels, B., Steg, G. and Ungerstedt, U., A method for mechanographical recording of muscle tone in the rat: the effect of some antiparkinsonian drugs on rigidity induced by reserpine. Brain Res., 140, 177-181 (1978)

26. Klawans, H., The pharmacology at extrapyramidal disorders. Karger, Basel (1973).

27. Klemme, R.M., Surgical treatment of dystonia, paralysis agitans and athetosis. Arch. Neurol. Psychiatr. 44, 926 (1940).

28. Kornhuber, H.H., Motor functions of the cerebellum and basal ganglia. The cerebello saccadic (ballistic) clock, the cerebellonuclear hold regulator, and the basal ganglia ramp (voluntary speed smooth movement) generator. Kybernetik 8, 157 (1971).

29. Laursen, A.M., Corpus Striatum. Acta physiol. scand., 59, Suppl. 211, (1963).

30. Martin, J.P., The Basal Ganglia and Posture. Pitman Med. Publ., London, (1967).

31. McLennan, H., The release of dopamine from the putamen. Experentia, 21, 725 (1965).

32. Meyers, R. The modification of alternating tremors, rigidity and festination by surgery of the basal ganglia. Res. Publ. Ass. Nerv. Ment. Dis., 21, 602-665 (1942).

33. Olmstead, C.E., Villablanca, J.R., Marcus, R.J. and Avery, D.L., Effects of caudate nuclei or frontal cortex ablations in cats. IV. Bar pressing and maze learning and performance. Exp. Neurol. 52, 389-420 (1976).

34. Péchadre, J.C., Larochelle, L. and Poirier, L.J., Parkinsonian akinesia, rigidity and tremor in the monkey. J. neurol. Sci., 28, 147-157 (1976).

35. Pijnenburg, A.J.J. and van Rossum, J.M., Stimulation of locomotor activity following injection of dopamine into the nucleus accumbens. J. Pharm. Pharmacol., 25, 1003-1005 (1973).

36. Pollock, L.J. and Davis, L., Muscle tone in Parkinsonian states. Arch. Neurol. Psychiat. 23, 303-317 (1930).

37. Purpura, D.P., Physiological organization of the basal ganglia. In The Basal Ganglia, Raven Press, N.Y., 91-114 (1976).

38. Siegfried, J., Advances in Parkinsonism, Editiones "Roche", Basel 332, 19, (1975).

39. Steg, G., Efferent muscle innervation and rigidity. Acta physiol. scand. 61, Suppl. 225 (1964).

40. Stern, J. and Ward, A.A., Supraspinal and drug modulation of the alfa motor system. A.M.A. Arch. neurol., 6, 404-413 (1962).

41. Tanaka, R., Reciprocal Ia inhibition during voluntary movements in man. Exp. Brain Res., 21, 529-540 (1974).

42. Trétiakoff, C., Contribution à l'étude de l'anatomic pathologique du locus niger de soemmering: Thèse de Paris, no 293, Paris (1919).

43. Ungerstedt, U., Adipoia and aphagia after 6-hydroxydopamine induced degeneration of the nigro-striatal dopamine system. Acta physiol. scand. Suppl. 367, 95-122 (1971).

44. Ungerstedt, U., Postsynaptic supersensitivity after 6-hydroxydopamine induced degeneration of the nigro-striatal dopamine system. Acta physiol. scand., Suppl. 367, 89-93 (1971).

45. Ungerstedt, U., Ljungberg, T. and Ranje, C., Dopamine neurotransmission and the control of behaviour. In Psychobiology of the striatum, Elsevier, Amsterdam, 85-97 (1977).

46. Ungerstedt, U., Ljungberg, T. and Steg, G., Behavioural physiological and neurochemical changes after 6-hydroxydopamine-induced degeneration of the nigropstriatal dopamine neurons. In Advances in Neurology, Vol. 5. Raven, N.Y., 421-426 (1974).

47. Vallbo, Å., Muscle spindle afferent discharge from resting and contracting muscles in normal human subjects. In New Developments in Electromyography and Clinical Neurophysiology. Karger, Basel (1973).

48. van Buren, J.M., Evidence regarding a more precise localization of the posterior frontal-caudate arrest response in man. J. Neurosurg., 24, 416-417 (1966).

49. Villablanca, J.R. and Marcus, R.J., The basal ganglia. A brief review and and interpretation. Acta neurol. lat.-amer., 21, 157-183 (1975).

50. Wallin, B.G., Hongell, A. and Hagbarth, K.-E., Recordings from muscle afferents in Parkinsonian rigidity. In New developments in electromyography, 3, 263, Karger, Basel. (1973).

51. Ward, A.A., McCulloch, W.S. and Magoun, H.W., Production of an alternating tremor at rest in monkeys. J. Neurophysiol., 11, 317-330, (1948).

52. York, D.H., Dopamine receptor blockade - A central action of chlorpromazine on striatal neurons. Brain Res., 37, 91 (1972).

Inhibitory Functions in the Neostriatum

José M. R. Delgado

Departamento de Investigación, Centro Ramon y Cajal
Ctra. de Colmenar km. 9, Madrid 34, Spain

INTRODUCTION

Inhibition was defined long ago by Brunton (12) as the arrest of function of one neuronal structure by the activity of another. Inhibitory processes seem as important and ubiquitous as excitation for the normal function of the nervous system (51,97) and range from altered neuronal discharges to modified behavioral manifestations. Studies on inhibition have been particularly abundant at the synaptic level (5,48,51,57,60,125,160), and also in relation to conditioning (8,68,96,116, 124). Much less is known about the inhibition of spontaneous and evoked behavior (3,38,148) where its role is poorly understood. Behavioral inhibition is neglected in most textbooks of physiology and psychology. In the recent symposium on "Psychobiology of the Striatum" (23), no attention was paid to behavioral inhibition. The lack of information and interest is partly due to methodological difficulties in the investigation of neuronal functions in the awake, unrestrained animals, and partly to the fact that study of inhibition must be related to some manifestations present or elicited during the experiment.

The demonstration of pre- and postsynaptic inhibition, the existence of nerve cells with specific inhibitory properties, the presence of inhibitory neurotransmitters such as GABA, glycine, and β-alanine, and recently developed methods for radio stimulation and telemetry provide some of the experimental bases for investigation of inhibitory functions and mechanisms.

In agreement with Delgado (34), Konorski (97), and Sherrington (139), we may accept the following types of inhibition:

(1) Reciprocal inhibition, as exemplified by flexion of a limb which induces diminished activity of the extensors due to reciprocal functional relations between the neuronal structures involved.

(2) Antagonic inhibition, such as inhibition of the scratch reflex by nociceptive stimulation. Other examples are the inhibition of hunger, sexual interest, somnolence, and other similar states by the influence of pain and fear. In these cases there is no established relation between the considered phenomena but a general blocking of drives induced by nociceptive stimuli.

(3) Unidirectional inhibition; for example, descerebrate rigidity and sham rage, involving a release phenomenon and a set of manifestations which appear by the removal of controlling inhibitory influences.

(4) Lateral inhibition: This phenomenon is produced by a special type of inhibitory neuron with short, ramified axons which influence the neighboring long axon neurons. Konorski (96) has emphasized its importance in perceptual pro-

cesses at the highest levels of the nervous hierarchy.

(5) General inhibition, as described by Sechenov (135) who observed that stimulation of the brain stem of the frog caused widespread inhibition of reflex activity. According to Thomas (147), a general inhibition may be identified when a conditioning volley to a particular structure will cause diminution in responding to test stimuli in a large variety of other structures. General inhibition of behavior induced by electrical stimulation of the brain may be identified (38) by its appearance at the onset of stimulation, affecting all behavior, reducing awareness, decreasing muscular tone, and persisting in spite of sensory stimulation.

(6) Specific inhibition of behavior induced by brain stimulation is characterized by its sudden start, affecting exclusively a determined activity such as food intake or aggressiveness (120,131).

(7) Sleep is a special type of inhibition which may be spontaneous and may also be induced by brain stimulation. In the second case it is characterized by a slow onset affecting all behavior; it is easily interrupted by sensory stimulation and depends primarily on environmental conditions (73). More experimental work is necessary to analyze the stages of induced sleep and the brain areas responsible for this effect.

(8) Conditioned inhibition has a long tradition (96,116,124). The inhibitory stimulus is defined as a multidimensional environmental event that, as a result of conditioning, develops the capacity to decrease performance below the level occurring when that stimulus is absent (68).

(9) The existence of tonic inhibition is demonstrated by the release of manifestations after the blocking or injury of a particular structure. For example, in monkeys, ablation of area 13 of the orbital cortex causes restlessness and hypermotility with a peak at noon and without affecting normal night sleep (100).

The above list of different types of inhibition is only provisional and reflects current terminology. In spite of its non-uniform criteria and overlaps, this list may be useful for evaluation and clarification of available data. For example, unidirectional inhibition of descerebrate rigidity could be considered a release phenomenon of tonic inhibition. In any case, it is essential to specity what is inhibited, and to realize that some activity must exist prior to being diminished, such as a neuronal discharge or motor performance.

CONDITIONING TEST AND MAPPING

Experimental design should distinguish neuronal inhibition (including pre- and postsynaptic, mono- and polysynaptic, and also electrical activity of neuronal fields) from behavioral inhibition (including spontaneous, induced, instrumental, social, and conditioned behavior). Cerebral inhibition should be related to both single neurons and total behavior, as discussed by Deadwyler et al., (27) who investigated the caudate-cortical circuits possibly involved at both levels of neuronal and behavioral inhibition.

One of the simplest methods is the conditioning test: first an electrical stimulation is applied to a neuronal structure with suitable parameters to produce a standard response (electrophysiological or behavioral). Then another suitable stimulation (conditioning volley) is applied with a known temporal sequence (pre-, post-, or simultaneous with the test volley) to an anatomical target in order to evaluate its inhibitory effects on the standard response. This method was used in the classic studies of Eccles et al. (52) to demonstrate the depression of monosynaptic EPSPs determined by stimulation of the innervation of antagonistic muscles.

In more complex experiments, we may study the inhibition of reliable responses induced by sensory stimulation. Positive and negative reinforcement, conditioning, motivation, and other phenomena may be investigated in this way. Another method is to analyze the inhibition of spontaneous activities such as mobility. In all these studies we may use mechanical, electrical, or chemical manipulation of specific areas of the brain.

Mapping of inhibitory points within the primate brain has been performed by our group (38,39,131). Arrays of needle guides were implanted stereotaxically in the skull and several days later, with the unanesthetized animal restrained in a chair, each guide was explored with roving electrodes introduced at 1mm steps. The following combination of behavior identified inhibition: (a) Cessation of spontaneous activity; (b) Hypomimic staring with closed mouth and placid expression; (c) Slow lowering of the head; (d) Lack of interest in offered food; (e) Lack of reaction to slight striking of the face; (f) Slight contralateral head turning was often a concommitant effect. In our studies, inhibitory responses were found in the striatum in all animals explored, but only about 7% of all striatal points investigated were inhibitory. These points were always bilateral but not symmetrical, and no systematic anatomical distribution was found.

NEUROCHEMICAL APPROACH

Behavioral inhibition in the striatum must have supporting neurochemical mechanisms about which little is known, in spite of extensive literature on the prominent role of the striatum dopaminergic system (6,20) and on the functional importance of the acetylcholine and dopamine balance (141). Recent findings about the high concentration of enkephalin in the striatum (80) could be related to the analgesic effects of caudate stimulation, inhibiting the response to peripheral painful shocks.

One useful technique is the chemitrode system (36,46) which consists of intracerebral push-pull cannulae (61) with arrays of multilead electrodes, permitting simultaneous delivery and collection of both electrical and chemical information. With this procedure we have shown that, in monkeys, during perfusion of the caudate nucleus with 14_C glucose, newly formed GABA, aspartate, glutamate a-alanine glutamine, and glycine may be collected (28). This method should allow the testing of metabolic correlates of behavior and also the study of chemical bases of inhibitory phenomena.

In other experiments, it was demonstrated that dopamine and noradrenaline can be synthetized locally in the caudate nucleus from their precursors tyrosine and Dopa (130). Other investigators (21) using push-pull cannulae in "encephale isolé" cats, have demonstrated the synthesis of 3H-dopamine from 3H tyrosine in the caudate nucleus.

According to Neill (113), there is a dopamine acetylcholine antagonism in the central anterior region of the corpus striatum whose net output modulates behavioral inhibition. This is supported by the fact that, in rats, direct application of crystalline dopamine, D-amphetamine, or scopolamine to this region increased their response to food in reinforcement experiments, while anticholinesterase eserine depressed responding.

In rats, unilateral destruction of the dopaminergic nigro-neostriatal bundle by injection of 6-hydroxidopamine produced a general disruption of bar pressing behavior and a loss of self-stimulation in caudate-putamen (119).

INHIBITORY PHENOMENA IN THE NEOSTRIATUM

Analysis of inhibition in the neostriatum must take into consideration the interrelations among cerebral structures which possess inhibitory properties including the orbitofrontal cortex, septum, caudate nucleus, putamen, pallidum, internal capsule, amygdala, hypothalamus, rostral thalamus, cingulate gyrus, and other structures (11,16,34,87,107,111,142,149). Electrically evoked inhibition may be related to blocking of cortico-caudate pathways (99), to fornix mediated inhibition of the lateral hypothalamus, or to disturbances in caudate-pallidum and caudate-nigra connections (146). Inhibition induced by septal and caudate stimulation may depend on activation of the amygdala and hippocampus (63) or the intralaminar nucleus of the thalamus (25). The dorsal raphe nucleus projecting to the substantia nigra and striatum may regulate the interaction between serotonergic and dopaminergic mechanisms, modulating in this way neostriatal excitability and functions (56,115). Excitability is considerably higher in the amygdala and hippocampus than in the septum and caudate, and in monkeys, electrical stimulation of the caudate may induce afterdischarges in the hippocampus without changing the spontaneous activity of the stimulated site (43).

For a clearer understanding of striatal inhibition, more information is needed concerning the functional interrelations among different structures with inhibitory properties. Their possible agonistic or antagonistic effects should be tested keeping in mind that inhibition from a tonic inhibition may result in the release of a function, manifested as active performance.

NEURONAL INHIBITION

As demonstrated by Peacock and Hodes (117), stimulation of the caudate nucleus inhibits the monosynaptic spinal reflex. Studying unitary activity in the cat, Segundo and Machne (136) demonstrated cessation of firing during sensory stimulation, mainly from the skin and muscles. Activation as well as inhibition of caudate and putamen neurons was found, with different patterns of temporal response according to the type or location of stimuli. About two thirds of the putamen neurons responded to sensory stimuli and very often there was a convergence of somatosensory and vestibular effects on single neurons, suggesting a striatal participation in the sensory motor regulation of complex behavior; see also "Motor Responses."

Inhibitory post-synaptic potentials have been induced in the anterior sigmoid gyrus of the cat by single shock applied to the head of the caudate nucleus (81). Buchwald *et al.* (14), working with cats, demonstrated that the inhibition of unitary activity induced by caudate stimulation is paralleled by behavioral inhibition. A study by Yoshida and Precht (162) indicates that caudate stimulation blocks the firing rate of the substantia nigra neurons and this finding may be related with the Gabaergic strionigral pathway and with the regulation of GABA-mediated inhibition of the dopaminergic nigrostriatal system (123). We also know that cholinergic neurons in the caudate are innervated by inhibitory dopaminergic synapses which may be blocked by haloperidol (105).

Most caudate nucleus neurons are inhibited by iontophoretic application of dopamine, including the cells which were activated by the administration of aminoacids (7,22,70,71,106). This inhibitory effect is also produced by dopamine release in the caudate induced by stimulation of the substantia nigra (155,156). In other experiments, however, stimulation of the substantia nigra could have inhibitory as well as excitatory effects on caudate neuron firing (15,82).

In cats under chloralose anesthesia, Kitano (92) demonstrated that most extralemniscal thalamic neurons are inhibited by conditioning caudate or hippocampal stimulation. This inhibitory effect was completely blocked by intravenous administration of strychnine and not by picrotoxin. These results suggest participation of both hippocampus and caudate in the perception of pain.

CORTICAL EXCITABILITY

The caudate nucleus may have an inhibitory influence on cortical excitability, and as described by Umbach(150), seizure activity of several brain structures can be inhibited by caudate stimulation. During seizures, when caudate excitability increased and high voltage potentials appear, there is usually a simultaneous inhibition of the tonic phase. During the clonic phase, the rhythmic slow waves and silent periods appear first in the caudate, suggesting a special role of the caudate nucleus in the mechanism of clonic convulsions, perhaps through intermittent inhibition (87). Destruction of the striatum increases the duration and severity of epileptogenic activity (91).

According to Buchwald et al. (16), there is a caudate loop which receives input from different thalamic nuclei, feeding back modified impulses to the nucleus ventralis anterior and then to the cortex. This system would be antagonistic to the reticular arousal system, although under some conditions it could also be synergic.

The functional relations between the neostriatum and intralaminar thalamic nuclei seem to be reciprocal because slow waves and spindling may be induced in the thalamus by caudate stimulation and recruiting responses have been recorded in the caudate nucleus after stimulation of the non-specific thalamic nuclei (1). Also, the usual low voltage EEG in the caudate and putamen change to high voltage when their cortical connections are interrupted.

Liles and Davis (100) reported that caudate stimulation may inhibit or facilitate cortically induced muscle contractions in the anesthetized cat. Inhibitory effects were localized in the anteroventral region of the caudate, while facilitatory responses were elicited from the posterodorsal area. The rostral poles of the thalamic reticular nuclei seemed to be the essential link mediating the inhibition of cortical motor response produced by caudate stimulation.

Working with anesthetized cats, Feltz et al. (54) showed that stimulation of the caudate nucleus inhibited the short latency responses to somatic stimulation in centrum medianum and dorsomedian nucleus of the thalamus without modifying subthalamic units or those of the lemniscal projection system. Also in cats, chemical activation of the caudate by direct microinjection of the potent neuroexcitatory agent N-methyl-al-aspartic acid resulted in a prolonged suppression of cortical evoked potentials to somatosensory stimulation (127). The secondary portion of cortical potential evoked by thalamic stimulation is inhibited by high frequency stimulation of the caudate (47), and the multisensory evoked potentials in the association cortex are inhibited by stimulation of the basal ganglia (54).

As indicated by Jung and Hassler (88), the caudate nucleus seems to restrain and control the level of excitability of the whole cerebral cortex, producing a selective inhibition of the sensory systems which are less significant for the behavioral performance of the moment. Also, the striatum seems crucial for the inhibitory control of cortical voluntary motor activity (18; see 100).

MOTOR RESPONSES

The early finding of Ferrier (55) that faradic stimulation of the corpus striatum produces contraversive turning of the head and body has been confirmed by many investigators. Stimulation of the caudate nucleus in unanesthetized, freely moving cats induces purposeful, contralateral turning of the head and body which often results in well coordinated circling (67). According to Forman and Ward (59), there is a somatotopic representation of turning movements with head turning and body bending located in the ventral part of the caudate nucleus, while neck and trunk movements are represented in the middle and dorsal caudate areas.

In apparent contradiction with evocation of motor effects, Mettler *et al.* (108) and Hodes *et al.* (79) described a suppressor effect that caudate stimulation exerted on motor responses evoked from the motor cortex. This finding, however, was not confirmed by later studies by Forman and Ward (59). There is also a peripheral inhibitory effect because, as described by Nagao *et al.* (112), caudate stimulation in cats inhibits the electromyographic response of the gastrocnemius elicited by dorsal root excitation. Caudate stimulation also affects the afferent somatic transmission at the level of the cuneate nucleus (85).

The existence of a continuous inhibitory tone of the caudate nucleus upon spontaneous walking is suggested by the effects of its destruction. The original studies of Lussana and Lemoigne (102), Magendie (104), Nothnagel (114), and Schiff (132) showed that bilateral destruction of the caudate nucleus produced an irresistible drive to run straight forward with well coordinated movements, regardless of obstacles. Mettler and Mettler (109) observed that in cats and monkeys, lesions smaller than 3mm in diameter failed to impair motor activity significantly,whereas after complete destruction of both caudate nuclei there was an incessant drive to walk straight ahead. After bilateral caudate ablation in cats, Villablanca *et al.* (153) described a "compulsory approaching syndrome," characterized by stereotyped and prolonged approaching and/or following persons, cats, or objects. In addition, there was exaggerated forelimb treading, hyperactivity, and other disturbances.

In our studies in dogs (Delgado, unpublished), we observed that after extensive lesions in both caudate nuclei, the animals had an irresistible urge to walk forward. When forceably restrained or when blocked by an obstacle, they continued to try and push ahead, pawing the floor without turning around or seeking to avoid the obstacle. This effect contrasted with stimulation of the motor cortex in unanesthetized animals which correctly processed sensory information and were able to modify the pattern of electrically evoked movement, avoiding obstacles and performing with adaptive efficiency (29).

The role of the striatum in regulating motor activity is well recognized and should be related to processing of sensory information with excitatory and inhibitory effects (9,50,134). Motor disturbances may appear after extensive bilateral lesions of the caudate nucleus or lenticular nuclei, including tremors in monkeys (53) and contralateral chorea in the chimpanzee (91). In rats, intrastriatal injection of GABA antagonists (Picrotoxin, bicuculline) may produce involuntary choreic-like movements, and this preparation has been proposed for the study of dyskinesias (143).

In human patients with electrodes implanted in the frontostriatal region, stimulation has produced arrest of voluntary movements and speech followed by amnesia and confusion (43,151).

SLEEP

In both barbiturate-induced and natural sleep, the caudate nucleus shows the earliest EEG changes. This fact has been demonstrated in patients (95,110) and in monkeys (78) with intracerebral electrodes, indicating the prominent role of the caudate in the mechanism of sleep (see also 133).

Direct electrical stimulation of the caudate nucleus in cats may produce a sleep-like effect, according to Akert and Andersson (2) and Hess *et al.* (76). In these experiments, the EEG patterns were similar to those of spontaneous sleep but behavior differed because the animals did not assume the balling position before sleep, muscle tone was preserved, and there was atypical behavior including awkward paw and ankle movements, suggesting disturbed propioceptive mechanisms. Stevens *et al.* (145) working also with cats, observed quietude, failure to respond to conditioned avoidance, and drowsiness when the caudate nucleus was stimulated electrically or chemically with carbachol. Similar inhibitory effects have been reported by Buchwald et al. (17) using low frequency stimulation, while arousal was obtained using high frequency or stronger intensities of stimulation (13).

A sleep-like effect in monkeys and also in one schizophrenic patient has been described, induced by stimulation of the head of the caudate nucleus (69). In our studies in awake, freely moving monkeys (128), intermittent stimulation of the septum for about 5 minutes decreased spontaneous activity of the animals, which started dozing and finally closed their eyes assuming a sleeping posture. Sensory stimulation produced arousal but the animals dozed again for about 15 minutes without further septal stimulation. In more recent studies in restrained monkeys, we have observed sleep-like effects induced by caudate stimulation (35,37). This subject, however, has been controversial and some investigators doubt that the inhibitory effects induced by caudate nucleus stimulation should be considered as true sleep (see discussion by Jasper, Akert, and Mettler in Heath and Hodes,69).

According to Hess (74), the caudate nucleus contributes to the regulation of somatomotor readiness and caudate stimulation activates only a part of the sleep mechanisms, affecting the stereoceptive and vegetative mechanisms less than the hypnogenic zone of the medial thalamus.

The role of the striatum in sleep mechanisms is also supported by the experiments of Corsi-Cabrera et al. (24) in which destruction of the medial portion of the caudate-putamen complex in rats increased the duration of paradoxical sleep 248% without changing slow wave sleep or wakefulness. However, bilateral removal of the caudate nuclei in cats (154) produced motor hyperactivity without EEG changes. Removal of most of the frontal tissue in front of stereotaxic plane A22 produced a significant permanent reduction of sleep, particularly REM sleep.

ARREST REACTION

One of the most spectacular inhibitory effects is the "arrest reaction," initially observed in cats by Hunter and Jasper (83) during stimulation of the intralaminar nucleus of the thalamus, and subsequently described by other investigators stimulating the striatum and other brain structures (see summary in 34). The effect consists of a sudden cessation of spontaneous motor activity, the animal remaining motionless with decreased responsiveness to sensory stimuli, with maintenance of postural tonus and reflexes, without autonomic changes, and resuming the interrupted spontaneous behavior as soon as the cerebral stimulation is over. The animals are "frozen" in the position held at the onset of stimulation, as if a moving picture had been halted; for example, immobilizing a cat between two steps of a staircase, or even with its tongue out of the mouth (29).

The arrest reaction should be differentiated from the "adynamia" described by Hess (72) in which inhibition of spontaneous movements is accompanied by loss of muscular tone and the animals are usually lying down on the floor with abnormal postures and diminished spontaneous activity, without falling asleep (34).

Arrest of motor activity has been produced in cats, dogs, and monkeys by different investigators stimulating the caudate nucleus, putamen, pallidum, cingulate gyrus, orbitofrontal cortex, and other structures; see Fig. 1 (2,13,19,29,39,40,42,59,62, 64,83,89,103,140,152,157).

Fig. 1 Radio stimulation of the caudate nucleus in the gibbon located at the left produced sudden inhibition of behavioral activities. Observe its passive attitude in contrast to the attentive interest of the other two gibbons.

The mechanism of arrest reaction is not well understood. It could be related to interference in the integrated normal function of the thalamic reticular system (86) or to inhibition of the whole sensorimotor system, with activation of part of the sleep mechanism (75). As similar types of arrest reaction are obtained from different areas, including striatum, thalamus, and limbic systems, further studies are necessary to determine synergisms, correlations, and functional equilibrium among excitatory and inhibitory structures of the brain. Grossman and Wang (66) and Grossman (65) have described a somatomotor inhibitory system in the thalamic reticular area and in the ventral part of the hypothalamus connected with the septal nuclei and lateral part of the head of the caudate nucleus. All these structures could be closely related in their inhibitory properties.

In monkeys trained to perform different types of arm movements to obtain food reward, Kitsikis (93) showed that caudate stimulation had a suppressor effect with different thresholds depending on the stages of the movement. Initiation of the motor action was easier to prevent than its actual performance. Taking up of food was also easier to suppress than carrying food to the mouth. These experiments suggest that caudate inhibition is one factor among several determinants with a dynamic equilibrium regulating motor performance.

FOOD INTAKE

The normal interest of monkeys in bananas and other food can be abruptly inhibited by electrical stimulation of the head of the caudate nucleus (30,131). This effect is characterized by (a) loss of interest in food offered to the animal; (b) dropping of pieces of food held in the hands; (c) immediate cessation of chewing movements; (d) spitting out of any food in the mouth; (e) active emptying of the mouth pouches if they were filled with food; (f) rejection of food, using the hands to push it away and averting the face; (g) walking away from the proximity of food; and (g) grimaces of distaste. All these effects started abruptly at the onset of stimulation and ended completely when stimulation was over, without leaving residual traces. The animals immediately resumed their interest in food on cessation of stimulation, grabbing it, chewing, and swallowing. The fact that the animals were well oriented, that motor coordination was well preserved, and that active and purposeful movements were performed to reject food indicate that inhibition was food-specific without having general inhibitory effects. In our studies, some caudate points had this specificity while others located only 3mm away could evoke general inhibition including diminution of postural tonus, decreased aggressiveness, and lack of interest in food.

Cessation of chewing and ejection of food have also been obtained in monkeys by electrical stimulation of septum, amygdala, and other structures, modifying only alimentary reactions without producing observable effects in the absence of food (126). Stimulation of the pyriform cortex in unanesthetized squirrel monkeys inhibited eating of a banana and the animal spit the fruit into the palm of its hand, peered at it with grimaces of distaste, and tried to escape from the testing stage (103).

Food intake seems to be regulated by a system involving the septum, caudate, and amygdala, correlated with but not dependent on the well known system located in the ventral-lateral hypothalamic area.

AGGRESSION

In rhesus monkeys restrained in a chair, a reliable aggressive response may be evoked by approaching the animal suddenly and waving a glove, stick, or bare hand. Characteristically the monkey reacts by opening its mouth, showing the teeth, moving the head forward, vocalizing with a typical low tone, and trying to grab, scratch, and bite the approaching object. These responses have been inhibited by electrical stimulation of the striatum, during which it was safe to touch the animal (34).

Threats toward monkeys free within a laboratory colony evoke this aggressive reaction plus pawing the floor, shaking the cage, and launching sudden attacks. All these manifestations have been inhibited by radio stimulation of the striatum, during which the animals were not depressed, arrested, adynamic, or stuporous, but had normal muscular tone, were well oriented, alert, able to follow moving objects with the eyes and to use the paws to push away any approaching objects. Inhibition of aggressive behavior lasted as long as the stimulation, after which the monkeys reverted immediately to their usual ferocious behavior.

In some experiments (38,120), it was demonstrated that programmed stimulations of inhibitory points of the putamen and caudate nucleus effectively blocked vocalizing, grimacing, threatening, and other offensive-defensive behavior without modifying food intake, walking, grooming, or instrumental responses to obtain pellets or avoid shock. We may conclude that stimulation of some cerebral points inhibits aggression without disturbing alimentary, locomotor, grooming, or learned instrumental behavior.

SOCIAL HIERARCHY

Hierarchical structure in monkey colonies is usually well defined with one animal dominating the group and the ranking order of the others determined by their reciprocal social relations, including territoriality, access to food, mounting, and offensive-defensive behavior. In colonies of five or six monkeys established in laboratory cages, the dominant animal is often alone, occupying over half of the territory. It is avoided by the rest of the group, and its stares or threats provoke signs of submissiveness such as cringing, presenting, and escaping.

In one of our experiments (31), programmed stimulation of the caudate nucleus in the boss monkey for one hour initially produced only a slight relaxation of his body. After 6 to 8 minutes, the other animals began to circulate more freely around the cage, often in proximity to the boss and occasionally crowding fearlessly around him. During caudate stimulation, the boss showed no territoriality or aggressive behavior and his top ranking position was lost. About 12 minutes after the end of stimulation, however, the boss resumed his aggressive behavior, recovering both territoriality and colony rank.

These changes in social status may be related to diminished aggressive reactivity or other factors to be clarified by further studies. Coyle and Kirkby (26) have reported that rats with bilateral lesions of the anterior caudate nucleus and tested in a straight-alley dominance situation won 91% of the encounters, demonstrating an increase in dominance related to changes in the level of aggression.

CEREBRAL HETEROSTIMULATION

Based on the finding that a boss monkey may lose his authority under caudate stimulation, other experiments were performed placing a lever inside the colony cage. In this case, lever pressing activated radio stimulation of the caudate nucleus in the boss. Without any cues, a submissive animal learned to press the lever in response to threats from the boss, thus inhibiting his aggressive behavior. The submissive monkey that learned this instrumental response did not become dominant but as long as the lever was connected, she was responsible for blocking many attacks against herself and for maintaining peaceful coexistence within the entire colony (32).

LEARNING, MEMORY, AND INSTRUMENTAL RESPONSES

The caudate nucleus seems to play a role in learning and performance because, in monkeys, small lesions disturb the delayed alternation task without altering visual discrimination (4,128,129). The punishing effect of electroshock to the feet is forgotten if it is immediately followed by stimulation of the caudate nucleus, and animals will persevere in the testing situation as if no shock had been received, while they learn quickly to avoid the shock in the absence of caudate activation (30,118,158,161). Less clear results were obtained in cats trained to eat from a metal dish and receiving aversive mouth shock from time to time, followed by thalamic or caudate stimulation (159). Tested 24 hours later, the animals stimulated in the thalamus had amnesia, while only half of the caudate stimulated cats had forgotten the aversive shock.

In rats, learning of maze problems was also disrupted by hippocampal and not by caudate stimulation (144). Apparently there is a time factor involved because, according to Wyers and Deadwyler (161), bilateral caudate nucleus stimulation in rats produced retrograde amnesia only after a 5 minute delay and not with delays of 15 minutes. Acquisition of conditioned emotional response by rats was also slower and extinction faster when single pulse caudate stimulation was delivered (100 msec) after each light flash, used as a conditioned stimulus (77).

According to Buchwald et al. (16), low frequency stimulation of the caudate in cats decreases lever pressing to obtain food, while high frequencies increase performance. In rats, disruption of normal caudate nucleus activity, induced by microinjections of potassium chloride, produced a marked impairment of the acquisition of a one-trial passive avoidance task, suggesting the involvement of caudate nucleus in the integration and storing of learned information (122).

In rhesus monkeys tested in pairs during lever pressing to obtain food pellets, radio stimulation of the caudate nucleus with a program of 5 seconds on, 5 seconds off clearly inhibited the instrumental response although there was a process of behavioral adaptation. The animal seemed aware of its handicap and often used the interstimulation time to obtain food by lever pressing at a faster rate, demonstrating that there were no residual or cumulative effects following cessation of inhibitory stimulation. During inhibitory stimulation of the dominant animal, there was a significant increase in the submissive partner's mobility and lever pressing for food (39).

In monkeys, the caudate nucleus and putamen have points with positive reinforcement properties, and therefore self-stimulation is possible (19,121).

Aversive conditioning in cats has been abolished by low frequency stimulation of the caudate (145), while higher frequencies produced contraversive rotation without modifying conditioning. Working with dogs, Shapovalova (137) has shown that stimulation of the head of the caudate nucleus inhibits electromyogram latency and performance of instrumental defensive reaction related to the maintenance of a certain posture. In rabbits, both septal and caudate stimulation inhibited previously learned avoidance and operant behavior, but these effects were obtained only with low frequencies of excitation (94).

In Boakes and Halliday's excellent book "Inhibition and Learning" (8), a review is presented of the role of inhibition in Pavlovian and instrumental conditioning. The role of the septum and importance of the limbic system are emphasized (49,84), but the possible role of caudate and putamen are not even mentioned in the index. The striatal areas, which have well defined inhibitory properties, probably participate in conditioned inhibition and this working hypothesis should be further investigated.

HEART RATE AND RESPIRATION

Behavioral inhibition induced by caudate stimulation does not usually affect autonomic functions. As described by Rubinstein and Delgado (131) in monkeys however, excitation of the caudate nucleus may produce cardiac and respiratory inhibition. Application of a single shock of 0.6mA slowed down the heart after a small delay: the first post-stimulation beat was not modified, while the second systole was completely abolished and the next three or four contractions were progressively less affected. These single shocks also produced a momentary arrest of respiration. When frequencies of 5 cps were used, cardiac and respiratory arrest could be evoked. Respiratory inhibition fatigued in seconds, while the decrease in heart rate persisted during as much as 30 seconds of continuous stimulation. Excitation at 100 cps stopped the heart for about four beats. Then escape occurred with gradual recovery of the initial rhythm.

These findings strongly suggest vagal activation. The fact that cardiac inhibition was induced with frequencies ranging from single shocks to 100 cps indicates that the triggered mechanism is independent of stimulus characteristics. The neighboring area 13 in the orbital surface of the frontal lobes is known to have important vagal connections which could go through the caudate nucleus. Another possibility would be a sympathetic inhibition, emanating from related structures, as suggested by Folkow et al., (58). In cats, injection of 200 µg of L-DOPA in the left lateral cerebral ventrical induced a period of unit silence in both caudates lasting from 30-50 minutes and coincident with a reduction of systolic blood pressure and heart rate (135). These findings support the possible role of the caudate nucleus in cardiovascular regulations.

FATIGABILITY

In our experience with monkeys (38,42), behavioral inhibition evoked by continuous striatal stimulation has always fatigued in a few minutes. Inhibition usually disappears within 60 seconds, and the animal resumes his aggressive behavior and interest in food in spite of the ongoing stimulation. The longest lasting evoked inhibition has been five minutes in a gibbon during stimulation of the head of the caudate nucleus.

In experiments with cats (42), we explored the fatigability of caudate stimulation. The tested effects were interruption of ongoing activities and contraversive head turning. Under continuous stimulation with 0.2-0.7mA, the motor response faded after 10 to 20 minutes. This fatigability seemed to be a local process because when intensity was increased, the motor response reappeared, and because when one point had been fatigued, thresholds for other points in the caudate nucleus and medial geniculate were not affected. Administration of reserpine (0.5mg/kg i.p.) and haloperidol (0.075 mg/kg i.m.) accelerated caudate fatigability while injection of imipramine (2 mg/kg i.m.) and cocaine (1 mg/kg i.m.) delayed fatigability, suggesting the involvement of catecholamines in the establishment of caudate fatigability.

FEEDBACK STIMULATION

Fatigability of inhibitory effects evoked by brain stimulation makes it difficult to obtain long term results. In order to prolong the duration of behavioral inhibition, two strategies have been tested: programmed and feedback stimulations (38, 42,44). From several programs used, the most effective has been that of 5 seconds on, 5 seconds off, with which inhibition could be prolonged for over an hour.This schedule, however, is rigid by design because stimulations are applied regardless of whether the animal is active or quiet.

In feedback experiments, the brain is stimulated only when the animal is active. For this purpose, monkeys were equipped with telemetric sensors of mobility (33, 98) which after every 25 counts, triggered a 1-second radio stimulation of a preselected inhibitory point in the brain (Fig. 2). In this way, stimulations were proportional and determined by the spontaneous mobility of the animal. With this feedback technique, the amount of stimulation necessary to inhibit spontaneous behavior was 25 to 30 times <u>less</u> than when fixed programs of 5 seconds on, 5 seconds off were used.

Contingency may be an essential element in the effectiveness of some feedback stimulations, as demonstrated in free chimpanzees equipped with telemetric recorders for depth EEG and brain stimulators (44). In this experiment, amygdala spindling was identified by a computer which automatically triggered radio stimulation of a negative reinforcing point located in the central gray. After two hours of this contingent stimulation, the rate of spontaneous amygdala spindling was reduced to 50% of control values. Following six days of 2 hour daily contingent stimulation cessions, the amygdala spindling decreased below 1%. In the absence of contingency however, the same amount of reticular stimulation did not modify amygdala spindling. With this type of feedback technology and with the introduction of the factor of contingency, the inhibitory effects of the striatum could be considerably increased and prolonged.

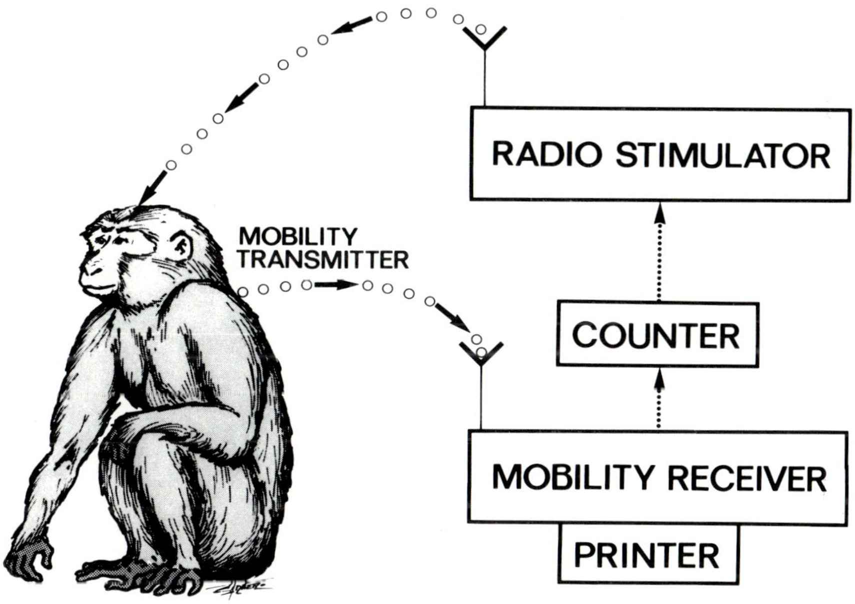

Fig. 2 To establish a feedback inhibition, the monkey was equipped with a telemetric sensor of spontaneous mobility which through a computer activated the radio stimulation of an inhibitory point in the pallidum. This feedback was 20-30 times more efficient in producing inhibition than the best automatically fired program of 5 sec on, 5 sec off.

PARADOXICAL INHIBITION

Testing the effects of caudate stimulation in monkeys during automatic delivery of food pellets (4 per minute) for 5 minutes, we observed a drastic reduction in the number of pellets taken and a great increase in walking (38). Identification of this caudate point as "inhibitory" or "excitatory" obviously depended on the type of behavior tested. This finding indicates the specificity of evoked effects with respect to behavioral categories and also to the quality of response, illustrating the need to explore multiplicity of responses.

The importance of social relations was evident during testing of pairs of monkeys, because when the stimulated animal was the dominant one, its inhibition of food intake was paralleled by a proportional increase in the number of pellets taken by the submissive partner, without any observable change in its walking activity (38).

DYNAMIC EQUILIBRIUM

The nervous system contains antagonistic subsystems such as the sympathetic and parasympathetic which regulate physiological activities at different levels of dynamic equilibrium. As our group has demonstrated in monkeys, gibbons, and chimpanzees (39,45), caudate nucleus inhibition can balance the behavioral activation of the central gray. In a typical experiment, a gibbon forming part of a peaceful laboratory colony had electrodes implanted in an inhibitory point of the caudate and an excitatory point of the central gray. Radio stimulation of the central gray evoked an offensive-defensive response characterized by showing of the teeth, vocalizations, and well directed attacks against a subordinate gibbon, causing unaccustomed disruptive fights. Although caudate stimulation inhibited this aggressive behavior, it has brief or no after effects, as reported earlier, while induced aggression usually outlasts brain stimulation by about 1 minute. It was therefore necessary to repeat the caudate stimulation several times in order to stop fighting evoked by excitation of the central gray.

If, however, caudate stimulation of appropriate intensity preceded central gray excitation, no offensive-defensive or other emotional effects were observed. Simultaneous stimulation of both structures could be maintained in equilibrium for about 1 minute without any social disturbance. After that time, the fatigable caudate inhibitory effect faded away and the gibbon resumed his aggressive behavior. This type of experiment consisting of simultaneous caudate and central gray stimulation seems appropriate for exploration of neurophysiological and neurochemical mechanisms of inhibition.

CLOSING COMMENTS

In the study of inhibitory functions in the neostriatum, there are two periods of productive scientific investigation: one in the 1950s when behavioral effects were investigated, and the other in more recent years when dopaminergic and other chemical systems have been considered. Published reports reveal the difficulty in establishing correlations between neuronal, behavioral, and chemical events.There appears to be great need to integrate the accumulated knowledge and to design multidisciplinary experiments in order to investigate, for example, the neuronal and neurochemical bases of arrest reaction, and the neuroanatomical and functional correlations among the different cerebral structures which play a role in the various aspects of inhibition.

REFERENCES

1. Ajmone-Marsan, C. and M. Dilworth, Non specific projection system of the thalamus and striate system. Electrophysiological study, EEG clin. Neurophysiol., Suppl. 3, 85 (1953).
2. Akert, K. and B. Andersson, Experimenteller Beitrag zur Physiologie des Nucleus caudatus, Acta physiol. scand., 22, 281-298 (1951).
3. Arnold, M.B. (Ed.), Feelings and Emotions, Academic Press, New York (1970).
4. Battig, K., H.E. Rosvold, and M. Mishkin, Comparison of the effects of frontal and caudate lesions on delayed response and alternation in monkeys, J. comp. physiol. Psychol., 53, 400-404 (1960).
5. Beritov, J.S., Neural Mechanisms of Higher Vertebrate Behavior, Trans. from Russian by W.T. Liberson, Little Brown, Boston (1965).

6. Birkmayer, W., K. Jellinger and Y. Riederer, Striatal and extrastriatal dopaminergic functions. In: A.R. Cools, A.H.M. Lohman, and J.H.L. van den Bercken (Eds.), Psychobiology of the Striatum, Elsevier/North Holland Biomedical Press, Amsterdam, New York, Oxford, 141-153 (1977).
7. Bloom, F.E., E. Costa, and G.C. Salmoiraghi, Anesthesia and the responsiveness of individual neurons of the caudate nucleus of the cat to acetylcholine, norepinephrine and dopamine administered by microelectrophoresis, J. Pharmacol. exp. Ther., 150, 244-252 (1965).
8. Boakes, R.A. and M.S. Halliday (Eds.), Inhibition and Learning, Academic Press, New York (1972).
9. Bossom, J., The effect of lesions in caudate nucleus and superior colliculus upon adaptation to prisms in monkey, Psychon. Sci., 26, 255-256 (1972).
10. Brady, J.V., A comparative approach to the evaluation of drug effects upon behavior. In: W.S. Fields (Ed.), Brain Mechanisms and Drug Action C.C Thomas, Springfield, Ill., 111-139 (1957).
11. Brayley, K.N. and D.J. Albert, Suppression of VMH-lesion induced reactivity and aggressiveness by electrical stimulation ventral to the anterior septum in the rat, Physiol. Behav., 18, 567-571 (1977).
12. Brunton, R.L., On the nature of inhibition, and the action of drugs upon it. Nature, 27, 419-422 (1883).
13. Buchwald, N.A. and F.R. Ervin, Evoked potentials and behavior. A study of responses to subcortical stimulation in the awake, unrestrained animal, EEG clin. Neurophysiol., 9, 477-496 (1957).
14. Buchwald, N.A., C.D. Hull, and M.C. Trachtenberg, Concomitant behavioral and neural inhibition and disinhibition in response to subcortical stimulation, Exp. Brain Res., 4, 58-72 (1967).
15. Buchwald, B.A., D.D. Price, L. Vernon, and C.D. Hull, Caudate intracellular response to thalamic and cortical inputs, Exp. Neurol., 38, 311-323 (1973).
16. Buchwald, N.A., E.J. Wyers, C.W. Lauprecht, and G. Heuser, The "caudate-spindle." IV. A behavioral index of caudate-induced inhibition, EEG clin. Neurophysiol., 13, 531-537 (1961b).
17. Buchwald, N.A., E.J. Wyers, T.Okuma, and G. Heuser, The "Caudate-spindle." I. Electrophysiological properties, EEG clin. Neurophysiol., 13, 509-518 (1961a).
18. Bucy, P.C., The neural mechanisms of athetosis and tremor. J. Neuropathol. exp. Neurol., 1, 224-239 (1942).
19. Bursten, B. and J.M.R. Delgado, Positive reinforcement induced by intracerebral stimulation in the monkey, J. comp. physiol. Psychol., 51, 6-10 (1958).
20. Carlsson, A., The occurrence, distribution and physiological role of catecholamines in the nervous system, Pharmacol. Rev., 11, 490-493 (1959).
21. Chéramy, A., A. Nieoullon, and J. Glowinski, Stimulating effects of γ-hydroxybutyrate on dopamine release from the caudate nucleus and the substantia nigra of the cat, J. Pharmacol. exp. Therapeut., 203, 283-293 (1977).
22. Connor, J.D., Caudate nucleus neurones: correlation of the effects of substantia nigra stimulation with iontophoretic dopamine, J. Physiol., Lond., 208, 691-703 (1970).
23. Cools, A.R., A.H.M. Lohman and J.H.L. van den Bercken (Eds.), Psychobiology of The Striatum, Elsevier/North Holland Biomedical Press, Amsterdam, New York, Oxford (1977).
24. Corsi-Cabrera, M., J. Grinberg-Zylberbaum, and L.S. Arditti, Caudate nucleus lesion selectively increases paradoxical sleep episodes in the rat, Physiol. Behav., 14, 7-11 (1975).
25. Cowan, W.M. and T.P.S. Powell, The projection of the midline and intralaminar nuclei of the thalamus of the rabbit, J. Neurol. Psychiat., 18, 266-279(1955).
26. Coyle, I.R. and R.J. Kirkby, Dominance behavior in the rat following lesions of the caudate nuclei, Physiol. Behav., 14, 75-80 (1975).
27. Deadwyler, S.A., N. Hirasuna and E.J. Wyers, Caudate stimulation and behavioral inhibition in the rat, Brain Res., 75, 59-70 (1974).

28. DeFeudis, F.V., J.M.R. Delgado and R.H. Roth, Content, synthesis and collectability of aminoacids in various structures of the brains of rhesus monkeys, Brain Res., 18, 15-23 (1970).
29. Delgado, J.M.R., Responses evoked in waking cat by electrical stimulation of motor cortex, Amer. J. Physiol., 171, 436-446 (1952).
30. Delgado, J.M.R., Brain stimulation in the monkey: technique and results. (Motion picture), Fed. Proc., 16, 29 (1957).
31. Delgado, J.M.R., Effect of brain stimulation on task-free situations, EEG clin. Neurophysiol., Suppl. 24, 260-280 (1963a).
32. Delgado, J.M.R., Cerebral heterostimulation in a monkey colony, Science, 141, 161-163 (1963b).
33. Delgado, J.M.R., Telemetry and telestimulation of the brain. In: L. Slater (Ed.), Bio-Telemetry, Pergamon Press, New York, 231-249 (1963c).
34. Delgado, J.M.R., Free behavior and brain stimulation. In: C.C. Pfeiffer and J.R. Smythies (Eds.), International Review of Neurobiology, Vol. VI, Academic Press, New York, 349-449 (1964).
35. Delgado, J.M.R., Radio control of behavior. Exhibit, XXIII Int. Congr. physiol. Sci., Tokyo, Sept., (1965).
36. Delgado, J.M.R., Intracerebral perfusion in awake monkeys, Arch. int. Pharmacodyn., 161, 442-462 (1966).
37. Delgado, J.M.R., Physical Control of the Mind: Toward a Psychocivilized Society, Vol. XLI, World Perspectives Series, Harper & Row, New York (1969).
38. Delgado, J.M.R., Inhibitory systems and emotions. In: L. Levi (Ed.), Emotions - Their Parameters and Measurement, Raven Press, New York, 183-204 (1975).
39. Delgado, J.M.R. and H. Bracchitta, Free and instrumental behavior in monkeys during radio stimulation of the caudate nucleus, Int. J. Psychobiol., 2, 233-248 (1972).
40. Delgado, J.M.R., J.M. Delgado-García, J.A. Amérigo, and C. Grau, Behavioral inhibition induced by pallidal stimulation in monkeys, Exp. Neurol., 49, 580-591 (1975).
41. Delgado, J.M.R., J.M. Delgado-García, M. Conde, and S. Sánchez-Robles, Fatigability of caudate nucleus stimulation in cats, Neuropsychologia, 14, 11-21 (1976a).
42. Delgado, J.M.R., J.M. Delgado-García and C. Grau, Mobility controlled by feedback cerebral stimulation in monkeys, Physiol. Behav., 16, 43-49 (1976b).
43. Delgado, J.M.R. and H. Hamlin, Spontaneous and evoked electrical seizures in animals and in humans. In: E.R. Ramey and D.S. O'Doherty (Eds.), Electrical Studies on the Unanesthetized Brain, Hoever, New York, 133-158 (1960).
44. Delgado, J.M.R., V.S. Johnston, J.D. Wallace, and R.J. Bradley, Operant conditioning of amygdala spindling in the free chimpanzee, Brain Res., 22, 347-362 (1970).
45. Delgado, J.M.R., A.M. Sanguinetti, and F. Mora, Aggressive behavior in gibbons modified by caudate and central gray stimulation, Int. Res. Comm. System, Sept., 16-2-32 (1973).
46. Delgado, J.M.R., P. Simhadri and J. Apelbaum, Chronic implantation of chemitrodes in the monkey brain, Proc. int. Union physiol. Sci., 2, 1090 (1962).
47. Demetrescu, M. and M. Demetrescu, The inhibitory action of the caudate nucleus in cortical primary receiving areas in the cat, EEG clin. Neurophysiol., 14, 37-52 (1962).
48. Diamond, S., R.S. Balvin, and F.R. Diamond, Inhibition and Choice. A Neurobehavioral Approach to Problems of Plasticity in Behavior, Harper & Row, New York (1963).
49. Dickinson, A., Septal damage and response output under frustrative non-reward. In: R.A. Boakes and M.S. Halliday (Eds.), Inhibition and Learning, Academic Press, New York, 461-496 (1972).
50. Diez-Martinez, O., M. Garcia Muñoz, G. Prieto, J.A. Roig, and H. Brust-Carmona, Sensory afferents to the caudate nucleus, Physiol.Behav.,16, 727-733 (1976).

51. Eccles, J.C., The Physiology of Synapses, Springer Verlag, Berlin (1964).
52. Eccles, J.C., R.M. Eccles and R. Magni, Central inhibitory action attributable to presynaptic depolarization produced by muscle afferent volleys, J. Physiol. Lond., 159, 147-166 (1961).
53. Edwards, D.J. and H.J. Bagg, Amer. J. Physiol., 65, 162 (1924).
54. Feltz, P., G. Krauthamer, and D. Albe-Fessard, Neurons of the medial diencephalon. I. Somatosensory responses and caudate inhibition, J. Neurophysiol., 30, 55-80 (1967).
55. Ferrier, D., Experimental research in cerebral physiology and pathology. West Riding Lunatic Asylum med. Rep., 3, 1-50 (1873).
56. Fibiger, H.C. and J.J. Miller, An anatomical and electrophysiological investigation of the serotonergic projection from the dorsal raphe nucleus to the substantia nigra in the rat, Neuroscience, 2, 975-987 (1977).
57. Florey, E. (Ed.), Nervous Inhibition, Pergamon Press, New York (1961).
58. Folkow, B., B. Johansson, and B. Oberg, A hypothalamic structure with a marked inhibitory effect on tonic sympathetic activity, Acta physiol. scand., 47, 262-270 (1959).
59. Forman, D. and J.W. Ward, Responses to electrical stimulation of caudate nucleus in cats in chronic experiments, J. Neurophysiol., 20, 230-244 (1957).
60. Frank, K., Basic mechanisms of synaptic transmission in the central nervous system, I.R.E. Trans. med. Electron., ME-6, 85-88 (1959).
61. Gaddum, J.H., Push-pull cannulae, J. Physiol., 155, 1P-2P (1961).
62. Gastaut, H., Corrélations entre le système nerveux végétatif et le système de la vie de relation dans le rhinencéphale, J. Physiol. Path. gén., 44, 431-470 (1952).
63. Gloor, P., Electrophysiological studies on the connections of the amygdaloid nucleus in the cat. II. The electrophysiological properties of the amygdaloid projection system, EEG clin. Neurophysiol., 7, 243-264 (1955).
64. Goldzband, M.G., S.E. Goldberg, and G. Clark, Cessation of walking elicited by stimulation of the forebrain of the unanesthetized dog. Amer. J. Physiol., 167, 127-133 (1951).
65. Grossman, R.G., Effects of stimulation of non-specific thalamic system on locomotor movements in cat, J. Neurophysiol., 21, 85-93 (1958).
66. Grossman, R. and S.C. Wang, Thalamic inhibition of subcortically induced locomotor movements, Fed. Proc., 15, 83 (1956).
67. Hassler, R., Die zentralen apparate der Wendebwegungen, Arch. Psychiat. Nervenkr., 194, 456-514 (1956).
68. Hearst, E., S. Besley, and G.W. Farthing, Inhibition and the stimulus control of operant behaviour, J. exp. Anal. Behav., 14, 373-409 (1970).
69. Heath, R.G. and R. Hodes, Induction of sleep by stimulation of the caudate nucleus in Macacus rhesus and man, Trans. Amer. neurol. Ass., 77, 204-210 (1952).
70. Herz, A. and W. Zieglgänsberger, Synaptic excitation in the corpus striatum inhibited by microelectrophoretically administered dopamine, Experientia, 22, 839-840 (1966).
71. Herz, A. and W. Zieglgänsberger, The influence of microelectrophoretically applied biogenic amines, cholinomimetics and procaine on synaptic excitation in the corpus striatum, Neuropharmacology, 7, 221-230 (1968).
72. Hess, W.R., Hypothalamische Adynamie, Helv. physiol. pharmacol. acta, 2, 137-147 (1944a).
73. Hess, W.R., Das Schlafsyndrom als Folge diencephaler Reizung, Helv. physiol. pharmacol. Acta, 2, 305-344 (1944b).
74. Hess, W.R., Das Zwischenhirn. Syndrome, Lokalizationen, Funktionen. 2. erweitert Aufl. Benno Schwabe, Basel, 2nd Ed. (1954).
75. Hess, W.R., The Functional Organization of the Diencephalon, Grune & Stratton, New York (1957).

76. Hess, R. Jr., W.P. Koella, and K. Akert, Cortical and subcortical recordings in natural and artificially induced sleep in cats, EEG clin. Neurophysiol., 5, 75-90 (1953).
77. Hirasuna, N., S.A. Deadwyler and E.J. Wyers, Disruption of the conditioned emotional response by caudate nucleus stimulation, Brain Res. Bull., 2, 173-179 (1977).
78. Hodes, R., R.G. Heath, and C.D. Hendley, Cortical and subcortical electrical activity in sleep, Trans. Amer. neurol. Ass., 77, 201-203 (1952).
79. Hodes, R., S.M. Peacock and R.G. Heath, Influence of the forebrain on somatomotor activity. I. Inhibition, J. comp. Neurol., 94, 381-408 (1951).
80. Hong, J.S., H.-Y.T. Yang, W. Fratta, and E. Costa, Determination of methionine enkephalin in discrete regions of rat brain, Brain Res., 134, 383-386(1977).
81. Hull, C.D., G. Bernardi, and N.A. Buchwald, Intracellular responses of caudate neurons to brain stem stimulation, Brain Res., 22, 163-179 (1970).
82. Hull, C.D., N.A. Buchwald, and J. Vieth, Cortical intracellular analyses of responses to inhibitory and disinhibitory stimuli, Brain Res., 6, 12-21 (1967).
83. Hunter, J. and H.H. Jasper, Effects of thalamic stimulation in unanesthetized animals. The arrest reaction and petit mal-like seizures, activation patterns and generalized convulsions. EEG clin. Neurophysiol.,1,305-324 (1949).
84. Isaccson, R.L., Neural systems of the limbic brain and behavioural inhibition. In: R.A. Boakes and M.S. Halliday (Eds.), Inhibition and Learning,Academic Press, New York, 497-528 (1972).
85. Jabbur, S.J., S.I. Harik and J.A. Hush, Caudate influence on transmission in the cuneate nucleus, Brain Res., 120, 559-563 (1976).
86. Jasper, H.H., Thalamic reticular system. In: D.E. Sheer (Ed.), Electrical Stimulation of the brain, Univ. Texas Press, Austin, 277-287 (1961).
87. Jung, R. Hirnelektrische Untersuchungen über den Elektrokrampf: Die Erregungsablaufe im corticalen und subcorticalen Hirnregionen bei Katze und Hund, Arch. Psychiat. Nervenkr., 183, 206-244 (1949).
88. Jung, R. and R. Hassler, The extrapyramidal motor system. In: J. Field, W. Magoun, and V.E. Hall (Eds.), Handbook of Physiology, Section 1: Neurophysiology, Vol. II, Williams & Wilkins, Baltimore, Md., 863-927 (1960).
89. Kaada, B.R., Somato-motor, autonomic and electrocorticographic responses to electrical stimulation of 'rhinencephalic' and other structures in primates, cat and dog, Acta physiol. scand., 24, Suppl. 83, 285 pp. (1951).
90. Kennard, D.M.A., Experimental analysis of the functions of the basal ganglia in monkeys and chimpanzees, J. Neurophysiol., 7, 127-148 (1944).
91. Kennard, D.M.A. and L.F. Nims, Effect on electroencephalogram of lesions of central cortex and basal ganglia in Macaca Mulata, J. Neurophysiol., 5, 335-348 (1942).
92. Kitano, K., Caudate and hippocampal inhibition of extralemniscal thalamic unitary responses and their blockade by strychnine in cats, Neuropharmacol., 15, 777-783 (1976).
93. Kitsikis, A., The suppression of arm movements in monkeys; threshold variations of caudate nucleus stimulation, Brain Res., 10, 460-462 (1968).
94. Klemm, W.R. and L.R. Dreyfus, Septal- and caudate-induced behavioral inhibition in relation to hippocampal EEG of rabbits, Physiol. Behav., 15, 561-567 (1975).
95. Knott, J., R.A. Hayne and R. Meyers, Physiology of sleep: Wave characteristics and temporal relations of human electroencephalograms simultaneously recorded from the thalamus, the corpus striatum and the surface of the scalp. Arch. Neurol. Psychiat., Chicago, 63, 526-527 (1950).
96. Konorski, J., Integrative Activity of the Brain. An Interdisciplinary Approach. Univ. Chicago Press, Chicago (1967).
97. Konorski, J., Some ideas concerning physiological mechanisms of so-called internal inhibition. In: R.A. Boakes and M.S. Halliday (Eds.) Inhibition and Learning, Academic Press, New York, 341-357 (1972).

98. Kupfer, D.J., T.P. Detre, G. Foster, G.J. Tucker, and J.M.R. Delgado, The application of Delgado's telemetric mobility recorder for human studies, Behav. Biol., 7, 585-590 (1972).
99. Laursen, A.M., Inhibition evoked from the region of the caudate nucleus in cats, Acta physiol. scand., 54, 185-190 (1962).
100. Liles, S.L. and G.D. Davis, Interrelation of caudate nucleus and thalamus in alteration of cortically induced movement, J. Neurophysiol., 32, 564-573 (1969).
101. Livingston, R.B., J.F. Fulton, J.M.R. Delgado, E. Sachs, Jr., S.J. Brendler, and G.D. Davis, Stimulation and regional ablation of orbital surface of frontal lobes, Res. Publ. Ass. nerv. ment. Dis., 27, 405-420 (1947).
102. Lussana, G. and A. Lemoigne, Arch. physiol., 117, 1877.
103. MacLean, P.D. and J.M.R. Delgado, Electrical and chemical stimulation of frontotemporal portion of limbic system in the waking animal, EEG clin. Neurophysiol., 5,91-100 (1953).
104. Magendie, F., Leçons sur les Fonçions et les Maladies du Système Nerveux. Paris, 1841.
105. Mao, C.C., D.L. Cheney, E. Marco, A. Revuelta, and E. Costa, Turnover times of gamma-aminobutyric acid and acetylcholine in nucleus caudatus, nucleus accumbens, globus pallidus and substantia nigra: effects of repeated administration of haloperidol, Brain Res., 132, 375-379 (1977).
106. McLennan, H. and D.H. York, The action of dopamine on neurons of the caudate nucleus, J. Physiol., Lond., 189, 393-402 (1967).
107. Mettler, F.A., Effects of bilateral simultaneous subcortical lesions in the primate, J. Neuropath. exp. Neurol., 4, 99-122 (1945).
108. Mettler, F.A., H.W. Ades, E. Lipman and E.A. Culler, The extrapyramidal system. An experimental demonstration of function, Arch. Neurol. Psychiat. Chicago, 41, 984-995 (1939).
109. Mettler, F.A. and C.C. Mettler, The effects of striatal injury, Brain, 65, 242-255 (1942).
110. Meyers, R., J.R. Knott, F.M. Skultety and R. Imler, On the question as to the existence of a "4s" suppressor mechanism, J. Neurosurg., 11, 7-23 (1954).
111. Moran, J.S. and E.M. Blass, Inhibition of drinking by septal stimulation in rats, Physiol. Behav., 17, 23-27 (1976).
112. Nagao, B., H. Kitano, K. Yamaga, and Y. Hori, Effect of caudate nucleus stimulation on repetitive evoked EMG in cat. In: Proc. 23rd Ann. Meeting, Amer. Assoc. Electromyography & Electrodiagnosis, San Diego, Nov. (1976).
113. Neill, D.B., Frontal-striatal control of behavioral inhibition in the rat. Brain Res., 105, 89-103 (1976).
114. Nothnagel, H., Arch. path. Anat., 57, 184 (1873).
115. Pasquier, D.A., T.L. Kemper, W.B. Forbes, and P.J. Morgane, Dorsal raphe, substantia nigra and locus coeruleus: interconnections with each other and the neostriatum, Brain Res. Bull., 2, 323-339 (1977).
116. Pavlov, I.P., Conditioned Reflexes. An Investigation of the Physiological Activity of the Cerebral Cortex. Trans. from Russian by G.V. Anrep, Oxford Univ. Press, London (1927).
117. Peacock, S.M. Jr., and R. Hodes, Influence of the forebrain on somato-motor activity. II: Facilitation, J. comp. Neurol., 94, 409-426 (1951).
118. Peeke, H.V.S. and M.J. Herz, Caudate nucleus stimulation retroactively impairs complex maze learning in the rat, Science, 173, 80-82 (1971).
119. Phillips, A.G., D.A. Carter, and H.C. Fibiger, Dopaminergic substrates of intracranial self-stimulation in the caudate-putamen, Brain Res., 104, 221-232 (1976).
120. Plotnik, R. and J.M.R. Delgado, Emotional responses in monkeys inhibited with electrical brain stimulation, Psychon. Sci., 18, 129-130 (1970).
121. Plotnik, R., D. Mir, and J.M.R. Delgado, Map of reinforcing sites in the rhesus monkey brain, Int. J. Psychobiol., 2, 1-21 (1972).

122. Prado-Alcalá, R.A., Z.J. Grinberg, Z.L. Arditti, M.M. García, H.G. Prieto, and H. Brust-Carmona, Learning deficits produced by chronic and reversible lesions of the corpus striatum in rats, Physiol. Behav., 15, 283-287 (1975).
123. Racagni, G., F. Bruno, F. Cattabeni, A. Maggi, A.M. Di Giulio, M. Parenti, and A. Groppetti, Functional interaction between rat substantia nigra and striatum: GABA and dopamine interrelation, Brain Res., 134, 353-358 (1977).
124. Rescorla, R.A., Pavlovian conditioned inhibition, Psychol. Bull., 72, 77-94 (1969).
125. Roberts, E. and C.F. Baxter (Eds.), Inhibition in the Nervous System and Gamma-aminobutyric Acid, Pergamon Press, New York (1960).
126. Robinson, B.W. and M. Mishkin, Alimentary responses to forebrain stimulation in monkeys, Exp. Brain Res., 4, 330-366 (1968).
127. Roemer, R.A., W.W. Baker, and D. Zivanovic, Intracaudate injection of N-methyl-dl-aspartic acid inhibits cortical evoked potentials, Physiol. Behav., 18, 225-229 (1977).
128. Rosvold, H.E. and J.M.R. Delgado, The effect on delayed-alternation test performance of stimulating or destroying electrically structures within the frontal lobes of the monkey's brain, J. comp physiol. Psychol., 49, 365-372 (1956).
129. Rosvold, H.E., M. Mishkin, and M.K. Szwarcbart, Effects of subcortical lesions in monkeys on visual-discrimination and single-alternation performance, J. comp. physiol. Psychol., 51, 437-444 (1958).
130. Roth, R.H., L. Allikmets, and J.M.R. Delgado, Formation and release of NA and DOPA, Arch. int. Pharmacodyn., 181, 273-282 (1969).
131. Rubinstein, E.H. and J.M.R. Delgado, Inhibition induced by forebrain stimulation in the monkey, Amer. J. Physiol., 205, 941-948 (1963).
132. Schiff, J.M., Lehrbuch der Physiologie des Menschen. I. Muskelund Nervenphysiologie, Schauenburg, Lahr, Vol. I (1858) Vol. II (1859).
133. Schneider, J., E. Woringer, G. Thomalske et G. Brogly, Bases électrophysiologiques des mécanismes d'action du pentothal chez le chat, Rev. neurol., 87, 433-467 (1952).
134. Schultze, M.J. and D.G. Stein, Recovery of function in the albino rat following either simultaneous or seriatim lesions of the caudate nucleus, Exp. Neurol., 46, 291-301 (1975).
135. Sechenov, I.M., Elements of Thought, In Russian, 1878. USSR Acad. Sci. Press Moscow (1943).
136. Segundo, J.P. and X. Machne, Unitary responses to afferent volleys in lenticular nucleus and claustrum, J. Neurophysiol., 19, 325-339 (1956).
137. Shapovalova, K.B., Instrumental defensive reaction during the stimulation of the head of the caudate nucleus in dogs, Neuroscience, 3, 143-146 (1978).
138. Shellenberger, K.M., D.N. Young, and M.S. Levine, Effects of intracerebroventricular L-DOPA on caudate unit firing, Brain Res. Bull., 2, 273-277 (1977).
139. Sherrington, C.S., The Integrative Action of the Nervous System, Cambridge Univ. Press, London (1947).
140. Sloan, N. and B.R. Kaada, Effects of anterior limbic stimulation on somatomotor and electrocortical activity, J. Neurophysiol., 16, 203-220 (1953).
141. Spehlmann, R., The effects of acetylcholine and dopamine on the caudate nucleus depleted of biogenic amines, Brain, 98, 219-230 (1975).
142. Spehlmann, R., O.D. Creutzfeldt and R. Jung, Neuronale Hemmung im motorischen Cortex nach elektrischer Reizung des Caudatum, Arch. Psychiat. Nervenkr., 201, 332-354 (1960).
143. Standefer, M.J. and R.E. Dill, The role of GABA in dyskinesias induced by chemical stimulation of the striatum, Life Sciences, 21, 1515-1520 (1977).
144. Stein, D.G. and S.L. Chorover, Effects of posttrial electrical stimulation of hippocampus and caudate nucleus on maze learning in the rat. Physiol. Behav., 3, 787-791 (1968).

145. Stevens, J.R., C. Kim, and P.D. MacLean, Stimulation of caudate nucleus. Behavioral effects of chemical and electrical excitation, Arch. Neurol., 4, 47-54 (1961).
146. Szabo, J., Projections from the body of the caudate nucleus in the rhesus monkey, Exp. Neurol., 27, 1015 (1970).
147. Thomas, E., Excitatory and inhibitory processes in hypothalamic conditioning. In: R.A. Boakes and M.S. Halliday (Eds.), Inhibition and Learning, Academic Press, New York, 359-380 (1972).
148. Tobach, E., (Chrmn.), Experimental Approaches to the Study of Emotional Behavior, Ann. N.Y. Acad. Sci., 159, Art. 3 (1969).
149. Tokizane, T., M. Kawakami, and E. Gellhorn, On the relation between the activating and recruiting systems, Arch. int. Physiol. Biochim., 65, 415-432 (1957).
150. Umbach, W., Zur elektrophysiologie des Caudatum der Katze, Arch. Psychiat. Nervenkr., 199, 553-572 (1959).
151. Van Buren, J.M., C.L. Li, and G.A. Ojemann, The fronto-striatal arrest response in man, EEG clin. Neurophysiol., 21, 114-130 (1966).
152. Vigouroux, R., H. Gastaut et M. Badier, Les formes expérimentales de l'épilepsie. Provocation des principales manifestations cliniques de l'épilepsie dite temporale par stimulation des structures rhinencéphaliques chez le chat non anesthésié, Rev. neurol., 85, 505-508 (1951).
153. Villablanca, J.R., R.J. Marcus, and C.E. Olmstead, Effects of caudate nuclei or frontal cortical ablations in cats, I. Neurology and gross behavior, Exp. Neurol., 52, 389-420 (1976a).
154. Villablanca, J.R., R.J. Marcus, and C.E. Olmstead, Effects of caudate nuclei or frontal cortex ablations in cats. II. Sleep-wakefulness, EEG, and motor activity, Exp. Neurol., 53, 31-50 (1976b).
155. Von Voigtlander, P.F. and K.E. Moore, The release of [^{3}H]-dopamine from cat brain following electrical stimulation of the substantia nigra and caudate nucleus, Neuropharmacology, 10, 733-741 (1971a).
156. Von Voigtlander, P.F. and K.E. Moore, Nigrostriatal pathway: stimulation-evoked release of [^{3}H]-dopamine from caudate nucleus, Brain Res., 35, 580-583 (1971b).
157. Ward, J.W. and V. Le Quire, Responses elicited by electrical stimulation of the gyrus cingulus in unanesthetized cats, Anat. Rec., 106, 256-257(1950).
158. Weiss, T.R., Influence of long-lasting stimulation of the caudate nuclei on sleep cycle in rats, Experientia, 23, 130-131 (1967).
159. Wilburn, M.W. and R.P. Kesner, Differential amnestic effects produced by electrical stimulation of the caudate nucleus and nonspecific thalamic system, Exp. Neurol., 34, 45-50 (1972).
160. Wolstenholme, G.E.W., and M. O'Connor (Eds.), The Nature of Sleep, Little Brown, Boston (1962).
161. Wyers, E.J. and S.A. Deadwyler, Duration and nature of retrograde amnesia produced by stimulation of caudate nucleus, Physiol. Behav., 6, 97-103 (1971).
162. Yoshida, Y. and W. Precht, Monosynaptic inhibition of neurons of the substantia nigra by caudate-nigral fibers, Brain Res., 32, 225-228 (1971).

Sensory Functions of the Neostriatum

George M. Krauthamer

Department of Anatomy, CMDNJ Rutgers Medical School
Piscataway, New Jersey 08854, U.S.A.

The basal ganglia, since Wilson's initial suggestion (207), are usually considered as an integral and major component of the extrapyramidal motor system. Support for this view was derived largely from clinicopathological observations which, understandably, focused attention on the most readily identifiable motor symptoms and paid scant attention to other, perhaps more subtle and complex changes in behavior (106, 187, 188). Experimental work has, similarly, emphasized the link between motor functions and the basal ganglia (12, 24, 26, 40, 41, 45, 53, 62, 87, 120, 125, 143, 196, 210, 211). Despite this wealth of information the nature of the neostriatal contribution to motor functions still remains quite unclear; indeed, the literature is replete with observation not readily incorporated into any motor schema (26, 75, 121, 122, 126, 134, 145, 179). In order to advance our understanding of the role of the neostriatum in terms of neuronal function and, eventually, in terms of the organization of behavior, it may be helpful to examine it from a different perspective.

The neostriatal relationship to efferent motor systems and afferent sensory systems is quite indirect if we properly limit these terms to mean centers with a relatively direct synaptic access to spinal and brainstem motor neurons and first or second order sensory neurons. The connections of the basal ganglia, that is, of the caudate nucleus and putamen, globus pallidus and substantia nigra, thus stand in marked contrast to the richness of cerebellar and motor cortical connections with motor and sensory neurons of the brainstem and spinal cord. Fundamental differences also exist in the intrinsic cytoarchitectural and histological organization and the neurochemical composition of the basal ganglia compared with centers possessing more direct connections to motor and sensory systems; major differences in the nature of information processing should therefore be expected.

The shortcomings inherent in the traditional concept of an extrapyramidal motor system and of the basal ganglia as a fundamentally motor structure have long been recognized and the term "sensorimotor" has gained acceptance as more accurately reflecting the neostriatal properties (21, 98, 141, 159). Unfortunately, this term begs the question since it can be applied with equal justification to almost all central structures of the brain. It would be helpful to establish more precisely the specific nature of the sensorimotor integration and to determine whether the neostriatal contribution leans toward the sensory or the motor component.

One important element in the sensorimotor equation is the reciprocal relationship between neostriatal functions and polysensory activities, a relationship which is direct, powerful, and dynamic. It may well constitute a key element in understanding the many varied and seemingly disparate behavioral manifestations attributed to the neostriatum.

Although the behavioral manifestations associated with neostriatal function are many and varied, a recurrent theme runs through many behavioral studies of the neostriatum regardless of the nature of the task and seemingly unrelated to the specific alteration of behavior. In a very broad sense, this common element may be viewed as a disturbance in some aspects of orienting behavior. It entails the ability to orient and then to attend to those environmental or internal stimuli which are or may become relevant at any given moment, to execute the situationally appropriate performance, including the requisite postural adjustments and movements, and to shift attention to other stimuli as these become situationally significant (18, 43, 49, 81, 127, 150, 187, 188, 197, 203).

The argument, to be present in this paper, will focus on recent physiological and anatomical findings, with the hope that they can be brought into consonance with the elaboration of those behavioral properties that one may categorize, albeit tentatively, as components of sequential orienting responses. At the very least, these findings point to the frequently underestimated role of the sensory component and thus may help us to reexamine the problem of neostriatal function.

NEOSTRIATAL RESPONSES TO SENSORY STIMULATION

The sensory responses of the neostriatum clearly indicate a close relation to polysensory systems.* Responses can be recorded to a wide range of sensory modalities in the manner typical of other polysensory structures, such as centrum medianum of the thalamus (3, 4), association cortex (5, 57, 190), reticular formation (13, 20, 48), and the deep layers of the superior colliculus (77, 88, 182). Typical polysensory neurons are characterized by partial or complete convergence of different sensory modalities onto a single unit, wide and often bilateral receptive fields, and an inability to follow repetitive sensory stimulation at rates exceeding 2 - 6 stimuli per sec. Unlike the activity of the modality specific lemniscal systems, polysensory activity is greatly depressed by barbiturates but it is unaffected or enhanced by chloralose, so that much of the information pertaining to polysensory systems has been obtained from animals anesthetized with α-chloralose. However, comparable results have been obtained from immobilized, locally anesthetized preparations and from unrestrained animals with chronically implanted electrodes. The use of chloralose anesthesia is, therefore, not a critical, determining factor for the demonstration of polysensory activities (89, 149, 189). The study of sensory convergence in the unanesthetized animal is complicated by the fact that polysensory activity varies as a function of arousal level (9, 10, 146, 168, 189). Thus, polysensory responses are maximal during the relaxed or drowsy state, but they are minimal when the animal has been alerted.

The polymodal nature of the sensory input to the neostriatum of the cat was first described in detail for the putamen by Segundo and Machne (176) and, for the caudate nucleus by Albe-Fessard et al. (7, 8). These initial observations have been confirmed repeatedly by subsequent studies in anesthetized and unanesthe-

*A number of comprehensive reviews of polysensory systems have been published (1, 2, 25).

tized preparations (7, 46, 109, 155, 175, 200). Convergence of somatosensory, visual, auditory, vestibular, splanchnic and vagal responses has been observed although there is no single study in which all of these stimulus modalities have been tested in order to determine precisely the degree of sensory convergence. Nor are there any systematic mapping studies extending beyond the head of the caudate nucleus to determine if a possible parcellation of the neostriatum into sectors optimally responsive to one or another sense modality exists. Most of the available information is, therefore, limited to data obtained from the head of the caudate nucleus. The occasional references to sensory convergence responses in the body of the caudate nucleus and in the putamen suggest a similar or identical organization of polysensory inputs.

Evoked Potentials

Polysensory evoked potentials are recorded readily in the head of the caudate nucleus (Fig. 1). When recorded with bipolar electrodes, the potentials appear abruptly upon penetration of the nucleus, and sharply demarcate the passage of the electrode from the overlying lateral ventricle into the neuropil. The amplitude of the evoked potentials increases rapidly, reaching a maximum of 200-500 μV in the dorsal third of the head of the caudate. As the recording electrode descends through the nucleus in a dorsoventral direction, the potentials change in amplitude and reverse their polarity, as is typical upon passage through a local potential field.

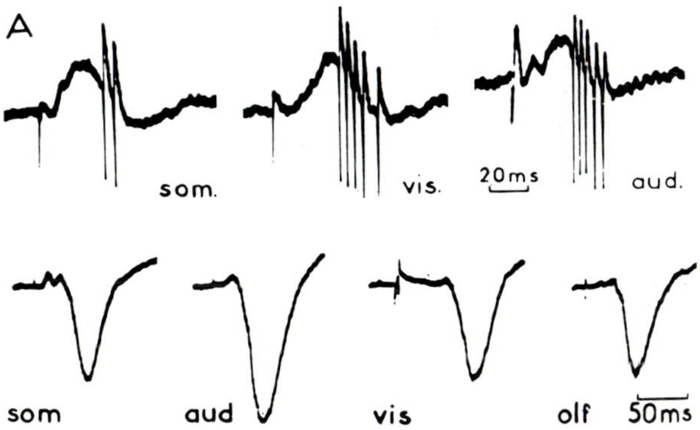

Fig. 1. Polysensory responses in caudate nucleus of cat. (A) Convergence of somatosensory, visual, and auditory inputs in a single unit. (Lower traces) Polysensory evoked potentials. From refs. 7 and 8.

In general, somatosensory evoked potentials are the largest and most stable, visually evoked potentials the smallest and least stable. Attempts to discern a differential pattern of regional distribution in terms of amplitude or latency have not been successful (7, 46). The ease with which potentials can be evoked by a given stimulus modality, and the differences in amplitude and latency which distinguish the responses to different stimuli appear to be variable and randomly distributed.

Single Units

Comparable results have been obtained in single unit studies (Fig. 1). Units with sensory convergence properties are distributed throughout the neostriatum but except for the report of Segundo and Machne (176), most studies have been limited to the head of the caudate nucleus. Partial or complete response convergence to somatosensory, visual, auditory, and visceral stimuli is generally observed. Although some units respond to only one sensory modality, their other response characteristics are typical of polysensory neurons (8, 176, 200).

Most caudate neurons display a very low level of spontaneous activity in the unanesthetized state and an even lower level when anesthetized (8, 22, 26, 40, 160, 175, 200). The reasons for this remarkable sluggishness of caudate neurons are unknown but this low activity level appears to be a fairly common state, whether or not the units have been identified as polysensory neurons.

Single units respond to sensory stimuli with one action potential or a brief discharge of 2 - 4 spikes at latencies ranging from 20 - 30 msec for somatosensory and auditory stimuli, and up to 60 msec for photic stimulation. Spontaneously active units respond initially with a brief excitatory or inhibitory phase followed by a prolonged rhythmic discharge pattern when stimuli are applied to peripheral nerves (175).

The Adequate Stimulus

As is the case for other polysensory structures, the physiologically adequate stimuli are light but sudden taps to the skin, bright flashes of light and brief sounds such as loud clicks or handclaps. Whether or not there is a proprioceptive input remains in dispute. Because of the potential importance of a proprioceptive input from muscle spindles and tendon organs for any structure directly involved in the control of movements, several authors have tested for such an input but the matter is still in doubt. Neither passive muscle stretch, selective stimulation of a muscle nerve (hamstrings) nor low intensity stimulation of group I fibers activates the polysensory units of the caudate nucleus (26, 134, 175). A similar absence of proprioceptive input to the polysensory centrum medianum-parafascicular complex of the cat thalamus has also been found by Mallart (133). However, Matsunami & Cohen (134) observed units which were excited by active movements induced by pyramidal tract stimulation, just as others have reported neostriatal unit responses temporally related to the execution of learned movements (26, 40, 41, 53, 167). The activation of other afferent fibers besides that of proprioceptive group I and II fibers can, of course, be at the origin of the movement associated responses. Hence, the caudate induced modulation of muscle spindles (196) need not reflect the operation of a proprioceptive feedback loop such as exists for the cerebellum and motor cortex.

The absence of modality specificity, the lack of a somatotopic, retinotopic, or phonotopic organization and the absence of intensity functions probably means that the sensory input to the neostriatum provides little analytical information about stimulus quality, place and modality. In any case, such information would be carried efficiently and rapidly by other channels, running in parallel, and projecting to modality specific subcortical and cortical areas and the cerebellum. Nevertheless, some aspects of this sensory information may be retained and encoded in a different manner by the neostriatum. Variations in latency, the number of discharging units and differences in their discharge pattern presumably carry some relevant information. Differences in the sensory convergence pattern have been thoroughly explored only for the polysensory or wide field neurons of the somatosensory cortex (15, 52, 56, 57, 148) but similar differences are probably equally demonstrable in the neostriatum, as the studies of Vinogradova & Sokolov (200) indicate.

Significance of the Polysensory Input

There is still little information available about the functional significance of the polysensory input. There is evidence that polysensory responses occur in several semi-independent systems (15, 193) although Thompson does not agree (165, 189, 191). It is quite likely that the polysensory input to the neostriatum bears some relationship to the alerting, signalling and orienting functions of the stimulus. Such a relationship has been demonstrated for the polysensory activities of the association cortex. Behavioral orienting responses and polysensory cortical evoked potentials covary and decrement in parallel during habituation (189). The fact that orienting behavior persists after ablation of the cortex indicates the importance of subcortical loci (180), and there is evidence that the organization of orienting responses is critically dependent upon the integrity of the hippocampus and caudate nucleus (199, 200).

A number of studies have analyzed the firing pattern of neostriatal units in relation to the execution of learned movements. Because these studies focused largely on motor functions, they generally did not include a systematic exploration of the sensory components associated with, or preceding the learned task performance (26, 40, 41, 53, 134, 167). However, incidental observations during these studies have led several authors to suggest a complex task and stimulus dependence of neostriatal firing patterns rather than a simple or direct relationship to limb and body movements (26, 53, 134).

Thalamic Origin of Polysensory Input

There is no evidence that the polysensory input to the neostriatum reflects the profuse and topographically organized projections from the cortex (16, 27, 76, 97, 101, 124, 177, 204, 205, 213). The variations in the relative responsiveness and response latency of polysensory units in the caudate nucleus bear no evident relation to the topography of the cortical projections from primary sensory or association cortex. Furthermore neostriatal polysensory responses persist following the blocking of cortical activity by KCl, acute ablation, or chronic suction ablation with consequent degeneration of corticofugal fibers (7, 108, 153, 166). Finally, the failure to find sensory responses in the neostriatum under barbiturate anesthesia, which does not depress the sensory specific cortical activity, provides additional evidence for the independence of the polysensory neostriatal activity from the modality specific lemniscal influx.

Cortical modulation. Although the polysensory input is not dependent on the cortex, it is subject to cortical modulation (59). The modulation is presumably exerted directly on spiny neurons of the neostriatum since these are the recipients of corticostriatal and thalamostriatal axodendritic synapses (100). Further, the initial effect of thalamic and cortical stimulation is excitatory (103) or, according to other investigators, excitatory - inhibitory but with predominance of cortical excitatory effects (22, 90). Corticostriatal inputs are thus in a position to affect the polysensory responsivity of the neostriatum.

The sensory convergence input apparently has its source in the medial thalamus. It is unlikely though not excluded, that either of the two other major subcortical projections, from substantia nigra (29, 63, 65, 99, 105, 123, 194, 195) or dorsal raphe (137, 138, 151) contributes significantly to the polysensory influx. For example, destruction of substantia nigra alters the firing pattern of polysensory neurons but does not eliminate the polysensory neostriatal input (114).

The intralaminar thalamus and centrum medianum project directly and massively to the entire neostriatum (33, 96, 119, 156, 157, 169) and the same thalamic zone is, in turn, the recipient of polysensory projections (129, 130, 135). As a consequence, a close functional and anatomical reciprocity joins the polysensory zones of the medial thalamus and the basal ganglia; it is highly probable that this plays a significant role in the regulation of neostriatal input and output.

NEOSTRIATAL MODULATION OF POLYSENSORY ACTIVITIES

The properties of this reciprocal relationship suggest the operation of a semiclosed loop between neostriatum and medial thalamus (Fig. 9). From the intralaminar thalamus and centrum medianum, the afferent polysensory influx is, in part, conveyed by direct anatomical projections to the neostriatum. The neostriatal output, in turn, is funneled back to the thalamus after further processing in substantia nigra and the internal segment of the globus pallidus or its feline homologue, the entopeduncular nucleus (79). One output component is directed towards nucleus ventralis anterior and nucleus ventralis lateralis, there to be integrated with ascending cerebellar and motor cortical functions (for current reviews see 28, 158, 159). The other component is channeled to the centrum medianum and adjacent zones. The polysensory input to the neostriatum is thus modulated at its source, in the medial thalamus, by the neostriatal output.

Extent of Modulation

The neostriatal modulation of polysensory activity is predominantly inhibitory (Fig. 3) but it is associated with a minor excitatory action (14, 66, 113, 116). Hence, it is more accurate to speak of modulation rather than inhibition when referring to the totality of the neostriatal action on the medial thalamus. The inhibition of polysensory activities (109) exerted by the caudate nucleus is widespread (Fig. 2). It affects the association cortex as well as the polysensory elements of the primary somatosensory cortex bilaterally (108, 109, 111), the contralateral caudate nucleus (109), the intralaminar thalamus and centrum medianum (66, 104, 107, 110, 131, 166), and the reticular formation as far caudally as nucleus gigantocellularis of the medulla (109, 112, 178).

Although the inhibition spreads over wide portions of the polysensory system (39, 64, 106), the intralaminar thalamus and centrum medianum are at the center of the inhibition (Figs. 3, 8). Since direct projections from the basal ganglia to cortex, reticular formation and contralateral neostriatum have not been demonstrated, the inhibitory effects found in these structures are presumably indirect reflections of the active inhibition exerted at the thalamic level.

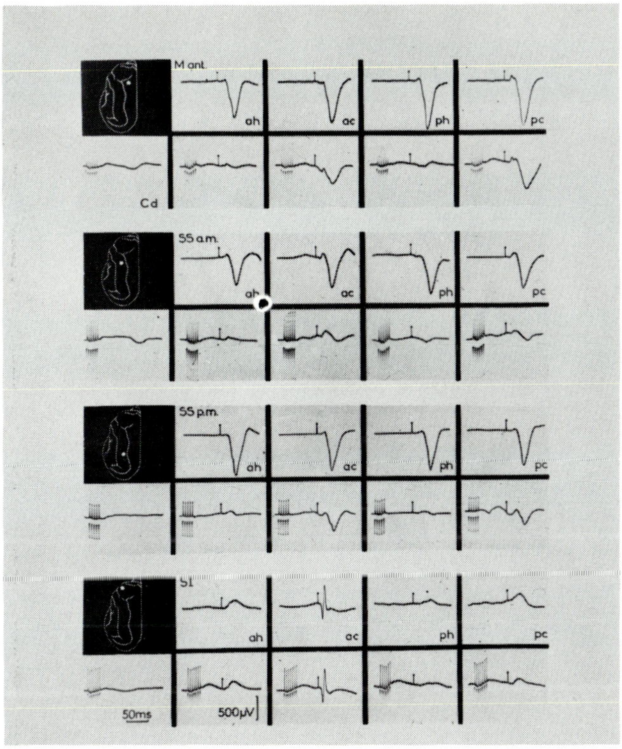

Fig. 2. Caudate inhibition of polysensory evoked potentials of suprasylvian association cortex (M ant, SS am, SS pm) and lack of inhibition of primary somatosensory cortical responses (SI). From ref. 108.

Since there are no projections from the basal ganglia to primary sensory specific structures, the inhibition observed in such structures following stimulation of the caudate nucleus (69, 93, 99), must be attributed to the activation of indirect pathways such as the diffuse intrathalamic and reticular projections of the medial thalamus. Single unit studies of typical modality and place specific lemniscal units in nucleus ventralis posterolateralis (66) and of primary evoked potentials in visual, auditory and somatosensory cortex (108) demonstrate the absence of any direct action on primary sensory projections (Figs. 2, 4).

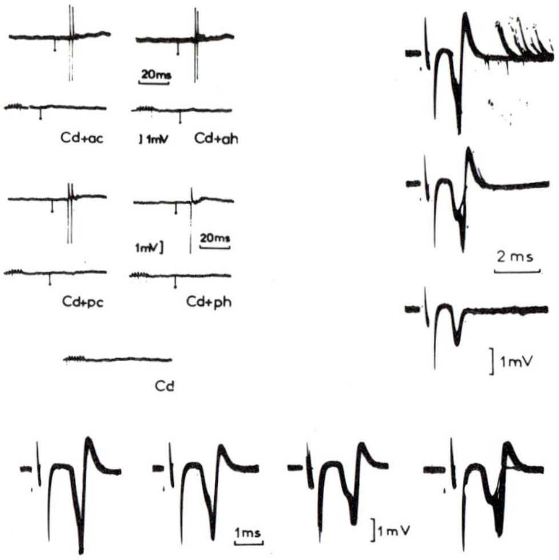

Fig. 3. Caudate inhibition of polysensory unit in centrum medianum and antidromic invasion. (Left panel) A brief pulse train applied to the caudate (Cd) inhibits the responses to stimulation of the four paws (ac, ah, pc, ph); no action potential is generated by Cd stimulation alone. (Right and bottom panels) Unit responding antidromically at fixed latency and orthodromically at variable latency to Cd stimulation. At increasing rates of stimulation the orthodromic spikes are first to drop out, followed by an exaggerated I-S break and final failure of S spike. From Refs. 66, 116.

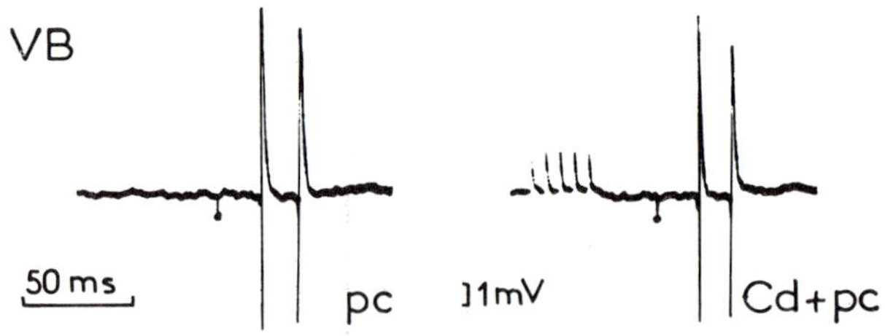

Fig. 4. Absence of inhibition of thalamic lemniscal unit in ventral posterolateral nucleus (pc, contralateral hindpaw). From ref. 108.

Inhibition of Polysensory Activities

Caudate stimulation leads to a powerful and prolonged inhibition of polysensory activities. The inhibition spreads throughout the polysensory system but it is seen most clearly in centrum medianum and adjacent portions of the intralaminar thalamus. It reaches its peak intensity 20 - 30 msec following caudate stimulation and lasts approximately 250-350 msec. The time course and intensity of caudate inhibition have been demonstrated by the usual method of applying a peripheral test stimulus at varying intervals after a prior caudate conditioning stimulus. The evolution of the inhibition can then be plotted as a function of evoked potential amplitude vs. the time interval between conditioning and test stimuli. A parallel curve is generated by plotting the probability of extracellularly recorded single unit spike discharges at increasing conditioning - test stimulus intervals. The similarity between these two curves indicates that the amplitude of the polysensory evoked potentials accurately reflects the level of neuronal excitability, as has been suggested by Fox and O'Brien (68). Intracellular studies reveal a corresponding evolution of the IPSPs generated by caudate stimulation (113).

The inhibition is very sensitive to small doses of strychnine, a glycine antagonist. A single intravenous dose of 0.1 mg strychnine completely blocks the inhibitory action of the caudate nucleus (Fig. 5) and, like the inhibition itself, the block is manifested throughout the polysensory system (107, 108, 115). Picrotoxin, a GABA antagonist, has no effect or only a slight effect on caudate inhibition but it does block the inhibition of polysensory responses produced by stimulation of the hippocampus (36, 104).

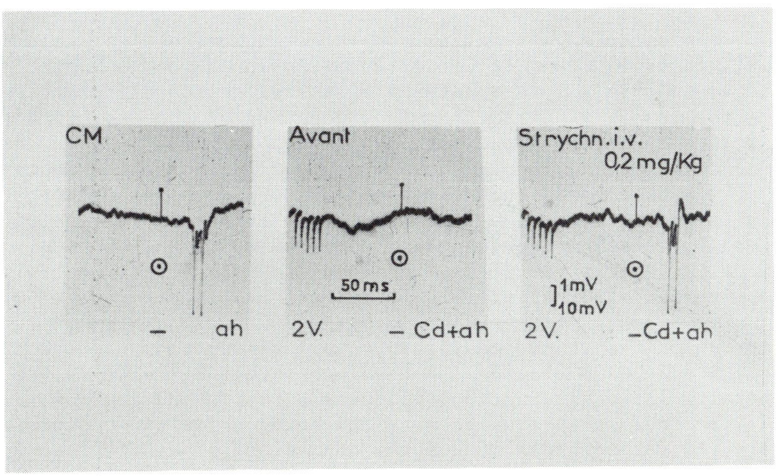

Fig. 5. Strychnine block of inhibition of polysensory unit in centrum medianum (ah, ipsilateral forepaw).

Differential modulation. The predominant inhibition of polysensory neurons after caudate or nigral stimulation is sometimes preceded by a brief excitation (14, 116), as also occurs in peripherally unidentified neurons of the medial thalamus after caudate stimulation (70, 71). Intracellular analysis of polysensory thalamic neurons has shown that the prolonged hyperpolarizing phase is always preceded by a brief depolarizing shift or EPSP (113). In most cases, the initial depolarization is minimal and does not reach firing threshold but in a small proportion of units the excitatory drive is sufficiently intense to generate an action potential prior to the onset of the inhibitory phase. This more intense excitatory drive is associated with stimulation of the dorsomedial head of the caudate nucleus while stimulation of the lateral half results in a merely subthreshold depolarization. The effects of nigral and sensory stimulation are analogous to those of dorsomedial caudate stimulation. An initial EPSP leading to the generation of one or more action potentials is followed by a prolonged IPSP (113).

The combination of an invariably occurring, prolonged hyperpolarizing phase preceded by a variable, brief period of depolarization, makes the temporal patterning of the various inputs a key element in the output of centrum medianum. It suggests that the reciprocal excitatory-inhibitory relationship linking neostriatum and medial thalamus is, in fact, a finely tuned mechanism in which temporal as well as spatial factors determine the precise composition of the excitatory-inhibitory balance.

Fractionation of the polysensory pool. An important factor in the modulation of polysensory activities is the site of initiation of the output within the neostriatum. An equally important factor is the differential action of the neostriatum on spatially distinct zones of the polysensory thalamic structures. Within the posterior medial thalamus neurons displaying sensory convergence properties are distributed throughout a wide zone including the centrum medianum, parafascicular nucleus, posterior portions of nucleus centralis lateralis and the adjacent paralaminar portion of nucleus medialis dorsalis and extending below the ill defined ventral margin of centrum medianum into the fields of Forel and zona incerta (66, 118, 171).

Although sensory convergence, partial or complete, is the rule for this extensive pool of neurons, sensory stimuli do not activate it in unison, for statistically significant differences in response latency characterize its different regions. Whether this is the result of intrathalamic processing of the sensory influx or whether it indicates different input pathways is not known. In any event, the result is a fractionation of the polysensory pool of the posterior medial thalamus into at least three different subpopulations, each responding at a significantly different time to the same sensory stimulus (66). Mean latencies for somatosensory stimulation are longest in centrum medianum, shortest for the scattered neurons in Forel's field, and intermediate for nucleus centralis lateralis and nucleus medialis dorsalis.

These three subpopulations of the polysensory pool are further differentiated by differences in caudate modulation. A high percentage of units is inhibited in nucleus centralis lateralis, nucleus medialis dorsalis and centrum medianum but only a small percentage in zona incerta and Forel's fields. Conversely, early excitatory responses which precede the inhibitory phase are rare in the dorsal zone but fairly common in centrum medianum and the ventral zone below it (Fig. 6).

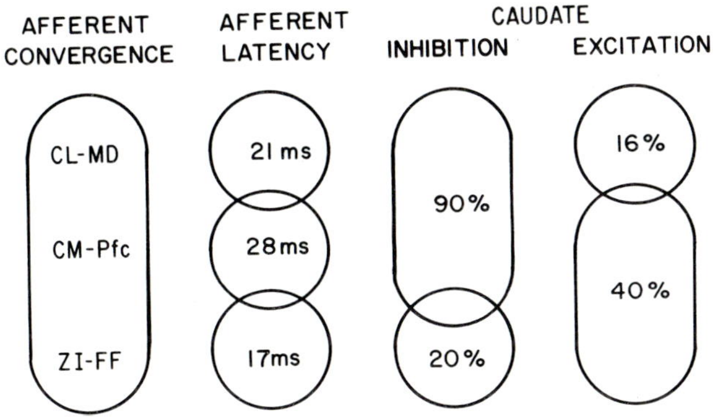

Fig. 6. Fractionation of the polysensory neuron pool of the medial thalamus by peripheral inputs and caudate modulation. The upper third of each diagram corresponds approximately to posterior nucleus centralis lateralis and paralaminar nucleus medialis dorsalis, the middle third to centrum medianum parafascicular complex, and the lower third to zona incerta and fields of Forel - see text for details.

Operation of the neostriatal-thalamic loop. Thus, there are at least three factors which contribute to the neostriatal modulation of polysensory activities: (1) The neostriatal zone whence the modulating output originates. This determines, in part, the intensity of the early excitatory drive preceding the inhibition. (2) The region of the medial thalamus to which the neostriatal output is directed. This determines, in part, the proportion of polysensory neurons brought to discharge prior to the inhibitory shut down. (3) The regional differences in the sensory response latency of medial thalamic polysensory neurons. This differentially sets the level of background excitability encountered by the neostriatal output at any given moment.

The output of the medial thalamus is largely fed back into the neostriatum, resulting in a highly dynamic, perhaps unstable, and constantly shifting differentiation and fractionation of the two nodal points of the semi-closed loop between neostriatum and medial thalamus. The loop is semi-closed because both nodal points, as well as the intervening nigral and pallidal way stations, are target and source of connections with other regions of the brain, so that the polysensory modulation must be viewed as a function of additional factors besides the three just listed (Fig. 9).

FUNCTIONAL DIFFERENTIATION WITHIN THE NEOSTRIATUM

The homogeneous distribution of the various cell types throughout the neostriatum (47,100) provides no indications for a regional differentiation of functions within the neostriatum. The existence of regional differences has been demonstrated, however, by physiological (61, 71, 108, 109, 111, 160, 166), behavioral (23, 35, 51, 94, 147, 208), and biochemical (67) observations.

Striatopetal Differentiation

There is no evidence that the polysensory input is selectively distributed since sensory convergence responses can be recorded from all parts of the caudate nucleus and putamen. Of course, it is possible that there exists a relative differentiation into regions optimally or preferentially responsive to one or another sensory modality but this remains to be investigated systematically.

The most likely source for input differentiation is certainly the corticostriatal projection (16, 76, 97, 101, 204, 205, 213) which can modulate the cortically independent polysensory input (59). Conceivably, a spatial and temporal differentiation is imposed on the neostriatal neuron population by the dual action of the cortex and the fractionated feedback from the medial thalamus.

Striatofugal Differentiation

Evidence supporting a differentiation of output functions is quite ample and contrasts with the dearth of evidence concerning a differentiation of input functions. Mapping studies and selective lesions have demonstrated regionally differentiated effects on reflex movements (125, 143) and on learned behavior (49, 106, 206, 208).

The inhibitory region. A marked regional differentiation underlies the neostriatal modulation of thalamic polysensory responses. Mapping studies and localized cooling within the caudate nucleus have outlined the intrastriatal organization of the inhibition. There are zones where stimulation gives rise to inhibition of polysensory activities and other zones which are silent in this regard. It is particularly striking that no inhibition can be obtained from the putamen. In the cat the most effective inhibitory zone occupies the dorsal and lateral two thirds of the head of the caudate nucleus. More caudally, the inhibitory zone is increasingly confined to the ventrolateral edge of the caudate nucleus and the subjacent internal capsule until it expands again within the substantia nigra and the entopeduncular nucleus where the lowest inhibitory thresholds have been found (108, 109).

Within the effective inhibitory zone there appears to be a further differentiation in terms of the intensity of the early excitatory drive. Moreover, the distribution of optimally effective points with the lowest stimulus thresholds is discontinuous, similar to the recently demonstrated discontinuity of the corticostriate projection pattern (76, 97, 213).

The extent of the caudate inhibitory zone has a considerable resemblance to the projection territory of the sensorimotor cortex and gyrus proreus. Indeed, in the cat, stimulation of these cortical fields, but not of suprasylvian association cortex, leads to the inhibition of polysensory responses analogous to that obtained by caudate stimulation (9, 91, 212).

It is conceivable that the corticostriate projection, which is largely independent of the corticospinal and corticobulbar ones (16, 103, 209), normally plays a role in the inhibitory process.

A close correspondence exists between the zone of caudate inhibition of polysensory activities and the course of the caudatofugal projections. From the head of the caudate nucleus these fibers appear to course intrastriatally in a posteroventral direction and gradually exit into the internal capsule. Their subsequent course coincides with that of the anatomically demonstrated projections to substantia nigra and entopeduncular nucleus (38, 78, 144, 186, 201).

THE ANATOMICAL SUBSTRATE FOR THE NEOSTRIATAL-THALAMIC LOOP

There are numerous, and extensive studies devoted to the anatomical connections of the neostriatum (for recent reviews see 28, 136) but comparatively little is known about the detailed organization of the thalamostriate projections (33, 96, 119, 156, 157, 169). Because of the importance of the thalamostriate pathway for the sensory functions of the neostriatum, we made use of the recently developed method of retrograde axoplasmic transport of horseradish peroxidase in order to trace precisely the projections to the limited, inhibitory zone of the caudate nucleus (119).

Projections From Centrum Medianum

In order to confine the enzyme to a very small portion of the caudate nucleus, horseradish peroxidase was ejected electrophoretically through micropipettes. This technique permits the concurrent recording of neuronal activity at the ejection site and constitutes an additional verification of pipette placement.

One disadvantage of small enzyme deposits is the corresponding reduction in the number of retrogradely labeleled neurons. When terminal branching covers a wide area, restricted injections of horseradish peroxidase will be taken up by only a fraction of the terminal branches of the axon leading to very weak labeling of some input sources or no labeling at all (96). This may produce false negative results. Conversely, false positive results may be the consequence of horseradish peroxidase uptake by injured fibers of passage (75). In the studies to be discussed, we have attempted to reduce this potential source of errors by varying the angle of approach to the ejection site.

When a small quantity of horseradish peroxidase is deposited in the head of the caudate nucleus, many neurons in centrum medianum are densely labeled with granular reaction product. In the case illustrated by Fig. 7, the labeled neurons clearly outline the centrum medianum and, depending upon one's definition of the anatomical boundaries extend, perhaps, into the most posterior and ventral portiontion of n. centralis lateralis (95, 135, 162). No retrogradely labeled cells are seen in the parafascicular nucleus or any other part of the intralaminar thalamus.

Scattered neurons are located in substantia nigra, pars compacta, near the oculomotor nerve, and a very few medium sized pyramidal cells are located in the 5th layer of the corresponding cortical projection zone. A comparison of the maps showing the distribution of inhibited thalamic polysensory neurons and neurons labeled retrogradely from the head of the caudate nucleus shows a complete overlap. It is apparent that the polysensory thalamic field most powerfully modulated by the caudate nucleus provides, in turn, the major thalamic input to the caudate trigger zone where the inhibition originates (Figs. 7, 8).

The existence of a heavy projection from centrum medianum to the caudate nucleus has also been demonstrated autoradiographically (169). The terminal distribution of the radioactive label in the caudate reveals that the centromedian projection is discontinuous, reminiscent of the patch-like corticostriatal pattern.

Evidently, the usual schema, whereby centrum medianum is said to project to the putamen and nucleus centralis lateralis to the caudate nucleus is misleading and needs to be revised, at least for the cat.

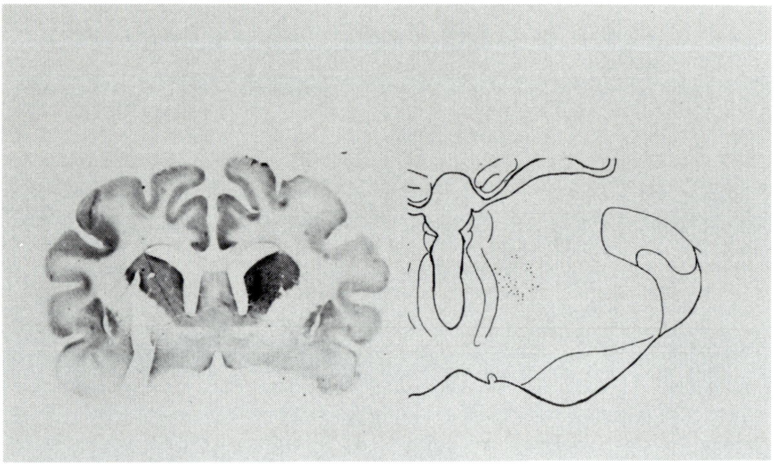

Fig. 7. HRP labeled neurons in centrum medianum following injection of HRP into caudate nucleus as shown. The dots indicate actual location of labeled neurons.

Whether or not the thalamic cells projecting to the caudate nucleus are identical to the polysensory neurons inhibited by the caudate, has not been established. There is some evidence for a very close synaptic relationship since some centrum medianum neurons can be activated antidromically as well as orthodromically from the caudate nucleus (116). Since the medial thalamic neurons have a dual projection, to the neostriatum and the cortex (60, 61, 96, 161), and the corticofugal neurons have an analogous dual projection to the neostriatum and the medial thalamus (209), the functional relationship between medial thalamus, cortex, and neostriatum is much more tightly coupled than heretofore believed.

Projections to Centrum Medianum

Given the intimate relationship between centrum medianum and caudate nucleus, it becomes important to determine precisely the origin of the several projections to centrum medianum (Fig. 8). This was accomplished by electrophoretically depositing small quantities of horseradish peroxidase in centrum medianum, the thalamic region maximally inhibited by caudate stimulation (130).

Projections from the basal ganglia. The major input source to centrum medianum, in terms of number and intensity of retrogradely labeled neurons, was the ipsilateral entopeduncular nucleus. Somewhat fewer labeled neurons were encountered in substantia nigra, pars reticulata. These results confirm previous findings (30, 31, 34, 80, 102); they also show that a small zone within the medial thalamus receives a dual input from the basal ganglia, one from substantia nigra and another from the entopeduncular nucleus.

Whether or not some of the functional differences associated with the caudate inhibition of polysensory responses can be linked to the dual nature of the projection pathway to centrum medianum, as has been suggested (72, 73), is not known. In any case, it is evident that the striatal influences on the medial thalamus cannot be attributed to one pathway exclusively.

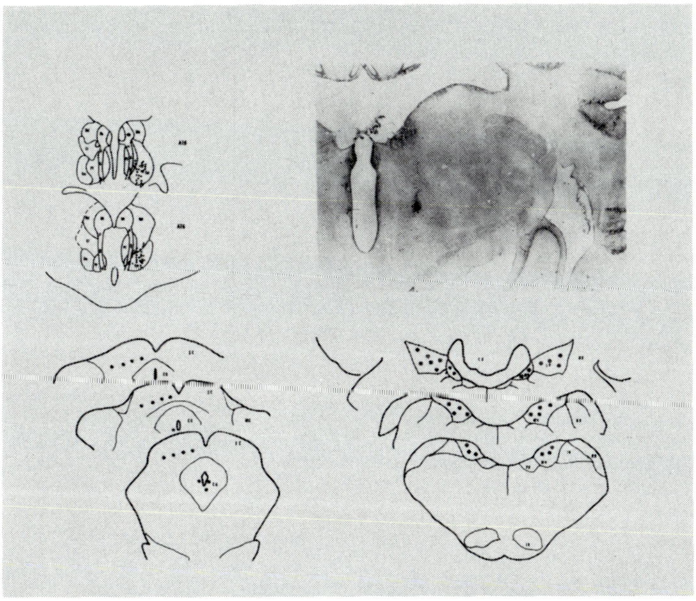

Fig. 8. HRP labeled neurons in the deep layers of superior colliculus, periaqueductal gray, and lateral and medial vestibular nuclei following injection of HRP in centrum medianum (upper right). The stars indicate the general location but not the actual sites of labeled neurons. The injection site lies within the area where most polysensory neurons inhibited by caudate stimulation are located. Two representative sections have been sketched (upper left). The left-hand side shows centrum medianum according to Mehler, the right-hand side according to a standard stereotaxic atlas (modified from Mehler, 135).

<u>Tectal and vestibular projections</u>. The most important, and quite unexpected result was the demonstration that, in addition to the basal ganglia, the major input sources to centrum medianum are the deep layers of the superior colliculus and the vestibular nuclei. The deep layers of the superior colliculus are known to receive a polysensory input and its neurons respond to auditory, visual, and somatosensory stimuli either selectively or by convergence (32, 55, 77, 182). Behaviorally, they have been closely identified with the performance of orienting movements although there remain considerable differences in interpretation of the underlying mechanism (181, 185).

A second important input source is derived from the lateral and medial vestibular nuclei; this projection is bilateral. While it is not known whether there are true polysensory neurons in the vestibular complex, the existence of neurons responding to proprioceptive and tactile stimuli has been established (170). The vestibular complex provides a first processing stage of sensory impulses leading to the organization of complex phasic and tonic eye-head-neck-body reflexes and postural adjustments. The convergence, in the medial thalamus, of vestibular and deep tectal projections constitutes a powerful argument that the centrum medianum, and its neostriatal link, represent major integrating centers for the elaboration of complex orienting responses. Support for this hypothesis is derived from studies which have demonstrated that the association cortex (128, 139, 184) and the intralaminar thalamus (152, 173) respond to situationally significant visual stimuli and that these regions are strongly implicated in the preparation for, and execution of target oriented eye and head movements (174).

Other input sources. Other, relatively minor inputs to centrum medianum originate from area 4 of the motor cortex, locus coeruleus, and scattered cells within the periaqueductal gray and the reticular formation (130). It is of some interest that the reticular input to centrum medianum is so small. Heretofore, it has been widely accepted that the polysensory responses of centrum medianum were primarily projected via a spino-reticulo-thalamic pathway, at least the somatosensory component, which has commanded most attention in the past (19, 20).

Projections of Nucleus Centralis Lateralis

Nucleus centralis lateralis also projects to the head of the caudate nucleus in the cat but less massively than centrum medianum (129). Many of its neurons are also modulated by the caudate nucleus (Fig. 6) but its afferent projections differ considerably from the projections to centrum medianum (58, 92, 163).

Recent studies with small electrophoretic deposits of horseradish peroxidase restricted to nucleus centralis lateralis (129) have shown retrogradely labeled neurons in the deep cerebellar, fastigial nucleus, the dorsal horn of the spinal cord, the pontine tegmental fields and the cuneiform nucleus, as well as the deep tectal layers. There appears to be no direct projection from the substantia nigra and only a minor one from the entopeduncular nucleus. The caudate inhibition of the polysensory neurons of this region (66) must, therefore, first engage intrathalamic neurons, presumably located in or near centrum medianum. The fact that nucleus centralis lateralis rather than centrum medianum receives a spinothalamic and spino-reticulo-thalamic projection has been emphasized by Mehler (17, 135) and underscores the marked anatomical and functional differentiations within the polysensory system. Since the caudate modulation of these polysensory fields is, itself, differentiated, the fallacy of treating the neostriatal - medial thalamic interaction as a unitary phenomenon becomes readily apparent.

CONCLUSIONS

Neostriatal - Centromedian Loop

The anatomical and physiological observations reviewed here suggest that the neostriatum functions in a close, finely tuned relationship with the polysensory medial thalamus. The neostriatum and centrum medianum are linked by a tightly organized, semi-closed loop which minimally may require no more than three synapses for its completion. Polysensory activities are channeled into the neostriatum from centrum medianum and adjacent portions of nucleus centralis lateralis and medialis dorsalis. In turn, the neostriatal output, passing through substantia nigra and globus pallidus (entopeduncular nucleus) is, in part, directed back to centrum medianum and, apparently less directly, to nucleus centralis lateralis (Fig. 9).

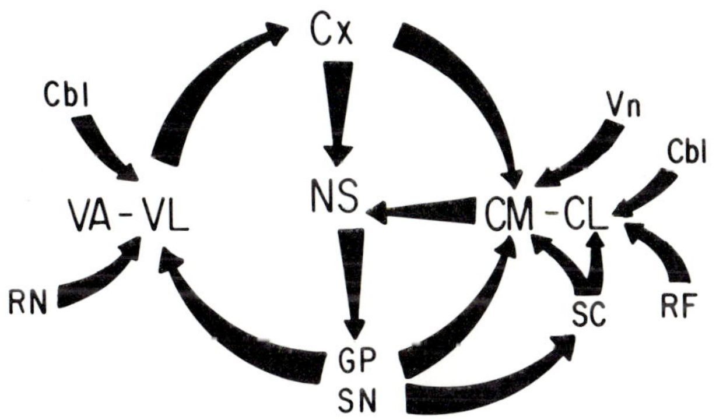

Fig. 9. The neostriatal-thalamic loop. The neostriatal output is funneled by globus pallidus-substantia nigra to the ventrolateral and medial thalamus. From VA-VL the major projections are to the cortex but from CM-CL the major projection is back to the neostriatum. The corticostriatal projections modulate the polysensory loop between neostriatum and medial thalamus. Major extrastriatal input sources to VA-VL and CM-CL are indicated; note the differences in the input to CM and CL. NS, neostriatum; GP, globus pallidus; SN, substantia nigra; CM, centrum medianum; CL, n. centralis lateralis; VA, n. ventralis anterior; VL, n. ventralis lateralis; Cbl, cerebellum; RN, red n.; Vn, vestibular n.; RF, reticular formation; SC, superior colliculus; Cx, cortex.

It is important to realize that the output of the medial thalamus is not only projected to the lateral thalamus and cortex (6, 44, 132, 172) but also, perhaps mostly, to the neostriatum. It can be said, therefore, that the basal ganglia are characterized by three subcortical loops. The connections between neostriatum and substantia nigra and between globus pallidus and subthalamic nucleus form two loops; the third loop is that between neostriatum and centrum medianum. These loops function quite independently of the more loosely organized circuit mediated by thalamocortical and corticostriatal projections to which the term "loop" is best not applied because of their profuse connections with other regions.

The centromedian-neostriatal loop probably plays an important role in the regulation of complex orienting responses. Schlag-Rey and Schlag have recently drawn attention to the association between orienting responses and sensory convergence structures (174). At the thalamic level, the polysensory input to centrum medianum is derived from the deep layers of the superior colliculus and the vestibular nuclei which are closely identified with the generation of orienting responses. Moreover, the tectal region is itself, a target for nigrofugal projections (11, 42, 86, 164). Similarly, at the cortical level, caudate inhibition is most clearly revealed on the anteromedial suprasylvian gyrus of the cat (108). This area corresponds to area 7 in the monkey (82) and in both species it has been linked to elements of the orienting response (128, 139, 184). In addition, some or perhaps all of these polysensory regions are subject to hippocampal modulation (104, 131) suggesting a close cooperation between parts of the limbic system and basal ganglia. Recent anatomical studies proposed a ventral extension of the neostriatum into the limbic basal forebrain area (83, 84) and other studies also provide evidence for very close links between limbic structures and basal ganglia (31, 37, 54, 140, 142). According to Vinogradova and Sokolov (200) the hippocampus and caudate nucleus provide the neuronal foundations for orienting reflexes (202).

The functional relationship between caudate nucleus and medial thalamus is most probably not a fixed one but depends upon cortical and other influences. These delineate and differentiate the extent of neostriatal activity as it is channeled towards the medial thalamus. Within the medial thalamus, the neostriatal zone of influence encompasses regions which differ in the origin of their polysensory input. Some, such as centrum medianum, are more closely linked with the deep tectal strata and the vestibular complex; others, such as nucleus centralis lateralis, reveal the predominance of deep tectal, cerebellar, reticular and spinothalamic inputs. Somatosensory, visual, vestibular and other sensory inputs thus converge in a differential manner upon the medial thalamus, there to be acted upon differentially by the neostriatum.

Significance

Many of the behavioral symptoms associated with neostriatal lesions and stimulation, such as hypoactivity and hyperactivity, arrest reactions and obstinate progression (183, 197, 203) can, perhaps, be interpreted as direct expressions of extrapyramidal motor disturbances. Alternatively, these symptoms and other manifestations of neostriatal dysfunction, such as head turning, circling, unilateral sensory neglect, including the neglect of noxious stimuli and its opposite, sensory overattention (43, 79, 81, 127, 214), may be viewed as different expressions of a disturbed modulation of sensory convergence mechanisms. Similar disturbances of sensory convergence may contribute to the impaired performance of various learned tasks, including delayed alternation and response and, in humans, the faulty orientation of one's body in space (18, 50, 117, 126, 154, 188, 206, 208).

It may, therefore, be more meaningful to treat these seemingly disparate symptoms of varying degrees of complexity, as the consequences of defective orienting behavior indicating, perhaps, the inability to maintain appropriate contact with a significant stimulus and the inability to shift to other stimuli at the critical moment. Of course, these represent only some of the elements of which orienting responses are constituted. Indeed, a rigorous definition of orienting is still lacking, especially if it is to include more than just the initial turning towards a sudden and unexpected stimulus. Here, one might simply suggest that orienting responses are a necessary aspect of sequentially organized behavior and that the polysensory nature of the neostriatal input and the striatofugal modulation of polysensory activities provide a key mechanism for their execution.

The feedback loop between centrum medianum and caudate nucleus enables the neostriatum to self-regulate its polysensory input at the level of the medial thalamus and to control polysensory activities from the medulla to the cortex. In the light of these findings, it seems plausible to consider the motor functions commonly attributed to the basal ganglia as derived largely from the regulation of polysensory activities by the neostriatum.

REFERENCES

1. Albe-Fessard, D. In: Contributions to Sensory Physiology, 2, W.D. Neff (Ed.) Academic Press, New York 101-167 (1967).

2. Albe-Fessard, D. & Fessard, A. In: Progr. Brain Res., G. Moruzzi, A. Fessard & H.H. Jasper (Eds.), Elsevier, Amsterdam, 115-148 (1963).

3. Albe-Fessard, D. & Gillett, E., Electroenceph. clin. Neurophysiol. 13, 257-269 (1961).

4. Albe-Fessard, D. & Kruger, L., J. Neurophysiol. 25, 3-20 (1962).

5. Albe-Fessard, D. and Rougeul, A., Electroenceph. clin. Neurophysiol. 10, 131-152 (1958).

6. Albe-Fessard, D., Levante, A. & Rokyta, R., Int. J. Neurosci, 1, 327-328 (1971).

7. Albe-Fessard, D., Oswaldo-Cruz, E., & Rocha-Miranda, C., Electroenceph. clin. Neurophysiol., 12, 405-420 (1960).

8. Albe-Fessard, D., Rocha-Miranda, C. & Oswaldo-Cruz, E., Electroenceph. clin. Neurophysiol., 12, 649-661 (1960).

9. Albe-Fessard, D., Besson, J.M., Guilbaud, G. & Levante, A. In: Corticothalamic Projections and Sensorimotor Activities, T.L. Frigyesi, E.R. Rinvik & M.D. Yahr (Eds.), Raven Press, New York 283-303 (1962).

10. Albe-Fessard, D., Massion, J., Hall, R.D. & Rosenblith, W.A. In: Brain Reflexes Progr. Brain Res. 22, E.A. Asratyan (Ed.), Elsevier, Amsterdam, 206-216 (1967).

11. Anderson, M. & Yoshida, M., Brain Res. 137, 361-364 (1977).

12. Aschoff, J.C. & Kornhuber, H.H. In: The Somatosensory System, H.H. Kornhuber (Ed.) Thieme, Stuttgart 145-157 (1975).

13. Bell, C., Sierra, G., Buendia, N. & Segundo, J.P., J. Neurophysiol, 27, 961-987 (1964).

14. Bendrups, A.P. & McKenzie, J.S., Brain Res. 80, 131-134 (1974).

15. Bignall, K.E., Exp. Neurol. 18, 56-67 (1967).

16. Blake, D.J., **Zarzecki, P.** & Somjen, G.G., J. Neurobiol, 7, 143-156 (1976).

17. Boivie, J., Exp. Brain Res. 12, 331-353 (1971).

18. Bowen, F.P. In: The Basal Ganglia. Res. Publ. Assoc. Res. Nerv. Ment. Dis. M.D. Yahr (Ed.), Raven Press, New York, 169-177 (1976).

19. Bowsher, D. In: The Thalamus, D.P. Purpura and M.D. Yahr (Eds.), Columbia Univ. Press, New York 99-108 (1966).

20. Bowsher, D., Mallart, A., Petit, D. & Albe-Fessard, D., J. Neurophysiol. 31, 288-300 (1968).

21. Brodal, A., Neurological Anatomy in Relation to Clinical Medicine, Oxford Univ. Press, New York (1969).

22. Buchwald, N.A., Price, D.D., Vernon, L. & Hull, C.D., Exp. Neurol. 38, 311-323 (1973).

23. Buerger, A.A., Gross, C.G., Rocha-Miranda, C.E., J. comp. physiol. Psychol. 86, 770-776 (1974).

24. Buser, P. In: The Thalamus, D.P. Purpura and M.D. Yahr (Eds.), Columbia Univ. Press, New York 323-347 (1966).

25. Buser, P. & Bignall, K.E., Int. Rev. Neurobiol. 10, 111-165 (1967).

26. Buser, P., Pouderoux, G. & Mereaux, J., Brain Res. 71, 337-344 (1974).

27. Carman, J.B., Cowan, W.M. & Powell, T.P.S., Brain Res. 86, 525-562 (1963).

28. Carpenter, M.B., Adv. Neurol 14, 7-47 (1976).

29. Carpenter, M.B. & Peter, P., J. Comp. Neurol. 144, 93-116 (1972).

30. Carpenter, M.B., Nakamo, U. & Kim, R., J. comp. Neurol. 165, 401-416 (1976)

31. Carter, D.A. & Fibiger, H.C., J. comp. Neurol. 177, 113-124 (1978).

32. Chalupa, I.M. & Rhoades, R.W., J. Physiol. 270, 595-626 (1977).

33. Chung, J.W., Hassler, R. & Wagner, A., Exp. Brain Res. 28, 345-361 (1977).

34. Clavier, R.M., Atmadja, S. & Fibiger, H.C., Brain Res. Bull. 1, 379-384 (1967).

35. Cohen, S.M., Acta Neurobiol. Exp. 32, 211-233 (1972).

36. Collins, R.J. & Simonton, V.R., Int. J. Neuropharmacol. 6, 349-356 (1967).

37. Costin, A., Gutman, J. & Bergmann, F., Electroenceph. clin. Neurophysiol. 15, 997-1005 (1963).

38. Cowan, W.M. & Powell, T.P.S., J. Neurol. Neurosurg. Psychiat. 29, 426-439 (1966).

39. Dafny, N. & Feldman, S., Electroenceph. clin. Neurophysiol. 23, 546-557 (1967).

40. DeLong, M.R., Brain Res. 40, 127-135 (1972).

41. DeLong, M.R. & Strick, P.L., Brain Res. 71, 327-335 (1974)/

42. Deniau, J.M., Hammond-LeGuyader, C., Feger, J. & McKenzie, J.S., Neurosci. Letters 5, 45-50 (1977).

43. Denny-Brown, D. & Yanagisawa, N. In: The Basal Ganglia. Res. Publ. Assoc. Nerv. Ment. Dis. 55, M.D. Yahr (Ed.), Raven Press, New York, 115-148 (1976).

44. Desiraju, T., Broggi, G., Prelevic, S., Santini, M. & Purpura, D.P., Brain Res. 15, 542-543 (1969).

45. Dieckmann, G. & Hassler, R., J. Hirnforsch 10, 187-225 (1968).

46. Diez-Martinez, O., Munoz, M.G., Prieto, G. & Brust-Carmona, H., Physiol. Behav. 16, 727-733 (1976).

47. DiFiglia, M., Pasik, P. & Pasik, T., Brain Res. 114, 245-256 (1976).

48. Dila, C., Brain Res. 25, 63-74 (1971).

49. Divac, I., Acta Neurobiol. Exp. 32, 461-477 (1972).

50. Divac, I., Physiol. Behav. 8, 519-522 (1972).

51. Divac, I., Rosvold, H.E., & Szwarcbart, M., J. comp. physiol. Psychol. 63, 184-190 (1967).

52. Doetsch, G.S., Escobar, N., Norman, H.L., Brain Res. 137, 277-289 (1971).

53. Dolbakyan, E., Hernandez-Mesa, N. & Bures, J., Neurosci. 2, 73-80 (1977).

54. Douglas, R.J., Psychol. Bull. 67, 416-442 (1967).

55. Drager, U.C., Hubel, D.H., J. Neurophysiol. 38, 690-713 (1975).

56. Dubner, R., Rutledge, L.T., Exp. Neurol. 12, 349-369 (1965).

57. Dubner, R., Rutledge, L.T., J. Neurophysiol. 27, 620-634 (1964).

58. Edwards, S.B. & de Olmos, J.S., J. comp. Neurol. 165, 417-432 (1976).

59. Encabo, H. & Buser, P., Electroenceph. clin. Neurophysiol. 17, 144-153 (1964).

60. Endo, K., Araki, T. & Ito, K., Brain Res. 132, 541-546 (1977).

61. Endo, K., Araki, T. & Ito, K., Brain Res. 132, 547-552 (1977).

62. Evarts, E.V., Canad. J. Physiol. Pharmacol. 53, 191-201 (1975).

63. Feger, J., Jacquemin, J. & Ohye, C., Exp. Neurol. 59, 351-360 (1978).

64. Feldman, S. & Dafny, N., Brain Res. 10, 402-417 (1968).

65. Feltz, P. & Albe-Fessard, D., Electroenceph. clin. Neurophysiol. 33, 179-193 (1972).

66. Feltz, P., Krauthamer, G. & Albe-Fessard, D., J. Neurophysiol. 30, 55-80 (1967).

67. Fonnum, F. In: The Neostriatum. Europ. Brain Behav. Soc. Workshop, I. Divac (Ed.), Pergamon Press, Oxford, The present volume.

68. Fox, S.S. & O'Brien, J.H., Science 147, 880-890 (1965).

69. Fox, S.S. & O'Brien, J.H., Science 137, 423-425 (1962).

70. Frigyesi, T.L. & Machek, J., Brain Res. 27, 59-78 (1971).

71. Frigyesi, T.L. & Machek, J., Brain Res. 20, 201-217 (1970).

72. Frigyesi, T.L. & Rabin, A., Brain Res. 35, 67-87 (1971).

73. Frigyesi, T.L. & Szabo, J., Exp. Neurol. 49, 123-139 (1975).

74. Frigyesi, T.L., Ige, A., Iulo, A., Schwartz, R., Exp. Neurol. 33, 78-87 (1971).

75. Fuster, J. & Alexander, C., Science 173, 652-654 (1971).

76. Goldman, P.S. & Nauta, W.J.H., J. comp. Neurol. 171, 369-386 (1976).

77. Gordon, B., J. Neurophysiol. 36, 157-178 (1973).

78. Goswell, M.J. & Sedgwick, E.M., Brain Res. 23, 112-116 (1970).

79. Grofová, I., Brain Res. 91, 286-291 (1975).

80. Grofová, I. & Rinvik, E. Anat. Embryol. 146, 113-132 (1974).

81. Gybels, J., Meulders, M., Callens, M. & Colle, J., Arch. inter. Physiol. Biochem. 75, 283-302 (1967).

82. Heath, C.J. and Jones, E.G., Erg. Anat. Entwicklungsgesch. 75, 6-64 (1972).

83. Heimer, L. & van Hoesen, G. In: The Neostriatum. Europ. Brain Behav. Soc. Workshop, J. Divac (Ed.), Pergamon Press, Oxford, The present volume.

84. Heimer, L. & Wilson, R.D. In: Golgi Centennial Symposium: Perspectives in Neurobiology, M. Santini (Ed.) Raven Press, New York, 177-193 (1975).

85. Herkenham, M. & Nauta, W.J.H., J. comp. Neurol. 173, 123-146 (1977).

86. Hopkins, D.A. & Niessen, L.W., Neurosci. Letters 2, 253-259 (1976).

87. Hore, J., Meyer-Lohmann, J. & Brooks, V.B., Science 195, 584-586 (1976).

88. Horn, G. & Hill, R.M., Exp. Neurol. 14, 199-223 (1966).

89. Huang, C.C. & Lindsley, D.B., Electroenceph. clin. Neurophysiol. 34, 265-280 (1973).

90. Hull, C.D., Bernardi, G., Price, D.D., & Buchwald, N.A., Exp. Neurol. 38, 324-336 (1973).

91. Ishida, Y., Neuropharmacol. 16, 163-170 (1977).

92. Itoh, K. & Mizuno, N., Exp. Brain Res. 30, 471-480 (1977).

93. Jabbur, S.J., Harik, S.I. & Hush, J.A., Brain Res. 120, 559-563 (1976).

94. Johnson, T.N., Rosvold, H.E. & Mishkin, M., Exp. Neurol. 21, 20-34 (1968).

95. Jones, E.G. & Burton, H., J. comp. Neurol. 154, 395-432 (1974).

96. Jones, E.G. and Leavitt, R.Y., J. comp. Neurol. 154, 349-378 (1974).

97. Jones, E.G., Coulter, J.D., Burton, H. & Porter, R., J. comp. Neurol. 173, 53-80 (1977).

98. Jung, R. & Hassler, R. In: Handbook of Physiology. Sect. 1 Neurophysiology, 2, J. Field, H.W. Magoun & V.E. Hall (Eds.), Amer. Physiol. Soc., Washington D.C., 863-928 (1960).

99. Kadobayashi, I. & Nakamura, M., Experientia 30, 260-262 (1974).

100. Kemp, J.M. & Powell, T.P.S., Phil. Trans. Roy. Soc. B 262, 383-401 (1971).

101. Kemp, J.M. & Powell, T.P.S., Brain Res. 93, 525-546 (1970).

102. Kim, R., Nakano, K., Jayaraman, A. & Carpenter, M.B., J. comp. Neurol. 169, 263-290 (1976).

103. Kitai, S.T., Kocsis, J.D. & Wood, J., Brain Res. 118, 137-141 (1976).

104. Kitano, K., Neuropharmacol. 15, 777-783 (1976).

105. Kocsis, J.D., Sugimori, M. & Kitai, S.T., Brain Res. 124, 403-413 (1977).

106. Krauthamer, G.M. In: Catecholamines and Behavior, 1, A.J. Friedhoff (Ed.) 59-87 (1975).

107. Krauthamer, G.M., Science 142, 1175-1176 (1963).

108. Krauthamer, G.M. & Albe-Fessard, D., J. Neurophysiol. 28, 100-124 (1965).

109. Krauthamer, G.M. & Albe-Fessard, D., Neuropsychol. 2, 73-83 (1964).

110. Krauthamer, G. & Albe-Fessard, D., C.R. Acad. Sci. 256, 265-267 (1963).

111. Krauthamer, G. & Albe-Fessard, D., C.R. Soc. Biol. 155, 1443-1450 (1961).

112. Krauthamer, G. & Bagshaw, M., J. Physiol. (Paris) 55, 274-275 (1963).

113. Krauthamer, G. & Dalsass, M., Brain Res. (in press).

114. Krauthamer, G. & Dalsass, M., Physiologist 14, 176 (1971).

115. Krauthamer, G. & Yamaguchi, Y., J. Physiol. (Paris) 56, 586 (1964).

116. Krauthamer, G., Feltz, P. & Albe-Fessard, D., J. Neurophysiol. 30, 81-97 (1967).

117. Krauthamer, G., Liebeskind, J. & Salmon-Legagneur, A., J. Physiol. 190, 18-19P (1967).

118. Kruger, L. & Albe-Fessard, D., Exp. Neurol. 2, 442-467 (1960).

119. Kuo, D., Pile, E., & Krauthamer, G., Soc. Neurosci. Absts. (in press)

120. Lamarche, M. & Buser, P., Experientia 24, 355-356 (1968).

121. Levine, M.S. & Schwartzbaum, J.S., J. comp. physiol. Psychol. 85, 615-635 (1973).

122. Lidsky, T.I., Buchwald, N.A., Hall, C.D. & Levine, M.S., Electroencyph. clin. Neurophysiol. 39, 79-84 (1975).

123. Liles, S.L., J. Neurophysiol. 37, 254-265 (1974).

124. Liles, S.L., Electroenceph. clin. Neurophysiol. 35, 277-285 (1973).

125. Liles, S.L. & Davis, G.D., J. Neurophysiol. 32, 574-582 (1969).

126. Livesey, P.J. & Rankine-Wilson, J., J. comp. physiol. Psychol. 88, 342-354 (1975).

127. Ljungberg, T. & Ungerstedt, U., Exp. Neurol. 53, 585-600 (1976).

128. Lynch, J.C., Mountcastle, V.B., Talbot, W.H. & Yin, T.C.T., J. Neurophysiol. 40, 362-389 (1977).

129. McGuinness, C. & Krauthamer, G., Soc. Neurosci. Abst. (in press).

130. McGuinness, C. Dalsass, M., Proshansky, E. & Krauthamer, G., Neurosci. Abst. 2, 67 (1976).

131. McKenzie, J.S. & Rogers, D.K., Brain Res. 27, 382-385 (1973).

132. Macchi, G. Bentivoglio, M., d'Atena, C., Rossini, P. & Tempesta, E., Neurosci. Letters 4, 121-126 (1977).

133. Mallart, A., J. Physiol. 194, 337-353 (1968).

134. Matsunami, K. & Cohen, B., Brain Res. 91, 140-146 (1975).

135. Mehler, W.R. In: The Thalamus, D.P. Purpura and M.D. Yahr (Eds.), Columbia Univ. Press, New York, 109-122 (1966).

136. Mehler, W.R. and Nauta, W.J.H., Confin. Neurol. 36, 205-222 (1974).

137. Miller, J.J., Richardson, T.L., Fibiger, H.C. & McLennan, H., Brain Res. 97, 133-138 (1975).

138. Moolenaar, G-M, Holloway, J.A. and Trouth, C.O., Exp. Neurol. 53, 304-313 (1976).

139. Mountcastle, V.B., Lynch, J.C., Georgopoulos, A., Sakata, H., & Acuna, C, J. Neurophysiol. 38, 871-908 (1975).

140. Nauta, H.J.W., J. comp. Neurol. 156, 19-28 (1974).

141. Nauta, W.J.H., Bain Res. 81, 319-341 (1958).

142. Nauta, W.J.H., & Mehler, W.R., Brain Res. 1, 3-42 (1966).

143. Newton, R.A. and Price, D.D., Brain Res. 85, 403-422 (1975).

144. Niimi, K., Ikeda, T., Kawamura, S. & Inoshita, H., Brain Res. 21, 327-343 (1970).

145. Niki, H., Sakai, M. & Kubota, K., Brain Res. 38, 343-353 (1972).

146. Noda, H. & Adey, W.R., Brain Res. 54, 243-259 (1973).

147. Öberg, R.G.E. & Divac, I., Acta Neurobiol. Exp. 35, 647-659 (1975).

148. O'Brien, J.H. & Fox, S.S., J. Neurophysiol. 32, 285-296 (1969).

149. O'Brien, J.H. & Rosenblum, S.M., Physiol. Behav. 15, 603-609 (1975).

150. Olmstead, C.E., Villablanca, J.R., Marcus, R.J. & Avery, D.L., Exp. Neurol. 53, 670-693 (1976).

151. Olpe, H.R. & Koella, W.P., Brain Res. 122, 357-360 (1977).

152. Orem, J., Schlag-Rey, M. & Schlag, J., Exp. Neurol. 40, 784-797 (1973).

153. Paderni, M.A. & Giammanco, S., Bol. Soc. ital. Biol. sper. 43, 1283-1286 (1967).

154. Potegal, M., Acta Neurobiol. Exp. 32, 479-494 (1972).

155. Potegal, M., Copack, P., deJong, J.M.V.B., Krauthamer, G., & Gilman, S., Exp. Neurol. 32, 448-465 (1971).

156. Powell, T.P.S. & Cowan, W.M., Brain Res. 79, 364-390 (1956).

157. Powell, T.P.S. & Cowan, W.M., J. Anat. 88, 307-319 (1954).

158. Purpura, D.P. In: The Basal Ganglia, Res. Publ. Ass. Res. Nerv. Ment. Dis., 55, M.D. Yahr (Ed.), Raven Press, New York, 91-114 (1976).

159. Purpura, D.P., Pharmacol. Therap. B 1, 17-38 (1975).

160. Purpura, D.P. & Malliani, A., Brain Res. 6, 325-340 (1967)

161. Rasminsky, M., Mauro, A.J. and Albe-Fessard, D., Brain Res. 61, 69-77 (1973).

162. Rinvik, E., Brain Res. 8, 237-254 (1968).

163. Rinvik, E., & Grofová, I., Anat. Embryol. 146, 95-111 (1974).

164. Rinvik, E., Grofova, I. & Ottersen, O.P., Brain Res. 112, 388-394 (1976).

165. Robertson, R.T. & Thompson, R.F., Physiol. Behav. 10, 245-252 (1973).

166. Rogers, D.K. & McKenzie, J.S., Brain Res. 56, 345-349 (1973).

167. Rolls, E.T. In: The Neostriatum. Europ. Brain Behav. Soc. Workshop, I. Divac (Ed.), Pergamon Press, Oxford, The present volume.

168. Rougeul, A. & Buser, P., J. Physiol. (Paris) 53, 466-467 (1961).

169. Royce, J., *Brain Res*. 146, 145-150 (1978).

170. Rubin, A.M., Liedgren, S.R.C., Milne, A.C., Young, J.A. & Frederickson, J.M. *Pflügers Arch*. 371, 155-160 (1977).

171. Rudomin, P., Malliani, A., Borlone, M. & Zanchetti, A., *Arch. ital. Biol*. 103, 60-89 (1965).

172. Scheibel, M.E. & Scheibel, A.B., *Brain, Behav. Evol*. 6, 332-358 (1972).

173. Schlag, J., Lehtinen, I. & Schlag-Rey, M., *J. Neurophysiol*. 37, 982-995 (1974).

174. Schlag-Rey, M. & Schlag, J., *J. Neurophysiol*. 40, 156-173 (1977).

175. Sedgwick, E.M. & Williams, T.D., *J. Physiol*. 189, 281-298 (1967).

176. Segundo, J.P., Machne, X., *J. Neurophysiol*. 19, 325-339 (1956).

177. Siegel, A., Sasso, L. & Tassoni, J.P., *Exp. Neurol*. 33, 130-146 (1971).

178. Siegel, J. & Lineberry, C.G., *Exp. Neurol*. 22, 444-463 (1968).

179. Soltysik, S., Hull, C.D., Buchwald, N.A. & Fekete, T., *Electroenceph. clin. Neurophysiol*. 39, 65-78 (1975).

180. Spear, P.D., *Brain Res*. 30, 218-222 (1971).

181. Sprague, J.M. & Meickle, T.H. Jr., *Exp. Neurol*. 11, 115-146 (1965).

182. Stein, B.E., Magalhaes-Castro, B. & Kruger, L., *J. Neurophysiol*. 39, 401-419 (1976).

183. Stevens, J.R., Kim, C. & MacLean, P.D., *Arch. Neurol*. 4, 47-54 (1961).

184. Straschill, M. & Schick, F., *Exp. Brain Res*. 19, 467-477 (1974).

185. Stryker, M.P. & Schiller, P.H., *Exp. Brain Res*. 23, 103-112 (1975).

186. Szabo, J., *Exp. Neurol*. 5, 21-36 (1962).

187. Teuber, H.L. In: *The Basal Ganglia. Res. Publ. Assoc. Nerv. Ment. Dis. 55* M.D. Yahr (Ed.), Raven Press, New York, 151-168 (1976).

188. Teuber, H.L. In: *Frontal Cortex and Behavior*, K. Akert & J.M. Warren (Eds.) McGraw-Hill, New York, 410-444 (1964).

189. Thompson, R.F. & Shaw, J.A., *J. comp. physiol. Psychol*. 60, 329-339 (1965).

190. Thompson, R.F., Johnson, R.H. & Hoopes, J.J., *J. Neurophysiol*. 26, 343-364 (1963).

191. Thompson, R.F., Smith, H.E. & Bliss, D., *J. Neurophysiol*. 26, 365-378 (1963).

192. Totibadze, N.K. & Moniava, E.S., *J. comp. Neurol*. 137, 347-360 (1969).

193. Towe, A.L., Tyner, C.F. & Nyquist, J.K., *Exp. Neurol*. 50, 734-756 (1976).

194. Ungerstedt, U., Acta physiol. scand. Suppl. 367, 1-48 (1971).

195. Usunoff, K.G., Hassler, R., Romansky, K., Usunova, P. & Wagner, A., J. Neurol. Sci. 28, 265-288 (1976).

196. Vedel, J.P., Exp. Brain Res. 22, 113-128 (1975).

197. Villablanca, J.R., Marcus, R.J. & Olmstead, C.E., Exp. Neurol. 52, 389-420 (1976).

198. Villablanca, J.R., Marcus, R.J., Olmstead, C.E. & Avery, D.C., Exp. Neurol. 53, 289-303 (1976).

199. Vinogradova, O.S. In: Neuronal Mechanisms of the Orienting Reflex, E.N. Sokolov & O.S. Vinogradova (Eds.), Erlbaum Assoc., Hillsdale, New Jersey, 128-154 (1975).

200. Vinogradova, O.S., Sokolov, E.N. In: Neuronal Mechanisms of the Orienting Reflex, E.N. Sokolov & O.S. Vinogradova (Eds.), Lawrence Erlbaum, Hillsdale, New Jersey 155-169 (1975).

201. Voneida, T.J., J. comp. Neurol. 115, 75-87 (1960).

202. Voromin, L.G., Leontiev, A.N., Luria, A.R., Sokolov, E.N. and Vinogradova, O.S., (Eds.), Orienting Reflex and Exploratory Behavior, Amer. Inst. Biol. Sci., Washington, D.C. (1965).

203. Wang, G.H. & Akert, K., Arch. ital. Biol. 100, 48-85 (1962).

204. Webster, K.E., J. Anat. 99, 329-337 (1965).

205. Webster, K.E., J. Anat. 95, 532-544 (1961).

206. Wilburn, M. & Kesner, R.P., Exp. Neurol. 45, 61-71 (1974).

207. Wilson, S.A.K., Brain 36, 427-492 (1914).

208. Winocur, G., J. comp. physiol. Psychol. 86, 432-439 (1974).

209. Wise, S.P. & Jones, E.G., J. comp. Neurol. 175, 129-158 (1977).

210. Yahr, M.D. (Ed.) The Basal Ganglia. Res. Publ. Assoc. Res. Nerv. Ment. Dis. 55, Raven Press, New York (1976).

211. Yahr, M.D. & Purpura, D.P. (Eds.) Neurophysiological Basis of Normal and Abnormal Motor Activities, Raven Press, New York (1967).

212. Yamaguchi, Y. & Krauthamer, G., C.R. Acad. Sci. 262, 1013-1015 (1966).

213. Yeterian, E.H. & van Hoesen, G.W., Brain Res. 139, 43-63 (1978).

214. Zimmerberg, B. & Glick, S.D., Brain Res. 86, 335-338 (1975).

"Cognitive" Functions of the Neostriatum

R. Gunilla E. Öberg* and Ivan Divac**

**Laboratory of Psychology, University of Copenhagen, Njalsgade 92, Copenhagen*
***Institute of Neurophysiology, University of Copenhagen, Juliane Maries Vej 36, Copenhagen*

The purpose of this chapter is to review the evidence for the involvement of the neostriatum (NS) in functions which do not seem to be usefully labelled as either sensory, inhibitory, or motor. We have chosen to term these remaining functions "cognitive" for a lack of a better alternative, and this label should only be regarded as a shorthand for processes which may be associative, mnemonic or complex perceptual, and for their expression in behavior. Other authors have proposed other terms, but these alternatives seem even less appropriate in the present context. Thus, the label "complex"[119] may refer to motor and sensory mechanisms as well as to "cognitive", and the term "sensorimotor integration"[65,124] is uninformative, since it describes the functions of the entire nervous system and can be applied to neural mechanisms at any level from tendon reflexes to cortical control of language. On the other hand, we cannot restrict our discussion to any more limited notion such as attention[49], memory consolidation[107], or corollary discharges[119]. These terms represent valid interpretations of some aspects of NS functioning, but in our view none of them can be stretched to cover all facets of the role of NS in functions which transcend sensation, movement, and inhibition.

During the last 15 years several reviewers have given a broadened perspective of NS functioning and have expressed views similar to those to be presented here[22,25,65,68,101,102,119,128]. This chapter attempts to give an exhaustive account of all sources of evidence for NS participation in cognitive functions.

ANATOMICAL EVIDENCE

The anatomy of NS is reviewed and interpreted in several chapters in this volume. Here, we will limit ourselves to the possible structural basis for NS' involvement in cognitive functions.

The NS is several synapses remote from both sense organs and muscles. Thus, this formation on one hand receives already processed information about the world, and on the other hand does not give final motor commands but submits its output to additional processing on the way to the final common path. If we were to follow stimulus information through its "vicissitudes" (referring to an influential demarcation of cognitive psychology[87]) we would find ourselves in the NS somewhere midway between the stimulus and the response.

The NS appears to be a well-informed structure. The entire cortical mantle sends fibers to NS, each cortical area finding both a primary target and auxilliary projection sites in this formation (Grofova, this volume; I. Divac and N.H. Diemer, ref. note 1). The projection from the motor cortex is substantial, and is sometimes taken as evidence for inclusion of the NS into the "motor system". However, the imput from sensory, cingulate and "association" areas of the cortex is at least as large (the afferents arriving from primary visual cortex appear to be the least extensive). The projection from the prefrontal cortex to the NS and the functional correlates of this connection stimulated Rosvold and his collaborators to include the NS into the "frontal lobe system"[101,102,105]. The involvement of NS in this "associational" system is much less widely accepted than its inclusion into the so-called extrapyramidal system, but the relation between NS and the prefrontal cortex is at least as well supported by hodology and, as we shall see, by behavioral studies. Furthermore, different sectors of NS may be included in still other systems , e.g. a "temporal lobe system" containing the temporal association cortex and related regions of NS.

Subcortical nuclei are likewise sources of information which may play a role for cognitive functions of the NS. Thus, multi-sensory input arrives to all parts of the NS from the intralaminar thalamic nuclei (Krauthamer, this volume), and the mediodorsal thalamic nucleus (an "association" nucleus with rich two-way connections with the frontal cortex[110]) probably projects to at least one region of the NS[8,25,108] (see also Grofova, this volume).

On the efferent side, NS can influence posture and movement through a minimum of four synapses. Obviously, every brain structure, regardless of its function, must be capable of influencing the behavior of the organism. Natural selection would not favour the evolution of any structure which is not "motor" in this broad sense. It is safe to assume that the biological function of cognitive capacities is behavioral adaptability.

EVIDENCE FROM ANIMAL BEHAVIOR

The behavioral roles of the non-human, mammalian NS have been studied with a variety of brain manipulation techniques: lesions of different kinds (Iversen, this volume), electrical stimulation (Phillips, this volume) and recording (Rolls, this volume), as well as pharmacological treatments (for example, refs. 36, 65, 124, and Iversen, this volume). Data favouring cognitive functions have been obtained with all of these methods. The following review is centered around clusters of behavioral tasks, rather than around presumed psychological processes. We are aware, however, that any behavioral classification can be disputed, and we do not claim that there are known cognitive mechanisms to suit the listed categories of behavior. With the risk of boring the well-informed, we have tried to make our presentation understandable for readers without much knowledge of experimental animal psychology.

It should be strongly emphasized that the main purpose of the reviewed studies of animal behavior has not been the elucidation of *how* the animal moves, but *where* or *when* it chooses to respond. In the following, therefore, an "impairment" subsequent to NS manipulation never refers to awkward movements and postures (in fact, obvious "motor" deficits are rarely observed in these animals), but to unadaptive (unrewarded or punished) choices among available response alternatives.

Studies of Spatial Choice

Spatial aspects of animal behaviour have been a frequent topic of investigation for comparative psychologists and learning theorists ever since the emergence of the discipline. There are several reasons for this interest. Spatial choice is obviously a fundamental component of adaptive behavior, e.g. wild animals are vitally dependent on their ability to find ways to food, shelter, etc., and many laboratory methods are developed as simplified models of such behavioral orientation in the environment. But animals' learning of spatial tasks has also become a test-case for theoretical disputes between "cognitively" oriented psychologists who emphasize the learning of stimulus relationships (e.g. internal "maps"), and "behaviorally" oriented psychologists who stress response-learning and stimulus-response associations. (For a glance into this many-facetted field of research, see, for example, refs. 9, 72, 82, 123). Organismic and environmental determinants of spatial choice have been studied in a wide variety of tasks, many of which have also been employed in studies of the functions of NS.

Repeated choices at one choice-point. A rather extensive body of research has concerned the contribution of NS in the control of an animal's behavior in situations where it is repeatedly exposed to the same response options and gets no guidance from external differential cues at the moment of choice. One behavioral situation of this kind is Hunter's delayed reaction task, first described 65 years ago[53] (see also ref. 38). This test was originally developed to investigate the presence of "ideas" in animals, and it has often been employed as a test of memory, although no-one has safely established what kind of ideas or memory animals rely on in this situation (see below). Other tasks in the category of "delayed response-type tests" are delayed spatial alternation, position reversals, spontaneous alternation, and spatial variants of time discrimination and delayed successive discrimination. All these tests depend on the functional integrity of the prefrontal cortex[25]. In terms of cognitive/behavioral requirements, these tasks seem to have three critical common denominators: the animal has to make spatial choices which are governed by some kind of organismic representation of a previous stimulus, and it has to make frequent changes of side of response (sometimes on every trial, as in delayed alternation, sometimes only after a sequence of responses, as in spatial reversals). In other words, this class of spatial tests combine an element of recent memory with a reversal feature.

Every task of this kind which has been studied was found to be sensitive to lesions or electrical stimulation in the anterior part of the NS in monkeys, rats, dogs, and cats (see Table 1). This high degree of replicability across species is combined with a remarkable generality across behavioral situations. The sensitive spatial-choice tasks span a wide spectrum of response modes: in some of the tests, the animals are trained to run down alleys (e.g. refs. 24, 33, 63, 74, 76, 129), in others to press bars or panels (e.g. refs. 1, 20, 46, 125), to lift plaques (e.g. refs. 7, 32), or to cross an open space (e.g. refs. 23, 26, 89). The animals may be required to choose between two (e.g. refs. 23, 24, 32) or more (e.g. ref. 76) loci. Furthermore, the employed tasks involve different sensory modalities: e.g. the predelay cue may be visual (e.g. ref. 103) or auditory (e.g. ref. 23). Yet animals with functional disruption in this area of NS are not generally debilitated, since they perform as proficiently as intact animals in a number of other tests, notably including learning tasks which involve spatial choice under guidance of differential cues (e.g. refs. 31, 32, 70).

To sum up, then, the NS seems to take part in mediating a phylogenetically widespread, supramodal function which is largely independent of response topography and which is operative under a rather well-defined range of circumstances. We shall now consider, in turn, the sources of the information to be processed by the NS, the nature of local NS processing, and psychological analysis of this innominate function.

Studies in several species have shown that only a limited portion of NS is of critical importance for delayed response-type tasks[23,32,69,70,89]. The focal NS

TABLE 1
SPATIAL "DELAYED RESPONSE-TYPE" TASKS
Effects of Manipulation of NS

Task	Treatment	Species	Reference
Delayed response	El.-lyt. or radio-freq. lesions	Monkey	6^{+}, 103^{+}, 111^{+}
		Dog	112^{+}
		Cat	23^{o+}
		Rat	76^{+}
	El. stimulation	Monkey	19^{+}, 103^{+}
Delayed alternation	El.-lyt. or radio-freq. lesion	Monkey	6^{+}, 32^{o+}, 104^{+}, 111^{+}
		Cat	26^{+}, 89^{o+}, 90^{+}
		Rat	46^{+}, 74^{+}, 89^{o+}, 129^{+}
	Kainic acid lesions	Rat	31^{o+}
	Cooling	Monkey	66^{+}
	El. stimulation	Cat	70^{o+}
	Scopolamine injection	Rat	96^{o+}
Spatial reversals	El.-lyt. lesions	Cat	90^{+}
		Rat	24^{+}, 63^{+}
	El. stimulation	Rat	69^{o+}
Delayed successive discrimination	El. stimulation	Monkey	20^{o+}
Spontaneous alternation	El.-lyt. lesion	Rat	33^{+}
Time discrimination	El.-lyt. lesion	Cat	100^{+}

$^{+}$Performance impaired. ^{o}Performance unimpaired.
$^{o+}$Performance impaired or unimpaired, depending on locus of treatment.

region is the recipient of information from that subdivision of the prefrontal cortex which is the neocortical "center" for the same class of behavior and probably also from the mediodorsal thalamic nucleus (see anatomical review above). These findings allocated one part of the NS to the "prefrontal system"[26,105], and recent experiments with the deoxyglucose technique confirmed this cortico-striatal relation: Increased neural activity in the prefrontal cortex of rats leads to a clearcut, localized enhancement of activity also in the anteromedial sector of NS (Divac and Diemer, ref. note 1). The mediodorsal thalamic nucleus was coactivated, as could be expected from its connections with both the prefrontal cortex and the implicated part of the NS, but it is not clear whether this source of afferents plays any role for delayed response-type behavior (see discussions in refs. 25 and 129).

The electrophysiological and neurochemical basis for NS' involvement in delayed response tasks is insufficiently investigated, and the available evidence does not particularly illuminate the fate, within the NS, of the information arriving from the frontal lobes. The average amplitude of EEG activity was found to increase both in the prefrontal cortex and in the head of the caudate nucleus during a delayed alternation task in monkeys[1], but the recording from the NS may in fact have picked up activity from fibers in transit. Niki and his collaborators likewise studied monkeys performing delayed alternation and encountered neurons with delay-related response characteristics in the prefrontal cortex, but not in the caudate nucleus; a small number of NS units were found to respond before the conditioned response in this task[88]. Pharmacological studies suggest that the performance of spatial tasks is sometimes affected by the level of dopamine in the NS, but the evidence is equivocal. Thus, Kitsikis *et al.* observed that intact cats which were poor learners of delayed responses tended to have depressed levels of dopamine in the NS, and that administration of L-Dopa improved their performance[62]. Unfortunately, the behavioral impairment and the subsequent treatment effect may actually have reflected a dopamine deficiency in non-striatal tissue, notably in the prefrontal cortex[29]. In rats which were trained to perform simple spatial discriminations (without delay and reversal features) in order to escape aversive stimulation, destruction of the nigro-striatal pathway severely impaired learning[99] or left learning intact[98], possibly depending on whether the animals were required to extinguish an innate stimulus preference or not[124] (see ref. 75 for a comparable effect of NS lesions); the deficit was presumably caused by dopamine depletion in the NS. Electrical stimulation of the substantia nigra disrupted spatial delayed responding in monkeys[19], possibly as a result of excessive release of dopamine in the NS. However, it is obviously unwarranted to assume that the substantia nigra and the nigro-striatal pathway affect behavior only by means of supply of neostriatal dopamine (Feger *et al.*, this volume). Furthermore, Sax *et al.* have shown that both delayed responses and alternation behavior can remain normal after administration of large doses of L-Dopa in otherwise intact monkeys; in this experiment, the pharmacological treatment caused motor hyperactivity without affecting error scores in any of the learning tasks[111].

If little is known about dopamine transmission and spatial behavior, even less is reported about the contribution of other transmitter mechanisms. One recent study on rats attempted to assess the role of cholinergic transmission for conditioned alternation[96]. The intriguing results suggest that cholinergic mechanisms in the anterior part of the NS may be important for effective retention of spatial alternation, but that over-training renders the same behavior independent of this neurochemical substrate. The possible involvement of other striatal neurochemicals (Fonnum and Walaas, this volume) is unexplored.

Finally, we shall shift focus to functional analysis of NS' involvement in delayed response behavior. The contribution of NS to this class of tasks has never been separately analyzed, the implicit assumption probably being that it does not differ importantly from the related function of the frontal cortex. Animals with prefrontal damage have been subjected to extensive and often ingenious experimentation and theorizing — all of which has not yet produced a generally accepted interpretation of their deficit. It is encouraging to note, however, that contemporary hypotheses of merit are either closely related or, at least, compatible. Goldman *et al.* have suggested that the implicated part of the frontal cortex is essential for a discrete spatial memory function[43]. Konorski likewise stressed memory processes, and postulated the existence of special "gnostic units" for kinesthesia in the frontal lobes[64], a view which has been shared and empirically supported by Stamm[117]. The latter formulation could be reconciled with Krauthamer's emphasis on orienting behavior (this volume). It does not seem to be possible to discriminate conclusively among these interpretations at present, and the particular importance of NS for any of these mechanisms remains undefined. Potegal has made the interesting suggestion that NS is a part of an "egocentric localization system", but this theory may not be relevant in the present context, since Potegal's experiments[95] compared to our own[89] point to different foci within the NS for tasks demanding "egocentric localization" and those requiring spatial delayed responses.

Pathfinding. Our understanding of the spatial functions of NS is hampered by lack of data on NS involvement in performance of more complex tasks. There are few studies of the role of NS in situations where the animal is obliged to make a series of choices between successive alternative paths (possible developing a "cognitive map"[82,123]), and the scarce data are not easy to synthesize: In one experiment, rats with rather extensive NS damage acquired a series of maze problems as proficiently as normal animals[46]. In another study, NS lesioned rats were deficient in learning and retention of the way through a maze which required alternate turns to the right and left, while maze problems not involving alternations did not reveal any difference between the lesioned and intact animals[11]. Electrical stimulation of the NS in rats retarded learning of a multiple alley maze, both when the stimulation was given after each choice point and when it was delivered only at the end of each completed run[93].

Whether the apparent discrepancies between these results should be attributed to differences in the behavioral tasks, or in the site and kind of brain manipulation, is not known. Moreover, any interpretation of the positive results is weakened by the fact that only rats have been studied in this way, since electrocoagulation or electrical stimulation of NS in rodents inevitably involves cortical connections (this problem will be further discussed below). Furthermore, there has been no attempt to assess whether any part of NS is essential for pathfinding, i.e. whether the same neural mechanisms are important for repeated choices at one choice path and for serial choices in a complex maze. Since clinical data suggest that NS may be involved in spatial orientation, including map-following, in humans (see below), pathfinding in animals seems to deserve further exploration.

Studies of Go, No-Go Choice

Obviously, not all behavioral adaptations of animals are best modelled as a choice between going to the right or to the left — "where-to-go"[101]. Frequently, the crucial decision rather concerns "whether-or-not", when, or under which circumstances to engage in one particular behavior. Psychologists have developed a large battery of learning tasks with which to study how animals emit, withhold, replace, and pattern responses. In some cases, the interest centers around the behavioral adjustment to the temporal distribution of appetitive stimuli (such as food or water for deprived animals). In other cases, the emphasis is on aversive control of behavior, e.g. animals are trained to avoid being subjected to noxious events. (In passive avoidance, the animal stops performing a punished response; in active avoidance, the animal acquires a response which leads to omission of punishment). Many tasks of both kinds have been employed in experiments aimed at elucidating the behavioral roles of NS.

Aversive conditioning. Animals with damage or functional disruption of the NS often show deficient behavioral adjustment to aversive stimuli. Examples of abnormalities of this kind are given in Table 2, which also illustrates that many different treatments of NS interfere with conditioned avoidance in rats. It seems safe to conclude, on the basis of this diverse evidence, that the NS is in some way important for avoidance behavior in this species.

The way in which NS subserves avoidance behavior is poorly understood. The NS seems to be regionally specialized also with respect to avoidance conditioning[80,86,132,133], but the structural heterogeneity of the NS has been regrettably disregarded in this line of research. Thus, the "dorsal" and "ventral" lesions which are commonly reported in this literature are rarely related to differential input to NS, to variations in its biochemical composition, or to any other structural basis for subdivision. Therefore the resulting data are neither easily compared to the effects of partial NS lesions in other behavioral situations, nor interpretable in

TABLE 2
AVERSIVE CONDITIONING ("GO, NO-GO")
Effects of Manipulation of NS

Task	Treatment	Species	Reference
Passive avoidance	El.-lyt. lesion	Cat	39^+
		Rat	80^+, 97^+, 114°, 132^+, 134^+
	Lesion by suction	Cat	90°
	Kainic acid lesion	Rat	109^+
	El. stimulation	Cat	131°
		Rat	42^+, 50^+, 135^+, 136^+
	Injection of:		
	carbachol	Rat	21^+
	scopolamine	Rat	50^+
	eserine	Rat	50°
	angiotensin	Rat	81^{o+}
	potassium chloride	Rat	97^+
	dopamine	Rat	59^+
Taste aversion	El.-lyt. lesion	Rat	30°
	El. stimulation	Rat	Phillips, this vol.^{o}
Sidman avoidance	Radiofreq. lesion	Rat	3^+, 4^+
Active avoidance, "one-way"	El.-lyt lesion	Rat*	61^+, 80^+, 132^{o+}, 134°
Active avoidance, "two-way"	El.-lyt. lesion	Rat*	2°, 45^+, 61^+, 80^+, 86^+, 132^+
		Cat	39°
	6-OHDA lesion	Rat	84^+
	Lesion by suction	Cat	121^{o+}
	Scopolamine injection	Rat	86^{o+}

$^+$Performance impaired. ^{o}Performance unimpaired.
$^{o+}$Performance impaired or unimpaired, depending on locus of treatment.

*Apparent discrepancies in the effects of NS lesions in rats may be attributed to differences in lesion site and/or experimental parameters (see discussion in refs. 115 and 128).

terms of differential information processing (e.g. differences in cortical afferents). Electrophysiological correlates of aversive conditioning have, to our knowledge, not been investigated, but a rather extensive pharmacological work is available (see Table 2). These studies suggest that both dopaminergic and cholinergic NS mechanisms may be involved in avoidance behavior, but the respective contribution of these two transmitter systems and the nature of their interaction remains open for speculation.

Several investigations have indicated that the nigro-striatal projection is of crucial importance for normal avoidance behavior in rats, but the evidence is not conclusive. Lesions of substantia nigra severely interfered with acquisition of both active[37,80] and passive[37] avoidance. Similarly, electrical stimulation of substantia nigra impaired passive avoidance[36,106]; unilateral stimulation was effective, but only when the ipsilateral nigrostriatal pathway was intact[36]. Surprisingly, however, bilateral 6-OHDA lesions of the nigrostriatal bundle, which caused a drastic reduction of intrastriatal dopamine, impaired neither acquisition nor retention of passive avoidance after a recovery period of about 2 weeks[36].

A limitation in current knowledge concerns the cross-species generality of the importance of NS for conditioned avoidance behavior. Data on primates are, as far as we know, lacking. The effects of caudate lesions in cats have been assessed in several studies; the results, although not quite consistent, suggest that cats can learn and retain avoidance responses without depending on NS mediation. Thus, Olmsead *et al.* found no deficits in passive avoidance in cats with much of the caudate but little other tissue removed[90]; a similar study by Fox *et al.* uncovered a marginal group effect, but some lesioned cats did not seem impaired[39]. Electrical stimulation of NS also did not convincingly disrupt learning of passive avoidance in cats[131]. Finally, active avoidance was unaffected by caudate damage[39], unless extensive lesions extending far beyond the NS were made[121,122]. Lesions aimed at the globus pallidus retarded learning of active avoidance[67], but the relevance of this finding for neostriatal mechanisms is uncertain, since the damage invaded the internal capsule. In short, there is no strong evidence for NS involvement in avoidance conditioning in species other than rats. Due to the difficulties inevitably involved in equating testing conditions across species it would, however, be premature to conclude that there is a genuine difference in the functions of NS.

For a functional analysis of the role of NS in aversive conditioning, it must be recognized that deficits in avoidance behavior can be caused by interference with many different basic mechanisms. Both active and passive avoidance obviously involve sensation/attention, motivation, learning (associative processes) and memory (consolidation, storage, and retrieval)[5]. In addition, active avoidance demands effective initiation and execution of movement ("go"), while passive avoidance requires withholding of responses ("no-go"). Not surprisingly, interpretations of NS' functions in aversive control of behavior have included behavioral initiation[98] or inhibition[39], regulation of fear[68] or arousal[61], sensory-motor integration[99], and memory consolidation[107]. We will argue that the first four interpretations have little merit; an explanation in terms of memory mechanisms seems most feasible, but it needs to be further qualified.

Lesions in some regions of NS as well as in substantia nigra produce impairments in both passive and active avoidance in rats[80,132]. A simultaneous deficit in both kinds of tasks means that the animal responds either too little (too late) or too much (too soon), depending on the situation. Such a syndrome does not easily lend itself to an interpretation in terms of disinhibition or motor deficits. Furthermore, rats with 6-OHDA lesions in substantia nigra showed good retention in an active avoidance task which was acquired preoperatively, while the same task could not be learned by naive animals with the same lesion[37]. Also this finding seems difficult to accommodate in any motor, inhibitory, or motivational interpretation. Most informative for a functional analysis are, however, the results of reversible experimental disruption of NS applied at different times in the training procedures. Injection of angiotensin into the dorsal part of NS[81] or of picrotoxin into the substantia nigra[58] caused clear deficits in passive avoidance when applied 5 minutes after the punishment (i.e. almost 24 hours prior to the retention test), but were ineffectual when applied 22 hours after this exposure (i.e. 2 hours before the test). Experiments on the effects of electrical stimulation of the NS similarly demonstrated that the disruptive agent is most effective when it is delivered shortly after the to-be-remembered event. Stimulation was found to be "amnesic" when applied as late as 5 minutes[135], occasionally even 15 minutes[42] after the punishment, but not thereafter. If these treatments affected avoidance by alterations on the efferent side, i.e. by disrupting motivation (fear), motor initiation or inhibition, or retrieval from memory (ability to "make use of experience"), they would be expected to be more, rather than must less, effective when applied close to the performance test. On the other hand, these treatments cannot have interfered with acquisition on the afferent side, i.e. by disrupting attention or sensation, since several minutes might elapse between the punishment and the effective manipulation of NS. Therefore, these results seem most compatible with the notion that NS plays an important role in the intermediate processing — quite conceivably in memory consolidation. We must hasten to add, however, that this presumed mnemonic function of the NS appears to be both situation-bound and anatomically fractionable. It has, for example, been demonstrated that small lesions in one part of NS in rats cause deficits in passive avoidance and two-way active avoidance without impairing a one-way avoidance task[132,134]. Lesions in another region may not interfere with any variant of avoidance conditioning[120].

Appetitive conditioning and extinction. In one of the earliest studies of this line of research, Bättig *et al.*[7] used a successive "go, no-go" discrimination task, and showed that monkeys with large lesions in the head of the caudate nucleus were impaired in learning *not* to make a manipulatory response in the presence of one stimulus, when the same kind of response paid off in the presence of another stimulus. It did not matter whether the discriminatory stimuli were auditory or visual: in either case, the monkeys with caudate damage tended to respond also on trials when they were not rewarded for doing so. In contrast, animals with such lesions were able to choose the correct visual stimulus whenever the two alternatives were presented at the same time[104], suggesting a deficiency in differential responding rather than a discriminatory problem *per se*. Support for this interpretation has since been obtained by Butters and Rosvold[17], who found that monkeys with large lesions in the head of the caudate nucleus showed abnormal persistence in performing a previously reinforced response which was no longer rewarded. Similar results have been obtained in rats, where electrical stimulation[52] as well as anterior NS lesions[40,63] lead to increased resistance to extinction of learned responses. In cats, large lesions of the caudate nucleus caused "perseverative responding which could take several hours of nonreinforcement to remove"[90].

Interference with NS tissue has also been found to alter the temporal patterning of behavior in free operant paradigms. In such situations, the animals have continuous access to the manipulandum (such as a lever), and the experimenter manipulates the delivery of reinforcements e.g. so that the animal can be rewarded only at certain, fixed, intervals, or is reinforced for spacing its responses in a specified way (so-called FI and DRL schedules). After a period of training, normal animals exhibit typical temporal variations in response rates in such tasks. Rats with damage to the NS, in contrast, showed maladaptive response-patterning (FI: ref. 48; DRL: refs. 85, 113). "Over-responding" in DRL was also induced by interference with either dopaminergic or cholinergic transmission in NS[83]. NS lesions produced a similar deficit in behavioral "timing" in an operant schedule of punishment in rats (Sidman avoidance, see refs. 3, 4).

All these findings suggest that the NS may be important for the ability to "put a brake" on learned behavior. It is interesting to recall, in this context, that Soltysik and his collaborators have found NS neurons which discharge in response to various aspects of a "waiting" task[116]. Such neurons may be contained in the essential substrate for temporal patterning of behavior (see also Rolls, this volume).

With respect to the present class of tasks, very little is known about regional variations of function within NS. The only strictly relevant evidence was obtained with a DRL schedule in rats[85]; in this study, over-responding was caused by "ventral", but not by "dorsal" lesions. The ablations performed in monkeys and cats have usually been large and always involved more than one functional subregion as derived from hodology and established with other behavioral methods. However, it seems reasonable to suggest, as Rosvold has done[101], that the deficits of "whether-to-go" mechanisms which were observed in monkeys with large lesions in the head of the caudate nucleus, are related to the very similar impairments which are found after orbitofrontal decortication in this species. Also in rats, extinction was impaired after either frontal cortical or anterior NS lesions[40,63]. It is to be predicted, then, that "go, no-go" tasks are relatively insensitive to functional disruption in other sectors of the NS.

It may be argued that over-responding in instrumental tasks is only an expression of a generalized behavioral disinhibition. However, the increased resistance to extinction in instrumental tasks in NS-damaged animals has been convincingly dissociated from other symptoms of disinhibition. Gade found, for instance, that rats with anteromedial NS lesions over-respond in the extinction phase of a VI operant schedule without showing any signs of locomotor hyperactivity in wheels or cages or in an open field[40]. Apparently, the disinhibition which is caused by this kind of NS inactivation is situation-bound; quite possibly, it is restricted to certain kinds of conditioned responses.

Discrimination learning and reversals

Evidence for NS involvement in behavioral responsiveness to stimuli is derived from different sources. In the present volume, Krauthamer summarizes anatomical and physiological evidence, and stresses the role of NS in regulation of polysensory activities. Removal of large portions of the caudate nucleus in cats caused an interesting syndrome of over-responsiveness to auditory, tactile, visual, and olfactory stimuli[125], while depletion of dopamine in the NS (and elsewhere in the forebrain) gave rise to a pronounced, multimodal sensory neglect[124]. While much remains to be clarified before these diverse observations form a unified picture, it already seems apparent that the NS is effectively informed about environmental events and takes part in controlling the organism's immediate reactions to them.

There are, however, many findings which indicate that the NS participates also in more "complex" forms of sensory processing, e.g. as expressed in different variants of discrimination learning. In the first study of this kind, published already in 1962, Buchwald and his collaborators demonstrated that strong electrical stimulation of the NS in monkeys can disrupt learning of a visual discrimination task[14]. A similar phenomenon was later observed in rats[51]. Most recently, Hirasuna *et al.*, also using rats, found behavioral effects of electrical stimulation which were interpreted as evidence for NS involvement in "visual memory"[54].

Perhaps the most instructive aspect of the research on NS participation in visual discrimination tasks concerns the coupling between neocortical and neostriatal mechanisms. The inferotemporal "association" cortex in monkeys, which is involved in processing of visual information[78], projects to the tail of the caudate nucleus and to the adjoining region of the ventral putamen (Grofova, this volume). Since the connection between the prefrontal cortex and the anterior part of the caudate nucleus had proven to be a good guide to a close functional relation in the frontal lobes, the temporal lobe was explored for a similar arrangement. And indeed, several studies in monkeys indicated that the NS projection field of the inferotemporal cortex is involved in visual processing: Lesions in either the tail of the caudate nucleus[32] or in the ventral putamen[15] impaired discrimination of simultaneously presented patterns, and electrical stimulation of the tail of the caudate interfered with a successive pattern discrimination task[20]. Of particular interest is the neat anatomical selectivity of these effects. The pattern discrimination task which was sensitive to damage in the posterior NS was unaffected by even larger lesions in the head of the caudate nucleus; these anterior lesions, on the other hand, impaired object discrimination reversals (another "visual" task which poses quite different stimulus-response requirements, and which depends on the integrity of the ventral part of the prefrontal cortex)[32]. Electrical stimulation of the tail of the caudate nucleus (and, in other cases, of the inferotemporal cortex) was disruptive only if delivered right before or during the presentation of the stimulus pattern, whereas stimulation of the anterodorsal NS region (or of the prefrontal cortex) impaired performance in the same test maximally when applied during the delay imposed between stimulus exposure and occasion to respond[20]. It seems, then, as if the posterior NS sector may be recruited by the posterior "association" cortex for perceptual processes, while the anterior NS region and the prefrontal cortex are more important for early consolidation and/or for preparation to respond on the basis of recent memory. The anterior regions of the neocortex and NS appear to be of joint importance for yet another aspect of visual guidance of behavior: behavioral adjustment to prisms which distort the visual world was retarded by lesions in either the frontal cortex or the head of the caudate nucleus in monkeys[12]. Additional support for yoked cortico-striatal functioning in visual-perceptual processes were provided recently by Jarvis *et al.*[57]. In this work, exposure to visual patterns in monkeys was shown to be followed by enhanced neural activity not only of the visual areas of the neocortex, but also in the associated parts of the caudate nucleus and putamen. The regional selectivity of the increased activity was in good agreement with the distribution of fibers from the temporal cortex to the NS, and with the neurobehavioral findings reviewed above (Mishkin, personal communication).

Also in rats, the NS seems to be functionally linked to cortical visual mechanisms. Electrical stimulation of the NS in this species impaired form discrimination reversals if applied to a posterior region, but not if applied to the anterior part of the nucleus[69]. Partial NS lesions were found to impair and not to impair visual tasks, depending on the location of the damage and different task requirements[31,60,75,115].

In contrast to the rather well-explored anatomical substrate for NS participation in visual discrimination tasks, almost nothing is known about the intrinsic physiological and chemical processes involved. No strictly relevant pharmacological work is available, and we know of no systematic neurophysiological recordings from NS during discrimination learning and performance. Rolls has, however, described NS neurons with interesting response characteristics to sensory stimulation (this volume).

This brief review has only dealt with the visual modality, reflecting a regrettable absence of studies on possible NS involvement in discrimination and reversal learning with non-visual stimuli. Since the NS is connected also to cortical areas which process auditory and somatosensory information (Grofova, this volume), as well as to polysensory thalamic nuclei (Krauthamer, this volume), and since NS neuronal activity is influenced through different modalities (Krauthamer, this volume; Rolls, this volume), it is conceivable that this formation may participate in mediation also of auditory, tactile, and kinesthetic aspects of discrimination learning and regulation of behavior. (As mentioned earlier in this chapter, NS lesions have been found to impair both delayed responding and "go, no-go" discrimination with auditory cues). In fact, any further substantial advancement of our understanding of the cognitive functions of NS seems unlikely until we get a more complete picture of its participation in complex sensory processes involving different modalities.

Discussion

A considerable body of data from animal behavior research suggests that NS is involved in mediation of cognitive functions. However, there are treacherous sources of artefacts inherent in most methods for manipulation of NS processing, and we have earlier cautioned that the effects of inactivation or stimulation of this structure may have been caused by unintended (but almost inevitable) interference with pathways leading to and from the cortex[25]. It seems necessary, therefore, to confront the data reviewed above with this alternative interpretation.

In the case of delayed response-type tasks, we no longer doubt a genuine involvement of NS neurons. The reasons for this confidence are several. First, cooling of the head of the caudate nucleus in monkeys to a temperature which is presumed to interfere with synaptic transmission without disrupting axonal conduction caused deficient performance of delayed alternation[66]. Secondly, rats which learnt delayed alternation after anterior decortication lost this ability after additional lesions were made in the NS; in this case, there was no cortex to be undercut by the second operation[130]. Somewhat similarly, in monkeys, anterior caudate lesions impaired delayed responding even at an age when the frontal cortex has not yet assumed any role for this kind of behavior[44]. Finally, injections of kainic acid into the NS of rats, which destroyed neurons of the treated region while apparently sparing most of the fibers of passage, severely impaired retention of delayed alternation[31].

Also with respect to conditioned avoidance (in particular passive avoidance) and visual discrimination tasks, NS mediation appears to be quite firmly established. A variety of pharmacological treatments, including lesioning with kainic acid (see Table 2), effectively impaired the performance of avoidance tasks, and the discovery of protein phosphorylation in NS after passive avoidance learning[35] lends additional credence to NS involvement. The importance of the tail of the caudate nucleus and the adjacent regions of putamen for visual discrimination in monkeys is now convincingly confirmed in studies of deoxyglucose consumption[57]. In contrast, the indicated involvement of NS in pathfinding, prism adaptation, and response modulation in different schedules of positive reinforcement is not yet unequivocally established. However, by analogy with the demonstrated cortico-striatal coupling in mediation of delayed responding and visual discrimination tasks, we predict that future research will confirm that NS is important at least for "frontal" tasks such as prism adaptation[12], "go, no-go" discrimination with auditory as well as visual cues[7], object reversals[32], and extinction of operant responses[16].

Somewhat unfortunately, different research groups with interest in NS mediation of cognitive processes have focussed on different classes of tasks as well as on different aspects of the anatomical substrate. Thus, as we have seen, research on avoidance conditioning has stressed the importance of the nigro-striatal bundle (and some authors talk about a nigro-striatal functional system), while neglecting the cortical input to the NS. In studies of, for example, spatial learning and different categories of visual discrimination tasks, on the other hand, the cortico-neostriatal connections have been used as guidance to telencephalic functional systems, while the nigro-striatal route has been almost totally disregarded. As a result of this arbitrary division of labour as well as perspective, it sometimes appears as if NS involvement in avoidance learning depended solely on transmission from the midbrain, while delayed responses and discrimination learning depended only on information from the cortical mantle. The role of the thalamo-neostriatal projection, in turn, seems to be overlooked by all.

Researchers using different kinds of behavioral tasks also tend to manipulate different parts of the NS. Thus, the "dorsal" and "ventral" lesions of one research tradition differ in locus from the partial lesions of another research tradition, leaving us at a loss when attempting to relate the different findings to each other in search for possible common mechanisms. As one obvious example of this unfortunate practice, no one seems to have employed animals with identical lesions in both "whether-to-go" tasks with appetitive reinforcement (e.g. extinction, DRL) and "go, no-go" tasks with aversive contingencies (e.g. passive avoidance, Sidman avoidance). Therefore, we do not know whether we are entitled to make the parsimonious assumption that one discrete NS mechanism is responsible for response suppression (response replacement) in both kinds of situations. (In fact, over-responding in appetitive and aversive tasks has occasionally be dissociated: cats with large lesions in the caudate nucleus showed abnormal resistance to extinction during shaping of a rewarded response, yet the same animals proficiently learnt a passive avoidance task[86]). If we are to avoid unfounded lumping or splitting of NS functions, it is essential that future behavioral studies are performed with more comparable localization of treatments. It is our belief that differential cortical projections to the NS gives a good basis for regional fractionation. (In stimulation or recording experiments, the implicated region may be identified by depositing horseradish peroxidase into NS for determining the associated cortical area). This approach to subdivision of NS has the great advantage of allowing both illumination of a possible linkage to cortical functional regions and cross-species comparisons of NS treatment effects. Studies of behavioral consequences of extensive (ideally total) inactivation of NS (e.g. refs. 90, 125) are, of course, a necessary complement for evaluation of the global behavioral role of the NS.

Two current notions about the functions of NS stress the importance of this formation for memory consolidation[107] and for "sensory-motor integration"[124]. The reviewed research on animal behavior suggests that the NS may indeed play a role both for memory formation *per se* and for perceptual as well as mnemonic guidance of behavior. Interestingly, however, these functions seem to be organized in a complex fashion so that different subregions of NS are involved, depending on different perceptual demands, different response-requirements (notably spatial choice *vs* "go, no-go"), and different temporal parameters (e.g. time between stimulus presentation and occasion to respond). Some other behavioral phenomena (e.g. resistance to extinction, maladaptive temporal patterning of behavior) may require other interpretations.

So far, there have been very few attempts to translate the rather contrived laboratory findings of NS functions to more naturalistic behavioral settings[33]. We would like to believe that NS is of importance also for solution of spatial problems as they occur in "real life" and for avoidance of aversive stimuli other than electrical shocks[30], but empirical support for these assumptions is really not at hand. This limitation of available knowledge does not favour the development of feasible ideas about the ecological advantage of the NS, i.e. of conceptions about the reasons for its evolution in phylogeny (Webster, this volume).

CLINICAL EVIDENCE

All neurological diseases which are known to involve pathological changes in the NS are characterized not only by abnormalities in posture and movement, but also by changes in intellectual functions[126]. Traditionally, the motor aspects of "basal ganglia diseases" were easier to detect and probably therefore more strongly emphasized than changes in intellectual and motivational processes. When the full range of behavioral manifestations is discussed, there is a tendency to ascribe disorders to movement and posture to the pathology of NS, and to blame non-striatal involvement for other symptoms. For the sake of argument, however, one could claim the opposite to be true: cognitive impairments may be rooted in NS malfunctioning, and motor deficits to originate from other structures. Obviously, it is not presently possible to localise with certainty any of the symptoms of, for example, Huntington's and Parkinson's disease in any particular structure. Nevertheless, the clinical evidence is crucial for the present thesis since the idea that NS takes part in mediation of cognitive processes cannot be upheld if such processes were in no way influenced by diseases involving the NS. Huntington's chorea and Parkinson's disease attack different neuron populations and cause different neurochemical changes in the NS as well as in other parts of the brain[126]. It is, therefore, no wonder that their behavioral manifestations, both "cognitive" and "motor", are different.

Huntington's chorea. As is well known, Huntington's chorea is one of the most seriously dementing neurological diseases. Interestingly, changes in mood and intellectual functioning often precede obvious motor symptoms, and even at a later stage patients sometimes complain that the disturbance of thought processes is at least as severe a handicap as the motor disabilities. One patient complained that "his thoughts jumped as much as his hands"[92].

The cognitive changes (which were clearly recognized already by Huntington[56]) have only recently been subjected to systematic research with application of methods and theories from current cognitive psychology. In one pioneering study, Oscar-Berman *et al.* tested aspects of concept formation, hypothesis-testing, and memory in patients with Huntington's disease[91]. The patients showed depressed performance on most measures, but seemed particularly impaired in ability to eliminate "irrelevant hypotheses", i.e. they insisted on behavioral strategies which were shown to be incorrect. More recently, Oscar-Berman has found that some Huntingtonian patients are impaired on a number of tasks which were previously employed in animal studies, among them spatial delayed responses and object reversal learning (personal communication).

During the last few years, there have also been several analyses of the memory deficits of Huntingtonian patients with use of techniques from experimental memory research. Two independent groups of investigators seem to agree that such patients are severely handicapped in encoding or storing new information[16,18,73]. These conclusions were drawn exclusively on the basis of tests of verbal memory, but application of broad test batteries have revealed subnormal performance on most measures of cognitive functioning, including tests of auditory memory for rhythms and tests of tactile memory[10]. Working in a quite different vein, finally, Potegal found that patients with Huntington's disease were impaired in spatial problems which required "compensation for self-produced movement", while showing rather normal performance on equally difficult tests which did not demand such compensation[94]; these observations are in agreement with Potegal's finding in rats with lesions in the mid-sector of the NS[95].

The pathological changes in the NS in Huntington's chorea are similar to the lesions induced with kainic acid in experimental animals, a finding which prompted a new theory of the pathogenesis of this disease (see, for example, ref. 27). As we have seen, such lesions produced severe learning impairments in both delayed alternation[31] and passive avoidance[109] in rats. These results suggest that the cognitive deterioration in Huntington's chorea patients could at least partly be due to the degeneration of NS neurons.

Parkinson's disease. The non-motor symptoms in Parkinsonism are less severe than in Huntington's chorea, and patients may remain unimpaired on several measures of "intelligence" even at quite advanced stages of the disease[13,79]. However, Parkinson recognized cognitive deterioration already in his original description of the syndrome, and careful testing reveals an array of subtle deficits. In the most extensive investigation to date, Bowen studied large groups of parkinsonian patients, and discovered, for example, (a) impairments of spatial orientation, involving both personal and extrapersonal space; (b) impoverished concept-formation; (c) difficulties in shifting from one behavioral strategy to another; (d) supramodal memory deficits, involving both verbal and non-verbal material[13]. In contrast, primary sensory functions were not found reduced in these patients. One interesting aspect of Bowen's work is that lateralized parkinsonian syndromes may involve both motor and cognitive symptoms: patients with primarily right-sided motor defects showed more severe decrements in tests of verbal cognitive processes than patients with predominantly left-sided motor disturbances. Bowen's investigations, as well as the study of McFie[79], also showed that parkinsonian patients are prone to exhibit impairments on many tasks which are known to be sensitive to damage to "association" areas of the human cerebral cortex. These behavioral indications of cortico-neostriatal functional relations in man are in encouraging concord with anatomical and neurobehavioral results obtained in animals. However, there are obviously many "cortical" cognitive symptoms, notably the aphasias, which are not seen in basal ganglia disease. It is also possible that trophic changes in the neocortex of Parkinsonian patients are the cause of some of their cognitive deficits.

The discovery of substantia nigra degeneration and dopamine depletion in the NS of parkinsonian patients[55] has stimulated the development of "animal models" of this disease (Steg, this volume, and ref. 34). Animals with pharmacologically or surgically induced "parkinsonism" usually have severe motor and motivational deficits which make both behavioral testing and interpretation difficult. However, as reviewed above, interference with dopaminergic mechanisms in the nigro-striatal system is sometimes associated with pronounced impairments in different forms of aversive conditioning, possibly indicative of defective memory consolidation. Scattered evidence suggests that also spatial tasks and temporal patterning of behavior may depend on adequate levels of dopamine. In most of these studies, the biochemical changes were not convincingly restricted to the NS, but at least the animal analogues of parkinsonism do not contradict the possibility that some cognitive symptoms of this disease may be caused by dopaminergic deficiency in the NS. It should be noted, however, that Bowen's carefully controlled investigations uncovered no improvements in cognitive functioning during L-Dopa treatment of parkinsonian patients. (The influence of L-Dopa on cognitive processes remains controversial, since several authors have reported beneficial effects; see, for example, refs. 47, 71). It should also be kept in mind that the NS of parkinsonians undergoes changes other than dopamine depletion[55].

SUMMARY AND CONCLUSIONS

Anatomically, the neostriatum occupies an intermediate position between all parts of the neocortex and multisynaptic efferent routes. An array of cognitive deficits are found in animals after experimental manipulation of NS tissue as well as in patients with Huntington's and Parkinson's diseases. Broadly speaking, NS dysfunctioning in both animals and man tends to be associated with: (a) spatial impairments; (b) reduced ability to switch behavioral "sets", "strategies", or "hypotheses" (e.g. resistance to extinction); (c) supramodal memory deficits. In addition, several aspects of visual guidance of behavior were shown to depend on the integrity of NS in animals. Thus, the NS seems to be involved in remarkably wide aspects of behavioral adaptability. Since many forms of impairments are dissociable by selective lesions or stimulation of NS in animals, any search for one common psychological denominator for all these symptoms is likely to be futile.

The NS subserves different functions in its different subregions, apparently depending on information from the associated cortical areas. Many cognitive functions of the neocortex thus seem to be relayed through the NS as they travel towards the final common path for behavioral expression. However, some "higher cortical functions" appear to be independent on this efferent link, e.g. language seems to be resistant to basal ganglia disease. It is possible that the most recently evolved functions of the human brain do not involve the NS, but follow other routes to efferent mechanisms.

The significance of the nigro-striatal pathway for cognitive processes is unclear. This connection has been implicated in some forms of learning in animals, but contradictory data are reported. The importance of dopamine for cognitive functions in man is similarly controversial. Our inadequate insights into the role of dopamine is combined with even greater ignorance about the involvement of other transmitters and the physiological properties of the implicated local circuitry.

EPILOGUE

Considering the manifold evidence for neostriatal involvement in processes which are not helpfully described as either sensory, inhibitory, or motor, it may appear rather puzzling that this aspect of NS functioning has not been more widely accepted. We have speculated on the reasons for this bias, and it may be of some value to provide a brief discussion of some possible explanations. (The following discussion does not claim to be exhaustive).

The historical inheritance is weighty; for many decades the NS was regarded almost exclusively as a "motor" structure, and generations of clinicians and experimentalists were educated in this tradition. In order to replace or supplement a dominating and seemingly coherent theory, the scientific community usually (and wisely) demands alternatives which are not only better supported by empirical evidence, but which also seem cohesive and which can be formulated in palatable concepts.

As we have seen, the cognitive role of NS is not easily presented with coherence and conceptual elegance. First of all, there is not *one* "cognitive function", but apparently many dissociable "functions", differently focused within the structure. Hence, an acceptance of NS involvement in cognitive mechanisms also demands acceptance of functional heterogeneity of NS, and this is a break with still another historically deeply rooted tradition. Instead of greater cohesion, the present view of NS functions at first sight offers dissection.

Secondly, the implicated cognitive processes have so far eluded attractive labelling. Selective impairments in, for example, delayed-response type tasks or conditioned avoidance may appear excessively esoteric and far less tangible than tremor and rigidity (although, on closer inspection, the latter symptoms turn out to be of a rather complex nature as well). Psychologists have invested a great deal of ingenious work into behavioral analysis of these deficits (see, for example, ref. 77), but there is still no consensus about classification and naming of the disturbed basic mechanisms. It is important to keep in mind, however, that neither everyday language, nor the terminology of academic psychology, evolved to suit the parcellation of behavioral and cognitive processes which emerges from clinical and experimental neurology. Behavioural neurology has inherited a conceptual apparatus which may be more appropriate for the behavioral and phenomenological observations of intact people, and it is probably unrealistic to expect that the same concepts apply in interpretation of neurological syndromes. The concept of "language" has broken down into unexpected and fascinating categories as a result of studies of aphasic patients[41], and "memory" is no longer regarded as a unitary process, but has emerged both fractionated and material-specific, for example from the research on amnesic patients and brain-manipulated animals[5]. Selective inability to learn delayed alternation may seem as bewildering as alexia without agraphia[41], but it is just as real. Furthermore, experience and anatomy indicate that we are likely to find numerous new examples of hard-to-label "cognitive" functions of the NS as we look for them with the curiosity that they deserve. In this search, as in the investigation of, for example, language[41] and visual processes[74], neuroanatomy may be a better initial guide and coherent theoretical framework than are traditional psychological categories. New concepts are likely to emerge in due time — to the mutual enrichment of neurology and psychology.

Finally, there may be an *a priori* resistance against housing both cognition and motor control in one and the same structure. But cognition often finds expression in behavior (or, to start at the other end, movements often reflect cognitive processes) and the exact borderline between cognitive and motor mechanisms is not always obvious. Anatomy indicates that the NS is an important first link between the computations of the neocortex and multisynaptic efferent routes. It does not seem very surprising that inactivation or dysfunctioning of an intermediate processing station such as the NS can give rise to both cognitive deficits and complex motor disturbances (Divac & Öberg, this volume) as a result of disconnection and/or aberrant information flow. This simple connectionist view seems capable of reconciling the apparently discordant observations of impairments of orientation, learning, and memory on one hand and altered patterns of movement on the other, and thus of bringing some cohesion to the heterogeneous behavioral functions which apparently can be ascribed to the NS. (For a more detailed account of this model, see ref. 29).

The NS may not be the interpreter in communication between the body and the soul, as suggested by Emanuel Swedenborg, but it is in a good position to serve as a dispatcher of the traffic going from the neocortex towards the final common path.

REFERENCES

1. Abplanalp, J. and Mirsky, A., *J. Comp. Physiol. Psychol.*, 85, 123-131 (1973).

2. Albert, M. and Bignami, G., *Physiol. Behav.*, 3, 141-147 (1968).

3. Allen, J.D. and Davison, C.S., *Behav. Biol.*, 8, 239-250 (1973).

4. Allen, J.D., Mitcham, J.C. and Byrd, J.I., *Psychon. Sci.*, 27, 157-160 (1972).

5. Baddiley, A.D., *The Psychology of Memory*, Basic Books, New York (1976).

6. Bättig, K., Rosvold, H.E. and Mishkin, M., *J. Comp. Physiol. Psychol.*, 53, 400-404 (1960).

7. Bättig, K., Rosvold, H.E. and Mishkin, M., *J. Comp. Physiol. Psychol.*, 55, 458-463 (1962).

8. Benjamin, R.M., Jackson, J.C. and Golden, T.T., *Brain Res.*, 141, 251-265 (1978).

9. Beritoff, J.S., *Neural Mechanisms of Higher Vertebrate Behavior*, Little, Brown, Boston (1965).

10. Boll, T.J., Heaton, R. and Reitan, R.M., *J. Nerv. Ment. Dis.*, 158, 61-69 (1974).

11. Borst, A., Delacour, J. and Libouban, S., *Neuropsychol.*, 8, 89-101 (1970).

12. Bossom, J., *Psychon. Sci.*, 2, 45-46 (1965).

13. Bowen, F.P., in: *The Basal Ganglia*, M.D. Yahr (ed.), Raven Press, New York, pp. 169-177 (1976).

14. Buchwald, N.A., Rakic, L., Wyers, E.J., Hull, C. and Heuser, G., *Exp. Neurol.*, 5, 1-20 (1962).

15. Buerger, A.A., Gross, C.G., and Rocha-Miranda, C.E., *J. Comp. Physiol. Psychol.*, 86, 440-446 (1974).

16. Butters, N., Tarlow, S., Cermak, L. and Sax, D., *Cortex*, 12, 134-144 (1976).

17. Butters, N. and Rosvold, H.E., *J. Comp. Physiol. Psychol.*, 65, 397-403 (1968).

18. Caine, E.D., Ebert, M.H. and Weingartner, H., *Neurol.*, 27, 1087-1092 (1977).

19. Cianci, S.N., *Exp. Neurol.*, 11, 104-114 (1965).

20. Cohen, S.M., *Acta Neurobiol. Exp.*, 32, 211-233 (1972).

21. Deadwyler, S.A., Montgomery, D. and Wyers, E.J., *Physiol. Behav.*, 8, 631-635 (1972).

22. Divac, I., *Acta Biol. Exp.*, 28, 107-120 (1968).

23. Divac, I., *Acta Biol. Exp.*, 28, 149-167 (1968).

24. Divac, I., *Neuropsychol.*, 9, 175-183 (1971).

25. Divac, I., *Acta Neurobiol. Exp.*, 32, 461-478 (1972).

26. Divac, I., *Physiol. Behav.*, 8, 519-522 (1972).

27. Divac, I., *Acta Neurol. Scand.*, 56, 357-360 (1977).

28. Divac, I., in: *The Psychobiology of the Striatum*, A.R. Cools, A.H.M. Lohman and J.H.L. van den Bercken (eds.), Elsevier/North Holland, Amsterdam, pp. 21-30 (1977).

29. Divac, I., Björklund, A., Lindvall, O., and Passingham, R.E., *J. Comp. Neurol.*, 180, 59-71 (1978).

30. Divac, I., Gade, A., and Wikmark, R.G.E., *Physiol. Psychol.*, 3, 43-46 (1975).

31. Divac, I., Markowitsch, H.J. and Pritzel, M., *Brain Res.*, 151, 523-532 (1978).

32. Divac, I., Rosvold, H.E. and Szwarcbart, M., *J. Comp. Physiol. Psychol.*, 63, 184-190 (1967).

33. Divac, I., Wikmark, R.G.E. and Gade, A., *Physiol. Psychol.*, 3, 39-42 (1975).

34. Duvoisin, R.C., in: *The Basal Ganglia*, M.D. Yahr (ed.), Raven Press, New York, pp. 293-303 (1976).

35. Ehrlich, Y.H., Rabjohns, R.R. and Routtenberg, A., *Pharmacol. Biochem. & Behav.*, 6, 169-174 (1977).

36. Fibiger, H.C., in: *The Psychobiology of the Neostriatum*. A.R. Cools, A.H.M. Lohman and J.H.L. van den Bercken (eds.), Elsevier/North Holland, Amsterdam, pp. 73-84 (1977).

37. Fibiger, H.C., Phillips, A.G. and Zis, A.P., *Pharmac. Biochem. Behav.*, 2, 87-96 (1974).

38. Fletcher, H.J., in: *Behavior of Nonhuman Primates*, A.M. Schrier, H.F. Harlow, and F. Stollnitz (eds.), Academic Press, New York, pp. 129-167 (1965).

39. Fox, S.S., Kimble, D.P., Lickey, M.E., *J. Comp. Physiol. Psychol.*, 58, 380-386 (1964).

40. Gade, A., *Behavioral Studies of the Prefrontal Cortex and the Neostriatum in the Rat.* Master's thesis, Department of Psychology, Copenhagen (1974).

41. Geschwind, N., *Scientific American*, 226, 76-83 (1972).

42. Gold, P.E. and King, R.A., *Behav. Biol.*, 7, 709-715 (1972).

43. Goldman, P.S., Rosvold, H.E., Vest, B. and Galkin, T.W., *J. Comp. Physiol. Psychol.*, 77, 212-220 (1971).

44. Goldman, P.S. and Rosvold, H.E., *Brain Res.*, 43, 53-66 (1972).

45. Green, R.H., Beatty, W.W. and Schwartzbaum, J.S., *J. Comp. Physiol. Psychol.*, 64, 444-452 (1967).

46. Gross, C.G., Chorover, S.L. and Cohen, S.M., *Neuropsychol.*, 3, 53-68 (1965).

47. Halqin, R., Riklan, M. and Misiak, H., *J. Nerv. Ment. Dis.*, 164, 268-272 (1977).

48. Hansing, R.A., Schwartzbaum, J.S. and Thompson, J.B., *J. Comp. Physiol. Psychol.*, 66, 378-388 (1968).

49. Hassler, R., *J. Neurol. Sci.*, 36, 187-224 (1978).

50. Haycock, J.W., Deadwyler, S.A., Sideroff, S.I. and McGaugh, J.L., *Exp. Neurol.*, 41, 201-213 (1973).

51. Herz, M.J. and Peeke, H.V.S., *Brain Res.*, 33, 519-522 (1971).

52. Herz, M.J., Marshall, K.E. and Peeke, H.V.S., *Physiol. Psychol.*, 2, 184-186 (1974).

53. Hunter, W.S., *Behav. Monogr.*, 2, No. 1 (1913).

54. Hirasuna, N., Deadwyler, S.A. and Wyers, E.J., *Brain Res. Bull.*, 2, 173-179 (1977).

55. Hornykiewicz, O., in: *The Neurobiology of Dopamine*, A.S. Horn, J. Korf and B. Westerink (eds.), Academic Press, New York (1978).

56. Huntington, G., in: *Advances in Neurology, Vol. 1: Huntington's Chorea 1872-1972.* A. Barbeau, T.N. Chase and G.W. Poulson (eds.), Raven Press, New York, pp. 33-35 (1972).

57. Jarvis, C.D., Mishkin, M., Shinohara, M., Sakurada, O., Miyaoka, M. and Kennedy, C., *Soc. Neurosci. Abstr.*, 4, 632 (1978).

58. Kim, H.J. and Routtenberg, A., *Brain Res.*, 113, 620-625 (1976).

59. Kim, H.J. and Routtenberg, A., *Soc. Neurosci. Abstr.*, 2, 445 (1976)

60. Kirkby, R.J., *Physiol. Behav.*, 4, 451-454 (1969).

61. Kirkby, R.J. and Polgar, S., *Physiol. Psychol.*, 2, 301-306 (1974).

62. Kitsikis, A., Roberge, A.G., Frenette, G., *Exp. Brain Res.*, 15, 305-317 (1972).

63. Kolb, B., *Physiol. Behav.*, 18, 237-244 (1977).

64. Konorski, J., *Integrative Activity of the Brain: An Interdisciplinary Approach.* Univ. Chicago Press, Chicago (1967).

65. Krauthamer, G.M., in: *Catecholamines and Behavior*, A.J. Friedhoff (ed.), Plenum Press, New York, pp. 59-87 (1975).

66. Krauthamer, G.M., Liebeskind, J. and Salmon-Legagneur, A., *J. Physiol. (Lond.)*, 190, 18-19 (1967).

67. Laursen, A.M., *Acts Physiol. Scand.*, 57, 81-89 (1963).

68. Laursen, A.M., *Acta Physiol. Scand.*, 59, Suppl. 211 (1963).

69. Livesey, P.J. and Muter, V., *J. Comp. Physiol. Psychol.*, 90, 203-211 (1976).

70. Livesey, P.J. and Rankine-Wilson, J., *J. Comp. Physiol. Psychol.*, 88, 342-354 (1975).

71. Meier, M.J. and Martin, W.E., *J. Am. Med. Ass.*, 213, 465-466 (1970).

72. Menzel, E.W., in: *Behavior of Nonhuman Primates*, Vol. 5, A.M. Schrier and F. Stollnitz (eds.), Academic Press, New York, pp. 83-153 (1974).

73. Mendell, P., Butters, N. and Montgomery, K., *Neuropsychol.*, 16, 507-510 (1978).

74. Mikulas, W.J., *Psychon. Sci.*, 5, 275-276 (1966).

75. Mikulas, W.J., *Psychon. Sci.*, 15, 132-133 (1969).

76. Mikulas, W.J. and Isaacson, R.L., *Psychon. Sci.*, 3, 485-486 (1966)

77. Mishkin, M., in: *The Frontal Granular Cortex and Behavior*, J.M. Warren and K. Akert (eds.), McGraw Hill, New York, 219-237 (1964).

78. Mishkin, M., in: *Brain and Human Behavior*, A.G. Karczmar and J.D. Eccles (eds.), Springer, Berlin, 187-208 (1972).

79. McFie, J., *J. Ment. Sci.*, 106, 1512-1517 (1960).

80. Mitcham, J.D. and Thomas, R.K. Jr., *J. Comp. Physiol. Psychol.*, 81, 101-107 (1972).

81. Morgan, J.M. and Routtenberg, A., *Science*, 196, 87-89 (1977).

82. Nadel, L. and O'Keefe, J., in: *Essays on the Nervous System*, R. Bellairs and E.G. Gray (eds.), Clarendon Press, Oxford, pp. 367-390 (1974).

83. Neill, D.B., *Brain Res.*, 105, 89-103 (1976).

84. Neill, D.B., Boggan, W.O. and Grossman, S.P., *Pharmac. Biochem. Behav.*, 2, 97-103 (1974).

85. Neill, D.B., Ross, J.F. and Grossman, S.P., *Physiol. Behav.*, 13, 297-305 (1974).

86. Neill, D.B. and Grossman, S.P., *J. Comp. Physiol. Psychol.*, 71, 311-317 (1970).

87. Neisser, U., *Cognitive Psychology*, Appleton-Century-Crofts, New York (1966).

88. Niki, H., Sakai, M. and Kubota, K., *Brain Res.*, 38, 343-353 (1972).

89. Öberg, R.G.E. and Divac, I., *Acta Neurobiol. Exp.*, 35, 645-659 (1975).

90. Olmstead, C.E., Villablanca, J.R., Marcus, R.J. and Avery, D.L., *Exp. Neurol.*, 53, 670-693 (1976).

91. Oscar-Berman, M., Sax, D.S. and Opoliner, L., in: *Advances in Neurology, Vol. 1: Huntington's Chorea 1872-1972*, A. Barbeau, T.N. Chase, and G.W. Paulson (eds.), Raven Press, New York, pp. 717-728 (1973).

92. Paulson, G.W., in: *The Basal Ganglia*, M.D. Yahr (ed.), Raven Press, New York, pp. 317-329 (1976).

93. Peeke, H.V.S. and Herz, M.J., *Science*, 173, 80-82 (1971).

94. Potegal, M., *Neuropsychol.*, 9, 233-235 (1971).

95. Potegal, M., *Acta Neurobiol. Exp.*, 32, 479-494 (1972).

96. Prado-Alcalá, R.A., Bermudez-Rattoni, F., Velázques-Martinez, D.N. and Gustavo-Bacha, M., *Life Sci.*, 23, 889-896 (1978).

97. Prado-Alcalá, R.A., Grinberg, Z.J., Arditti, Z.L., García, M.M., Prieto, H.G. and Brust-Carmona, H., *Physiol. Behav.*, 15, 283-287 (1975).

98. Price, M.T.C. and Fibiger, H.C., *Pharmacol. Biochem. Behav.*, 3, 285-290 (1975).

99. Ranje, C. and Ungerstedt, U., *Brain Res.*, 134, 95-111 (1977).

100. Rosenkilde, C.E. and Divac, I., *J. Comp. Physiol. Psychol.*, 90, 343-352 (1976).

101 Rosvold, H.E., in: *Mind as a Tissue*, C. Rupp (ed.), Harper & Row, New York, pp. 21-38 (1968).

102. Rosvold, H.E., *Acta Neurobiol. Exp.*, 32, 439-460 (1972).

103. Rosvold, H.E. and Delgado, J.M.R., *J. Comp. Physiol. Psychol.*, 49, 365-372 (1956).

104. Rosvold, H.E., Mishkin, M. and Szwarcbart, M.K., *J. Comp. Physiol. Psychol.*, 51, 437-444 (1958).

105. Rosvold, H.E. and Szwarcbart, M.K., in: *The Frontal Granular Cortex and Behavior*, J.M. Warren, K. Akert (eds.), McGraw Hill, New York, pp. 1-15 (1964).

106. Routtenberg, A. and Holzman, N., *Science*, 181, 83-86 (1973).

107. Routtenberg, A. and Kim, H.J., in: *Cholinergic-Monoaminergic Interactions in the Brain*, L.L. Butcher (ed.), Academic Press, New York (1978).

108. Royce, G.J., *Brain Res.*, 153, 465-475 (1978).

109. Sanberg, P.R., Lehman, J. and Fibiger, H.C., *Brain Res.*, 149, 546-551 (1978).

110. Sapawi, R.R. and Divac, I., *Neurosci. Lett.*, 7, 183-189 (1978).

111. Sax, D.S., Butters, N., Tomlinson, E.B. and Feldman, R.G., in: *Advances in Neurology, Vol. 1: Huntington's Chorea 1872-1972*, A. Barbeau, T.N. Chase and G.W. Paulson (eds.), Raven Press, New York, pp. 657-663 (1973).

112. Sidyakin, V.K., *Zh. Vyssh. Nerv. Deyat. im Pavlova*, 22, 1073-1076 (1972).

113. Schmaltz, L.W. and Isaacson, R.L., *J. Comp. Physiol. Psychol*, 65, 343-348 (1968).

114. Schultze, M.J. and Stein, D.G., *Exp. Neurol.*, 46, 291-301 (1975).

115. Schwartzbaum, J.S. and Donovick, P.J., *J. Comp. Physiol. Psychol.*, 65, 83-92 (1968).

116. Soltysik, S., Hull, C.D., Buchwald, N.A. and Fekete, T., *EEG Clin. Neurophysiol.*, 39, 65-78 (1975).

117. Stamm, J.S., *J. Comp. Physiol. Psychol.*, 70, 437-447 (1970).

118. Talland, G.A. and Schwab, R.S., *Neuropsychol.*, 2, 45-53 (1964).

119. Teuber, H.-L., in: *The Basal Ganglia*, M.D. Yahr (ed.), Raven Press, New York, pp. 151-168 (1976).

120. Thomas, R.K. and Hill, A.S., *Bull. Psychon. Soc.*, 1, 346-348 (1973).

121. Thompson, R.L., *J. Comp. Physiol. Psychol.*, 52, 650-659 (1959).

122. Thompson, R.L. and Mettler, F.A., *Amer. J. Ment. Defic.*, 67, 526-535 (1963).

123. Tolman, E.C., *Psychol. Rev.*, 55, 189-208 (1948).

124. Ungerstedt, U., Ljungberg, T. and Ranje, C., in: *Psychobiology of the Striatum*, A.R. Cools, A.H.M. Lohman and J.H.L. van den Bercken (eds.), Elsevier/North Holland, Amsterdam, pp. 85-97 (1977).

125. Villablanca, J.R., Olmstead, D.E. and de Andrés, I., *Exp. Neurol.*, 61, 635-649 (1978).

126. Vinken, P. and Bruyn, G. (eds.), *Handbook of Clinical Neurology*, Vol. 6, North Holland, Amsterdam (1968).

127. Weiskrantz, L. Mihailović, Lj. and Gross, C.G., *Brain*, 85, 487-504 (1962).

128. Webster, K.E., *Proc. Roy. Soc. Med.*, 68, 203-210 (1975).

129. Wikmark, R.G.E., Divac, I. and Weiss, R., *Brain, Behav. Evol.*, 8, 329-339 (1973).

130. Wikmark, R.G.E. and Divac, I., Paper presented at 81st Annual Convention of American Psychological Association, Montreal, Canada (1973).

131. Wilburn, M.W. and Kesner, R.P., *Exp. Neurol.*, 34, 45-50 (1972).

132. Winocur, G., *J. Comp. Physiol. Psychol.*, 86, 432-439 (1974).

133. Winocur, G., *Physiol. Psychol.*, 3, 255-256 (1975).

134. Winocur, G. and Mills, J.A., *J. Comp. Physiol. Psychol.*, 68, 552-557 (1969).

135. Wyers, E.J. and Deadwyler, S.A., *Physiol. Behav.*, 6, 97-103 (1971).

136. Wyers, E.J., Deadwyler, S.A., Hirasuna, N. and Montgomery, D., *Physiol. Behav.*, 11, 809-819 (1973).

137. Wyers, E.J., Peeke, H.V.S., Williston, J.S. and Herz, M.J., *Exp. Neurol.*, 22, 350-366 (1968).

Ref Note 1. Divac, I. and Diemar, N.H., *J. Comp. Neurol.*, in press

Is a Theory on Neostriatal Functions Now Possible?

Concluding Remarks on two sessions on physiological and behavioural explorations of the neostriatum

D. Albe-Fessard and P. Buser*

Laboratoire de Physiologie des Centres nerveux, 9, Quai Saint Bernard, 75005 Paris
**Laboratoire de Neurophysiologie comparée, 9, Quai Saint Bernard, 75005 Paris*

During this EBBS workshop on the Neostriatum (NS), two sessions were devoted to reviewing some of the salient features of striatal functions, as they may be deduced from physiological and behavioural tests. Since several important points were raised in both sessions-although often from different angles - it has been the common feeling of the respective chairpersons that a brief comment could be written under a single heading. Our aim here will be to try to sum up some of the suggested hypotheses, with the obvious question in mind of whether a unique function can now be ascribed to the neostriatum, based on data yet available.

As rightly pointed out by most, if not all speakers, the main emphasis for many decades has been placed on motor functions of the NS, particularly under the weight of clinical data. Lesion studies initially stressed the importance of the striatum as the highest extrapyramidal control system and credit should be given to JUNG and HASSLER for their thorough contribution to this functional aspect. That disturbed motor functions, including posture and/or locomotion and/or directed voluntary activity (dystonia and dyskinesia) may be associated with anatomical alterations in the NS and connected structures like substantia nigra (SN) is a well established fact. Animal models have to a certain extent contributed to our knowledge of the mechanisms involved. In fact, various major difficulties remain in trying to draw parallels between patients and animals, or to understand the underlying processes : (i) neostriatal lesions determine non-unequivocal effects in animals, i.e. hypoactivity or hyperactivity, or "arrest" reaction or obstinate progression ; (ii) such lesions do not always clearly reproduce the "extra-pyramidal" syndrome as observed in the human ; this specific point, often discussed at other meetings on the subject was not developed at length in this Symposium ; (iii) electrical stimulation of the NS can elicit some movements or dystonia, but this is certainly not a major nor easy way to explore NS functions. Everyone would agree that the caudate is certainly not a "motor" centre in the common sense. In connection to these aspects, KORNHUBER's speculative view, that NS mainly contributes to the control (or organization) of slow"ramp" movements as opposed to fast "ballistic" ones, was also brought up.

ROLLS' interesting review on unit studies in NS in connection to movement also brought out the difficulty to consider these structures as motor : with the exception of DE LONG's data, showing a clear acceleration of putamen cells when a monkey performs slow (ramp) but not fast (ballistic) movements (a finding which in a sense confirms KORNHUBER's view), most data (NIKI, ROLLS, BUSER) has been fairly disappointing and it can probably be assumed that caudate cells do not participate massively in the performance of a simple motor task.

It may of course well be that all the above authors (except DE LONG) were testing excessively rapid movements. The reason for this silence or "non-participation" may be more profound though.

Curiously enough, and despite these problems, there is excellent although less simple than initially suggested,correlation between extrapyramidal symptoms which are usually thought to be due to NS system dysfunctioning and neurochemical data. The well known facts on the parallelism of DA content and dyskinetic syndromes (Parkinsonism's hypokinesia and hypertonia) were recalled by STEG who also pointed out a reverse correlation between the "hyperkinetic-hypotonic" syndrome and DA : Huntington's chorea is increased by DOPA and decreased by DA receptor blocking agents like neuroleptics. Some other well known neurochemical correlates of motor effects of some substances were also reported : the effects of reserpine, of apomorphine, of amphetamine, that of 6OH DA, are all logically explained through their action on DA receptors in NS or in n. accumbens. That these correlations - in spite of their interest - only give a partial view on the NS functions in general is however likely. It was clearly indicated in other sessions of the Symposium that NS is both histologically (PASIK et al) and neurochemically (FONNUM) a heterogeneous structure with a complex system of afferent and efferent pathways (ALBE-FESSARD, FELTZ, FEGER, GROFOVA, DENIAU and many others). The DA mechanisms do not act in isolation,for other transmitter links are now clearly recognized (Ach, GABA, Glutamate, 5 HT, substance P, enkephalin). It is expected that more and more correlations will be found between transmitter contents in NS and related structures (SN) and motor disturbances. However, besides these correlations which consider the "motor" side of NS functions as the dependent variable, more and more other ones are now being brought up between neurochemical data and more complex behavioural changes, as shown in our meeting.

Not being a "motor" structure, is NS a sensory structure ? To answer that question, KRAUTHAMER gave a thorough summary of data essentially obtained through electrophysiological recording. Neither stimulation nor lesion studies could have predicted that the NS receives sensory projections.

It is now known that striate units can be activated (or sometimes inhibited) by a wide range of sensory modalities, thus placing NS within the set of polymodal or polysensory structures. Although the bulk of these data was first obtained under chloralose anaesthesia, enough information is now available to show that this polymodal property does not strictly depend upon the anaesthetic used. KRAUTHAMER reviewed numerous other facts, regarding the pathways for these projections, their thalamic origin and the influence of the neocortex which is not essential for their existence but merely acts as a gating system allowing sensory messages to reach the striatum via polysensory medial thalamic nuclei (intralaminar nuclei, centrum medianum CM). KRAUTHAMER than developed hypotheses trying to correlate some symptoms that derive from caudate lesions to the "sensory side" of the NS function : while suggesting that "the loop between caudate and CM may be of importance to modulate the polysensory activities at several sites in brain" and that "some manifestations of NS dysfunction may be viewed as expressions of a disturbed modulation of sensory convergence mechanisms" or "as reflecting the inability to maintain appropriate contact with a significant stimulus", KRAUTHAMER in this paper was very imaginative and indeed interesting, but not completely convincing.

Clearly, the bulk of data presented during these sessions dealt with striatal participation in higher brain operations which involve other processes, not just sensory nor simply motor. Here again, the four main approaches, lesions, electrical stimulations, neurochemical tests and electrical single recordings, have diversely contributed to build up theories - or at least hypotheses - on the participation of NS. PHILLIPS summarized data on behavioural effects of NS electrical stimulation. While purely "motor" symptoms are of minor importance - as indicated above - low frequency caudate stimulation is known to elicit behavioural "inhibition" with arrest (BUCHWALD, ROUGEUL etc.). But, interestingly enough, CD stimulation inhibits recently learned movements more easily than inborn or "natural" ones (animal stops pressing bar for food during stimulation but is still perfectly able to catch offered food with paw etc.). At this stage, the idea can be retained that CD acts as an "inhibitory" structure for a given set of activities belonging to the behavioural repertoire (an oversimplified hypothesis apparently first suggested by METTLER's lesion studies). Behavioural arrest, however, does not necessarily indicate that the stimulated structure is inhibitory, but rather that it intervenes in some way in the elaboration or "programming" of some kind of higher order integration, including a goal-directed task. It is not unreasonable to think that low frequency stimulation acts through a disruption of the normal functioning of this delicate network. In our opinion, this might be a sensible, though temporary conclusion that can be drawn from stimulation studies : they confirm the hypothesis of NS participation in higher integrations, but not much more can be deduced from them. It is true that other stimulation studies have placed emphasis on more particular effects of CD stimulation. PHILLIPS was right to report (i) that self-stimulation points were found in the caudate, and these were shown to depend upon a complex neurochemical mechanism, including both DA and 5HT systems ; (ii) that caudate or SN stimulation delivered after the trial, when passive avoidance is being learned, could elicit memory disruption. The latter effect was suggested as being due to an excess of DA release. ROLLS' unit recording studies were also indicative of NS subserving higher functions. He described neurons that were activated by the sight of interesting objects, which could not be considered as trivial visual stimuli ; these "objects" were related to the preparation of a goal-directed response, in particular toward food-reward. On the other hand, cells in the pallidum and in SN were more closely related, in their activation,to the movement itself, rather than to the sight of the reward and/or to the preparation of the goal-directed activity. It is true that these data, like those obtained from various structures throughout the upper brain (cat suprasylvian cortex, monkey infero-temporal, frontal and parietal areas, etc.) often remain anecdotal and would require more systematisation in the future. Nevertheless, their interest is great, in that the reactive pattern of each particular unit may reveal one aspect of the complex interplay of sensory, motor and integrative processes in the NS.

The next step was taken by S. IVERSEN and Ivan DIVAC who both reviewed facts and hypotheses, mainly derived from lesion and neurochemical NS studies, which could lead to some new concepts about its integrative functions. Two converging sets of data did in fact contribute to this new trend. Anatomically, it was shown that the whole neocortical mantle, including frontal and temporal lobes and cingulate cortex, projects in an orderly, topographical way on the neostriatum. Functionally, a new fact was brought to our knowledge when it was shown that various aspects of the "frontal syndrome" in monkeys could be replicated through lesions of anterodorsal parts of the CD nucleus : such lesions could, just like those of the dorsolateral and/or orbital prefrontal cortex, elicit deficits in various performances with spatial choice (delayed alternation, delayed reactions, conditional positional discrimination) or in those involving merely "inhibition" like go no-go tasks , bar pressing extinctions, and object discrimination reversal. The situation with avoidance reactions is less clear since those were impaired in rats, but no in cats. IVERSEN also reviewed facts which may allow to correlate these lesion data to some of the well known facts on neurochemical mechanisms. To sum up the view was reinforced that DA release in the striatum (from the SN-CD DA pathway) is essential for information processing in the striatal substrate. This, however, does not rule out a possible role of other transmitter substances.

These concepts were further developed by DIVAC, who presented a well documented review, first rightly pointing out that performance of a learned task (be it a spatial choice or a go no-go test) involves a number of different fundamental mechanisms among which attention-perception, memory functions, motivation etc.. Effective CD lesions (or even stimulations) could thus impair a broad category of functions, depending on the employed test. Impairment in spatial tasks, lack of inhibition (i.e. of a "break"), and deficits in discrimination learning or retention may each reveal one particular facet of the "memory function" of the caudate, which controls various sets of brain operations that could be grouped as "cognitive". To support this cognitive function, the cortico-striatal links may be of major importance.

The NS certainly remains one of the most mysterious structures in the CNS. Therefore, syntheses like those proposed during the sessions deserve considerable interest. In this respect, all data, discussions and hypotheses brought up at this meeting will undoubtedly contribute to our understanding of the striatal functions. However, in spite of these valuable suggestions, the final solution is probably far from being reached. It may well be that the caudate really controls "cognitive" functions. In our opinion, it may also be that what was synthesized under "cognitive" is only one side of a more general domain of integrative operations which remain to be characterized.

Index

The page numbers refer to the first page of the article in which the index term appears.